AF539963

CHEMICAL PROCESSES IN ORGANIC SYNTHESIS

CHEMICAL PROCESSES IN ORGANIC SYNTHESIS

By

R K Bansal

2016

SBS Publishers & Distributors Pvt. Ltd.
New Delhi

All rights reserved. No part of this publication may be reproduced, stored in a retrieval system, or transmitted in any form or by any means, electronic, mechanical, photocopying, recording or otherwise, without the prior written permission of the publisher and the copyright holder.

ISBN 13 : 9789380090795

First Published in 2016

© Reserved

Published by:

SBS PUBLISHERS & DISTRIBUTORS PVT. LTD.

2/9, Ground Floor, Ansari Road, Darya Ganj,

New Delhi - 110002,

INDIA

Tel: 0091.11.23289119 / 41563911

Email: mail@sbspublishers.com

www.sbspublishers.com

Preface

Organic synthesis is a special branch of chemical synthesis and is concerned with the construction of organic compounds via organic reactions. Organic molecules often contain a higher level of complexity than purely inorganic compounds, so that the synthesis of organic compounds has developed into one of the most important branches of organic chemistry. There are several main areas of research within the general area of organic synthesis: total synthesis, semi-synthesis, and methodology. Chemical synthesis is a purposeful execution of chemical reactions to obtain a product, or several products. This happens by physical and chemical manipulations usually involving one or more reactions. In modern laboratory usage, this tends to imply that the process is reproducible, reliable, and established to work in multiple laboratories. A chemical synthesis begins by selection of compounds that are known as reagents or reactants. Various reaction types can be applied to these to synthesize the product, or an intermediate product. This requires mixing the compounds in a reaction vessel such as a chemical reactor or a simple round-bottom flask. Many reactions require some form of work-up procedure before the final product is isolated. Synthesis can start with very simple compounds and work toward a complex one, or it can start with natural products from plants in which various portions of the desired end-product molecules are already conveniently assembled.

Editor

Contents

Chapter 1

FROM POLYMER TO SMALL ORGANIC MOLECULES: A TIGHT RELATIONSHIP BETWEEN RADICAL CHEMISTRY AND SOLID-PHASE ORGANIC SYNTHESIS

Danilo Mirizzi and Maurizio Pulici

Department of Chemical Core Technologies, Nerviano Medical Sciences S.r.l., V.le Pasteur, 10, 20014, Nerviano (MI), Italy

ABSTRACT

Since Gomberg's discovery of radicals as chemical entities, the interest around them has increased through the years. Nowadays, radical chemistry is used in the synthesis of 75% of all polymers, inevitably establishing a close relationship with Solid-Phase Organic Synthesis. More recently, the interest of organic chemists has shifted towards the application of usual "in-solution" radical chemistry to the solid-phase, ranging from the use of supported reagents for radical reactions, to the development of methodologies for the synthesis of small molecules or potential libraries. The aim of this review is to put in perspective radical chemistry, moving it away from its origin as a synthetic means for solid supports, to becoming a useful tool for the synthesis of small molecules.

ABBREVIATIONS

AIBN: Azo-bis-isobutyronitrile, AMBN: azobismethylisobutyronitrile, ATRC: Transition-metal-catalyzed atom transfer radical cyclization, 9-BBN: 9-Borabicyclo[3.3.1]nonane, CAN: Ceric ammonium nitrate, DAST: Diethylaminosulfur trifluoride, DCE: 1,2-Dichloroethane, DCM: Dichloromethane, DIC: Diisopropyl carbodiimide, DIPAD: Diisopropyl azadicarboxylate, DIPEA: diisopropylethylamine, DLP: lauroylperoxide, DMA: N,N-Dimethylacetamide, DMAP: 4-dimethylaminopyridine, DMF: N,N-Dimethylaminoformamide, EPHP: N-ethylpiperidine hypophosphite, FRET: Fluorescence Resonance Energy Transfer, HASC: α-Hetero-Atom Substituted Carbonyl HBTU: 2-(1H-Benzotriazole-1-yl)-1,1,3,3-tetramethyluronium hexafluorophosphate, HMPA: Hexamethylphosphoramide, IBX: 2-Iodoxybenzoic acid, LA: Lewis acid, NMM: N-methyl morpholine, NMP: N-methyl pyrrolidine, NMQ: N-methylquinolinium, PMHS: Polymethylhydrosiloxane, RCM: Ring-closing metathesis, SET: single electron transfer, SP: Solid-phase, SPS: Solid-phase synthesis, TBGH: Tributylgermanium hydride, TBTH: Tributyltin hydride, TEMPO: Tetramethyl pentahydropyridine oxide, TFA: Trifluoroacetic acid, THF: Tetrahydrofuran, TON: Turnover number, TTF: Tetrathiofulvalene, TTMSS: Tris(trimethylsilyl)silane, Troc: 2,2,2-Trichloroethoxycarbonyl

INTRODUCTION

Since Gomberg's discovery of free radicals as chemical entities [1], the interest around this electron-deficient species has increased throughout the years. Their success arrived during WWII, when radical chemistry took a primary role in the manufacturing of synthetic substitutes of natural rubber. Nowadays, radical processes are used to prepare 75% of all polymers, including the solid supports employed in organic synthesis.

An additional and interesting application of radical chemistry involves post polymerization modifications, which transforms the surface of the polymer for several purposes [2-4]. The change of polymers' surface characteristics allows the acquisition of new properties of the solid phase (SP) and, although this represents an interesting scientific aspect of the subject, touching the field of organic synthesis as well, it is also somewhat out of the purpose of this review and, for those interested, it would be more appropriate to refer to more specific literature.

Inevitably, all of these roles of radical chemistry have allowed creating a thread with SP synthesis. SP synthesis is an important feature in modern organic chemistry, however, the role of free radicals includes more than the methodologies just used for polymerization. The real boost came after physical organic chemists begun to determine absolute rates of radical reactions [5,6], together with the advent of new spectroscopic techniques (ESR) to detect them. At that point, radical chemistry became a more and more common methodology for day-to-day organic synthesis. With a wider range of available radical reactions, several applications in the synthesis of natural compounds, drug-like molecules and functional groups interconversion strategies were identified. It would have been just a matter of time to find extensions of those methodologies to SP approaches.

This review focuses on the applications of radical chemistry in solid-phase synthesis of single or classes of compounds, which includes also resin loading and cleavage and the use of solid-supported reagents. This subject has been last reviewed a few years ago [7]. The aim of this work is not to offer a comprehensive overview of the applications of radical chemistry on SP, but rather to emphasize the methodologies and developments in a field that might be useful in daily organic chemistry.

However, since classical radical chemistry is governed by kinetic considerations, before going through the examples that the literature offers on the subject it is fundamental to understand how radical reactions behave when they take place on the surface of a solid bead. In this respect it is important to mention Curran's work [8], dealing with the measurement of the kinetics of radical reactions on SPs. Curran calculated the rates of common reactions involving radical precursors anchored to an Ellman THP resin by the ratio of the possible products, in relation to the same reaction run in solution.

He found that the rate constant of H-abstraction from Bu3SnH by an alkyl radical linked to SP is probably not so different from the solution-phase one. This means that reactions of polymer-bound radicals, at least those mediated by tin hydride, can be planned as if they were performed in solution.

On the other hand, experiments performed with the generation of aryl radicals demonstrated that Ellman's resin behaves like a solvent with relatively reactive C-H bonds. This means that H-abstraction from the polymer backbone is a feasible process, implicating that, at least in the case of the studied aryl radical, it is difficult to set up reactions whose kinetics are slow. Accordingly, for resin-bound aryl radicals there is a limitation to the fastest intramolecular classes, while slower intra- and bimolecular reactions will suffer from competing hydrogen transfer from the backbone and linker C-H bonds.

Solid-Supported Reagents, Resins and Linkers

Among the various relationships between radical chemistry and SP, two interesting aspects are the use of special resins or linkers cleavable under radical conditions and solid-supported reagents designed to assist radical reactions in solution. In this regard, this section will provide a review of the literature.

Linkers and Resins

In the quest for reliable, clean and flexible linkers for SP synthesis, an important role is assigned to those cleavable via radical mechanisms. In particular, photolabile moieties represent the great majority of the type and several resins with appropriately designed linkers are commercially available. Since this specific subject covers more than 20 years of research and has been accurately review in the past, we refer readers to more appropriate publications [9,10].

Beside light-sensitive linkers, a number of studies have treated innovative linkages based on cleavage under radical conditions. Radical cleavage of linker units usually enables the introduction of either a hydrogen atom at the cleavage site (traceless linker [11]), or, which is of even more relevance in the context of diversity oriented synthesis [12], of an additional diversity element (diversity linker unit [13]). In this section of the review, the attention will be shifted among the "key" atom in the linker.

Sulfur-Based Linkers

In the need of new methods for SP synthesis with C-H bond formation after cleavage (traceless linker strategy), a linking strategy consisting in the presence of a sulfur atom, in different oxidation states susceptible of radical cleavage, was developed. The work performed by Janda [14] falls under this category (Scheme 1)

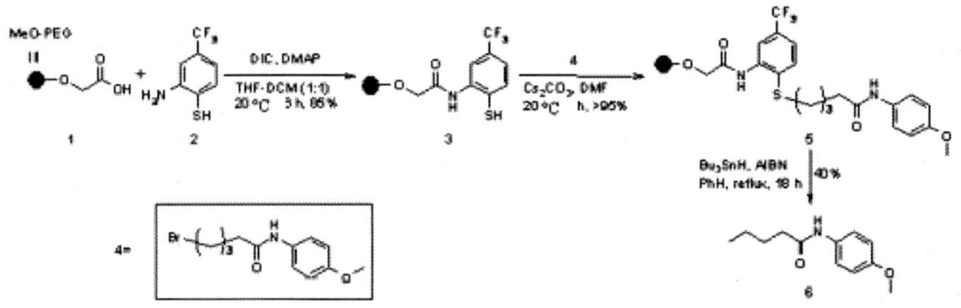

Scheme 1. Janda's sulfur-based linker.

Loading of test compound 4 was performed on a soluble PEG-polymer using the pre-loaded linker 3. Reductive cleavage with C-H bond formation was tried under classical radical conditions; however, hydrogenolysis (Ni-Raney) gave better yield (94%) after shorter reaction time (3 h).

In the following example, Winssinger and coworkers [15] completed the synthesis of a series of derivatives of the natural product aigialomycin D, in order to explore the potential activities of this class of compounds (Scheme 2). Polymer-supported 8 was generated from 7 through functionalization of the benzyl carbon and subsequent RCM. Cleavage from the resin under radical conditions gave reduced products 9

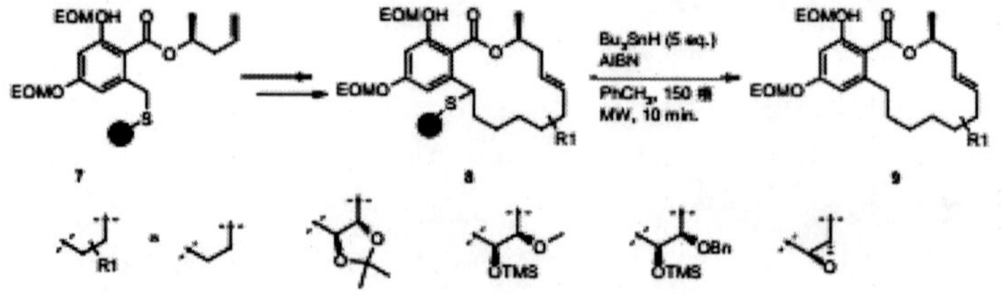

Scheme 2. Winssinger's sulfur-based linker.

Radical desulfonation is a known reaction in solution-phase synthesis [16]. Based on this reaction, Luo and Huang [17] studied the potentiality of sulfonamide moieties as traceless linkers cleaved by radical mechanism in order to generate a library of secondary amides (Scheme 3).

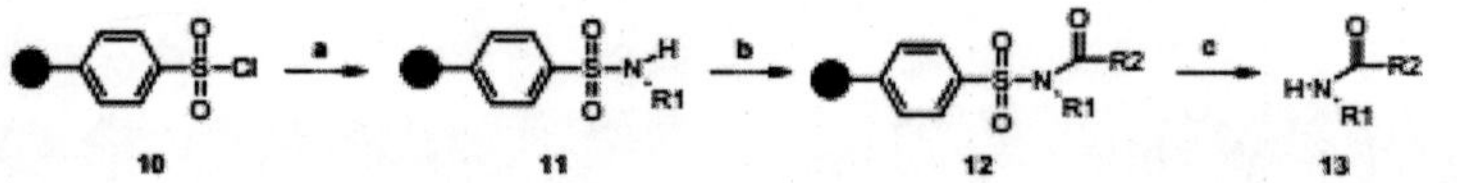

a) R^1NH_2, Py, 80 °C 2 h; b) Py, R^2COCl, r.t., 24 h or i) NaH/Py, ii) $(R^2CO)_2O$; c) i: $TiCl_4/Zn$, THF, refluxing, 24 h, ii: 3% HCl

R^1: *p*-CH_3Ph; Bn; *p*-CH_3Bn; *p*-CH_3OBn; *m*-CH_3OBn; *o*-CH_3OBn; *o*-ClBn; *n*-Bu; EtO_2CCH_2
R^2: Ph; *p*-CH_3Ph; CH_3; *p*-ClPh; *p*-O_2NPh; *p*-AcHNPh

Scheme 3. Huang's sulfur-based linker.

The results of the final cleavage were deeply dependent on the nature of the R1 group. Generally speaking, substituents containing electron-donating groups (such as a m-methoxyphenyl group) were cleaved successfully, while those containing electron-withdrawing moieties, such as a nitrophenyl group, gave no cleavage at all. In the former case, the presence of o,p-methoxyphenyl groups gave by-products from a competitive radical cascade, while, in the latter, reduction and acetylation of the resulting amino group were necessary to perform radical cleavage. The proposed mechanism (Scheme 4) gives an idea about how the nature of R1 can influence the radical fragmentation. In fact, if R1 contains a carbon atom capable to form a stable radical, the direction of the fragmentation can vary from the one illustrated, yielding unwanted side-products.

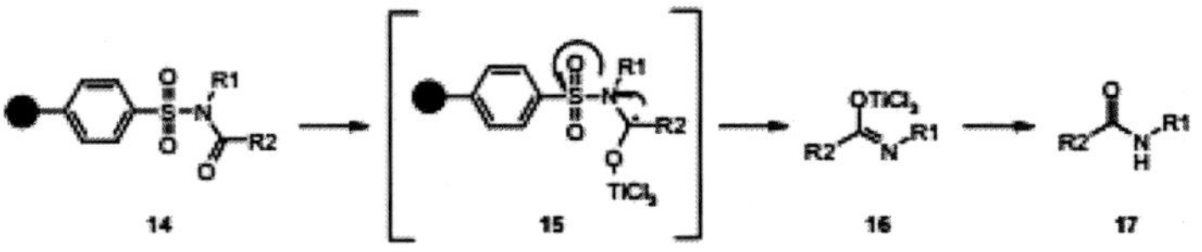

Scheme 4. Huang's proposed mechanism.

D'Herde and De Clercq [18] used the single-electron reductive property of samarium(II) for the cleavage of PS-β-benzolyloxysulfones in a Julia-Lythgoe olefination, where the SP-sulfone partly works as a traceless linker (Scheme 5). The best E:Z ratio (19:20 = 94:6) was obtained when R = Bz and 1,3-dimethyl-3,4,5,6-tetrahydro-2(1H)-pyrimidinone (DMPU) was used as an additive, although isolated yields were rather low (around 25%).

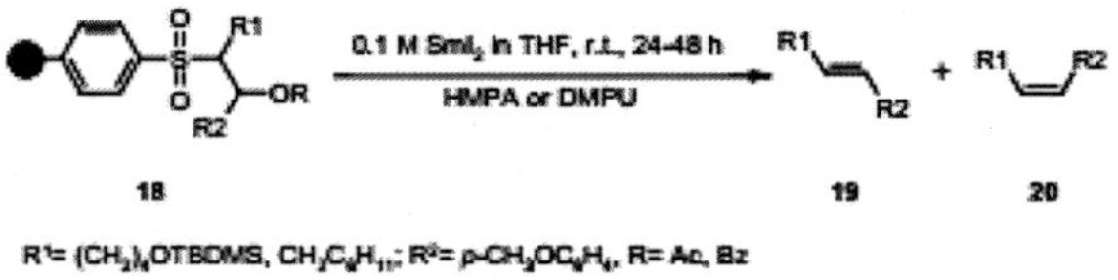

Scheme 5. De Clercq's proposed mechanism.

Radical cyclization to prepare azoles has been explored in solid-phase synthesis using a thiophenyl linker [19], using benzimidazole as a probe heterocycle. To prove the validity of the methodology, the same reactions were also carried out in solution (Scheme 6). Radical precursors 21a-c were loaded on the resins through the acid moiety on the arylthiol.

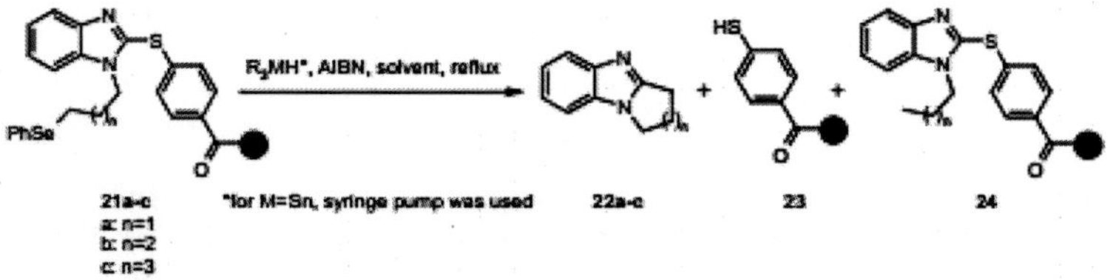

Scheme 6. Bowman's sulfur-based linker.

For n = 1, under the best conditions, only a 11% of desired product 22a was observed, in line with the results obtained in solution, indicating that this 5-exo-trig cyclization is not strongly favored. For n = 2 the reaction was smoother and use of TBTH gave a 60% of isolated 22b. When TBGH was used, a lower yield of 22b was isolated (22%), while TTMSS gave similar results (20%). For n = 3, again poor yields were obtained (4%), just as the reaction run in solution with TBTH. Three resins were used for the scope, however, if Wang and Rink resins gave similar results, amino-Merrifield proved to be unsuitable, as only traces of desired products were found (for n = 1). The rationale behind the use of 2-thioaryl derivative is summarized in the proposed mechanism (Scheme 7).

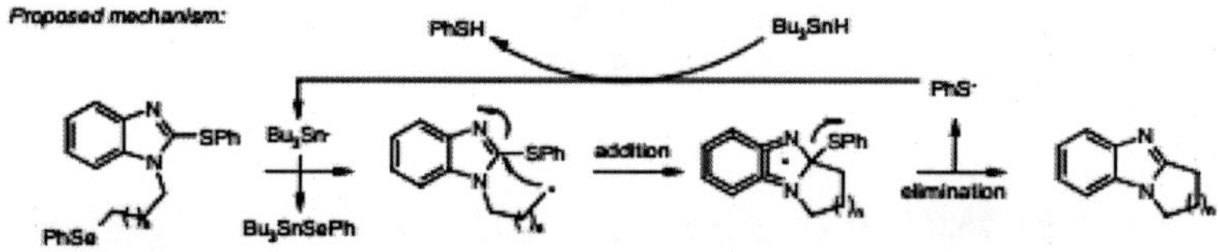

Scheme 7. Bowman's proposed mechanism.

The 2-thiophenyl group behaves as a radical leaving group to achieve rearomatization and, subsequently, as radical carrier to abstract a hydrogen atom from the metal hydride, propagating the chain [20]. Following the mechanism, it is possible to note that the desired cyclized product 22 is released after formation, while side-products, such as non-cyclized, reduced 24 and unreacted starting material are left on the resin.

In order to shorten the reaction times, focused microwave irradiation was also used. As a test case radical precursor 21b (n = 2) on Wang resin was taken. Under optimized conditions (using AMBN) a 52% yield was obtained, very close to the 60% obtained under conventional heating, but with dramatically shortened reaction times (20 min vs. 7 h). The reaction was also run on 21a (n = 1) on amino-Merrifield resin, giving poor yields (3% of 22).

Among the new SP techniques, the α-Hetero-Atom Substituted Carbonyl (HASC) traceless linker studied by Procter, occupies an important role. It exploits the SET characteristics of Sm(II) for reductive cleavage of an X-C bond. The general idea of the methodology starts with the attachment of α-halo-carbonyl compounds to an appropriate resin, followed by further transformation of the resin-linked compounds, ending in the reductive cleavage from the resin (Scheme 8).

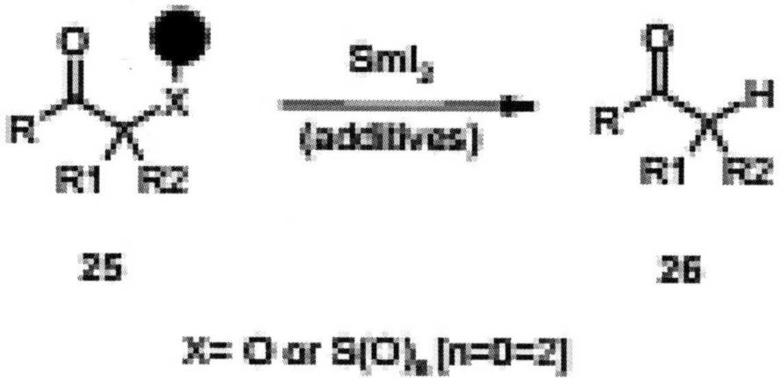

Scheme 8. Procter's HASC strategy.

Heterocycle synthesis was performed following this strategy using a sulfur-linkage [21-23]. Benzylthiol resin 27 was prepared from Merrifield resin, treated with the appropriate α-bromoacetamide and oxidized (Scheme 9). This activated substrate was then submitted to the Pummerer cyclization to yield oxindole 28, which could be cleaved using SmI2, giving the final products 29 in good yield, or alternatively, it could be further functionalized with allyl bromide prior to detachment (Scheme 10).

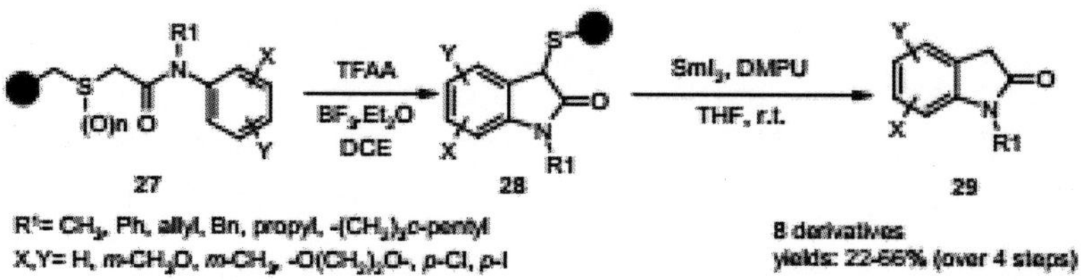

Scheme 9. Procter's use of a sulfur linker for the synthesis of oxindoles.

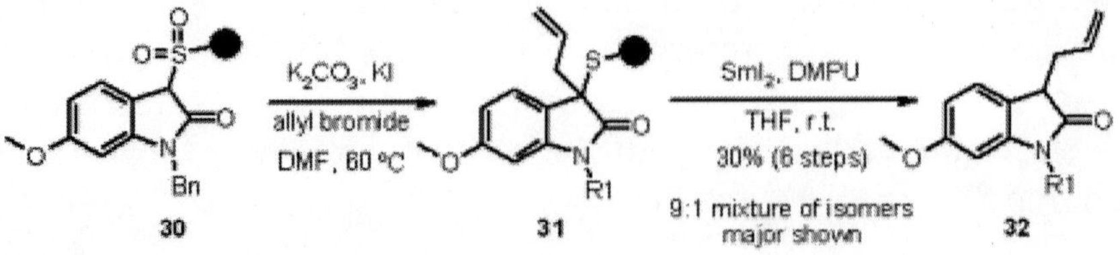

Scheme 10. Procter's synthesis of 3-allyl oxindoles.

Tetrahydroquinolones were prepared in a similar fashion, using a Heck reaction followed by sulfur oxidation and Michael addition to perform cyclization. A couple of examples are shown in Scheme 11.

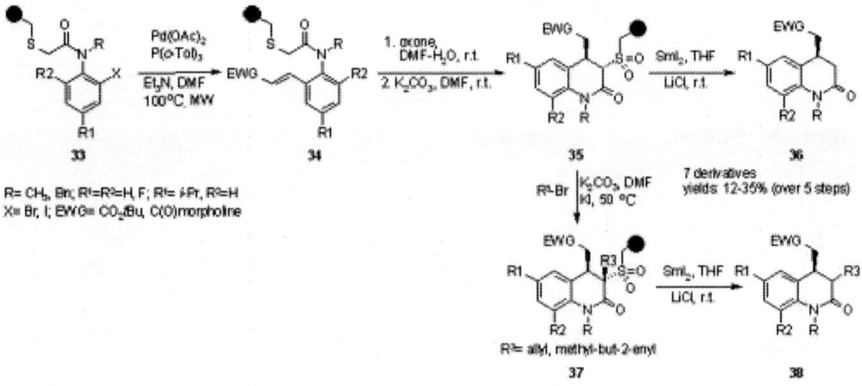

Scheme 11. Procter's use of a sulfur linker for the synthesis of tetrahydroquinolones.

Reductive cleavage from the resin was performed before further functionalization to give 36 or the sulfur-bearing carbon atom was derivatized and only subsequently cleaved off the resin to give 38.

In Scheme 12 an example of samarium iodide-mediated cyclative cleavage to give 40 through a radical or ionic process is shown.

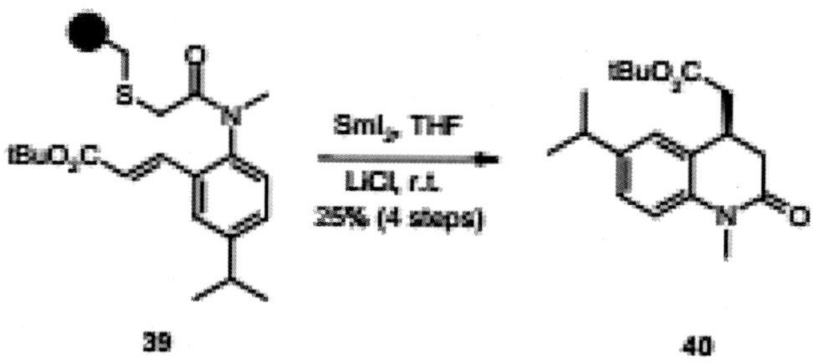

Scheme 12. Example of SmI2-mediated cyclative cleavage.

The general mechanism for samarium-induced reductive cleavage of α-heteroatom carbonyl compounds is displayed in the scheme below (Scheme 13).

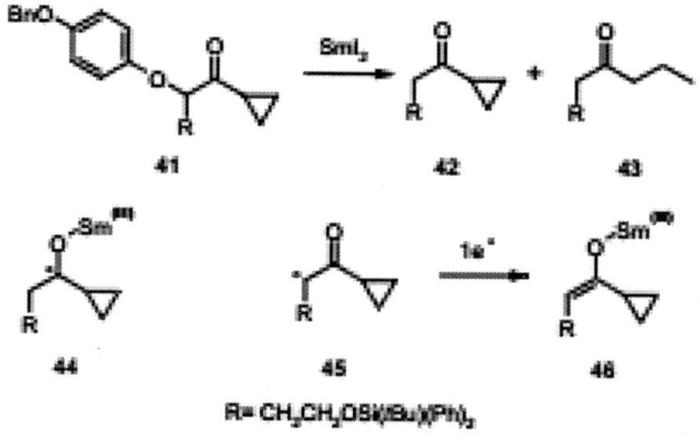

Scheme 13. Procter's proposed mechanism.

The cleavage mechanism was studied in solution for cyclopropyl derivative 41. The samarium SET reductive cleavage would account for a radical type 44, leading to cyclopropyl ring opening (compound 43) via a cyclopropylmethyl radical. However, since there was no trace of ring-opened derivative 43, the alternative and more probable explanation accounts for different intermediates such as 45-46. The presence of a samarium enolate of the kind of 46 was demonstrated by the capturing of electrophiles during cleavage.

Selenium (and Tellurium)-Based Linkers

A number of reports deal with the use of selenium linkers, which are known to be amenable to several cleavage strategies, and the topic has been reviewed recently [24]. The following section is simply meant to highlight the value of such types of linkers in contemporary SPS.

Ruhland and coworkers produced interesting works replacing sulfur with selenium, which proved to be a versatile traceless linker in both aliphatic C-H bond-forming reactions, taking advantage of easier homolytic cleavage, and alkene formation by selenium oxidation followed by β-elimination. In a first paper [25] bromopolystyrene 47 was obtained through thallium acetate-catalyzed bromination of commercially available polystyrene.

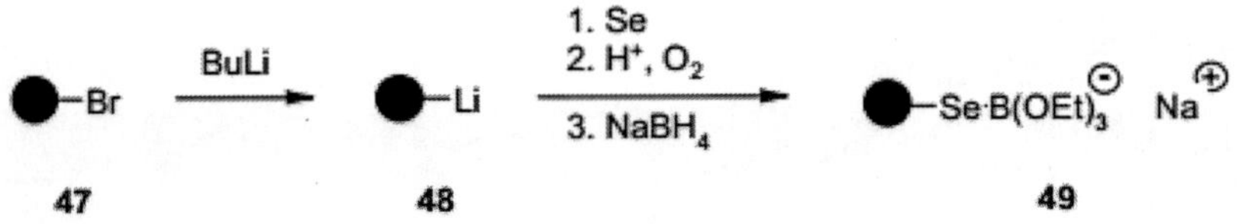

●–Br (47) $\xrightarrow{\text{BuLi}}$ ●–Li (48) $\xrightarrow[\text{3. NaBH}_4]{\text{1. Se; 2. H}^+\text{, O}_2}$ ●–Se·B(OEt)$_3^{\ominus}$ Na$^{\oplus}$ (49)

Scheme 14. Ruhland's synthesis of selenium-based resin.

After lithiation with BuLi, the resin was treated with selenium, in order to replace the metal anion. Reduction with NaBH4 was necessary because of the occurrence, during work-up in the presence of air (step 2) of Se-Se bridges. Reduction gave back the swelling properties to the resin, lost because of the high degree of cross-linking encountered in step 2. Loading of resin 49 resulted lower compared to the bromopolystyrene 47, probably due to incomplete reaction with selenium or partial availability of the metal in the resin itself. To prove the utility of resin 49, it was carried out the synthesis of an alkyl-aryl ether small library using the Mitsunobu reaction (Scheme 15).

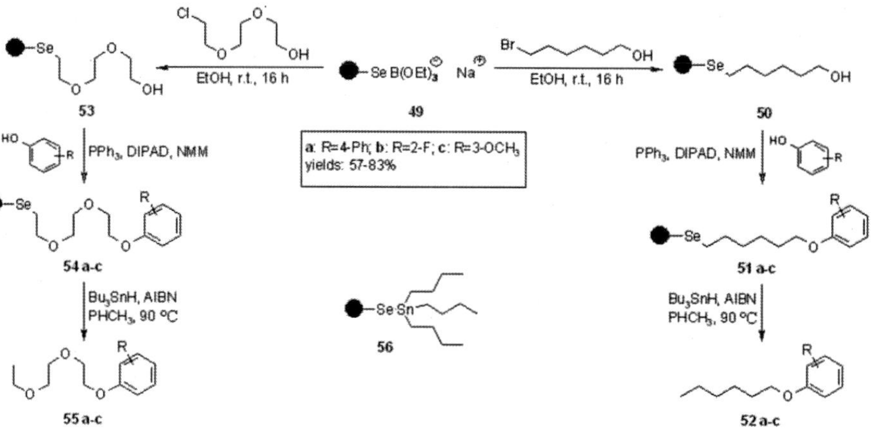

Scheme 15. Ruhland's synthesis of ethers library.

After application of the Mitsunobu protocol, cleavage was performed under radical conditions. Contrary to the reaction carried out in solution, the tributylstannyl phenyl selenide 56, often a problem for efficient purification, remained anchored on the SP and the only by-product of the cleavage was the excess Bu3SnH, which was efficiently separated by SP extraction.

Polymer-bound tellurium was next introduced by the same group [26], which was prepared exploiting a strategy similar to the above. Also in this case a small library based on the Mitsunobu reaction was produced, following tributyltin-mediated product detachment. Interestingly and contrary to expectations, it was found that the reactivity of the polystyrene-bound tellurium towards homolysis is lower in comparison to the polystyrene-bound selenium. This is possibly due to residual elemental tellurium (remaining within the resin) interfering with the radical chain process in the cleavage step.

The selectivity of bond cleavage offers an important general advantage when utilizing this SP methodology. However, contamination of the cleaved products by tin derivatives can hold back its use, especially for biologically interesting libraries. To address this issue the same group showed later that it is possible to use greener traceless cleavage, exploiting silanes or germanes as valid alternatives to tin-containing reagents. In fact, it was demonstrated [27] that among others especially tris (trimethylsilyl) silane, performs similarly to tributyltin hydride in homolytically cleaving alkyl-selenium derivatives. On the other hand, resin-bound tellurium appears to have a preference for tris(trimethylsilyl)germane (see Figure 16). In general, however, both yield and purity are superior for the polystyrene-bound selenium, though tellurium is somewhat less sensitive to the number of cleavage reagent equivalents.

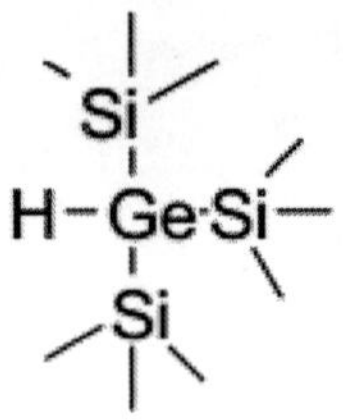

Figure 16. Tris(trimethylsilyl)germane.

A chiral selenium electrophile has been also developed and applied to the stereoselective selenenylation of alkenes. The linker has been generated on solid-phase by loading a methoxymethyl-protected selenophenol to a suitable polymeric support, and subsequently generating the electrophile by reaction with bromine at low temperature (Scheme 17). Indeed, mesoporous silica proved superior to other supports, although its loading is usually low. Polystyrene works better than TentaGel and it was predominantly used for the stereoselective additions. Selenenylation of alkenes in the presence of a suitable nucleophile, usually an alcohol, yielded up to 80% ee, making the polymer-supported reagent almost equally attractive as its soluble counterpart (Scheme 18).

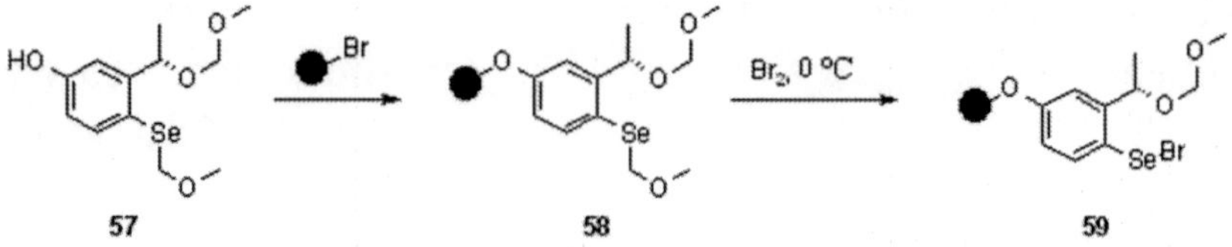

Scheme 17. Synthesis of asymmetric selenyl linker.

Different cleavage protocols were shown in the paper, including the classical oxidation followed by β-elimination, tributyltin hydride-mediated homolytical carbon-selenium bond cleavage and also the use of allyltributyltin, which led to the introduction of an allyl moiety in the product (Scheme 18).

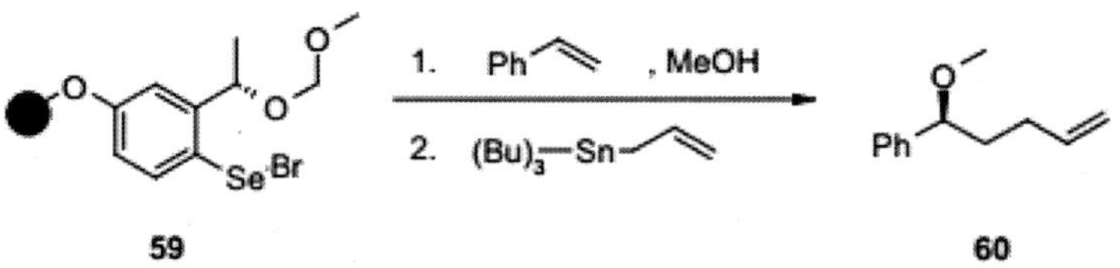

1. Ph, MeOH
2. $(Bu)_3$—Sn
59
60

Scheme 18. Application of asymmetric selenyl linker.

Supported selenyl bromide was prepared also by Nicolaou and coworkers [28] and showed its utility, beside that of a series of other selenium-containing resins, for the use in combinatorial chemistry as linkers (Scheme 19).

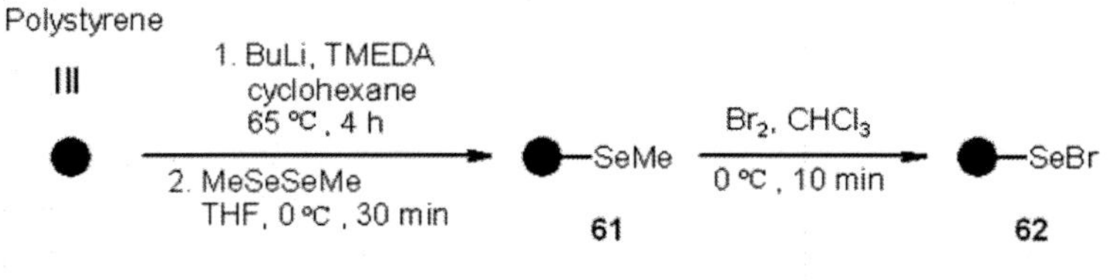

Polystyrene
1. BuLi, TMEDA cyclohexane 65 °C, 4 h
2. MeSeSeMe THF, 0 °C, 30 min
SeMe
61
Br_2, $CHCl_3$
0 °C, 10 min
SeBr
62

Scheme 19. Nicolaou's preparation of PS-selenyl bromide.

Compound 62 was exploited to load alkene 63, and subsequently cleaved under radical conditions (Scheme 20).

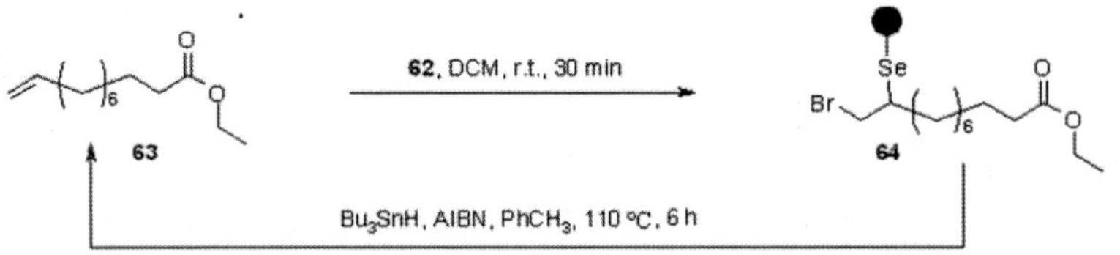

63
62, DCM, r.t., 30 min
64
Bu_3SnH, AIBN, $PhCH_3$, 110 °C, 6 h

Scheme 20. Nicolaou's SP alkene selenylation.

Compound 62 can also be reduced to resin 65 (Scheme 21), which can be reacted with halides to load the corresponding alkane. Standard cleavage under radical conditions gave reduction product 69 (Scheme 21).

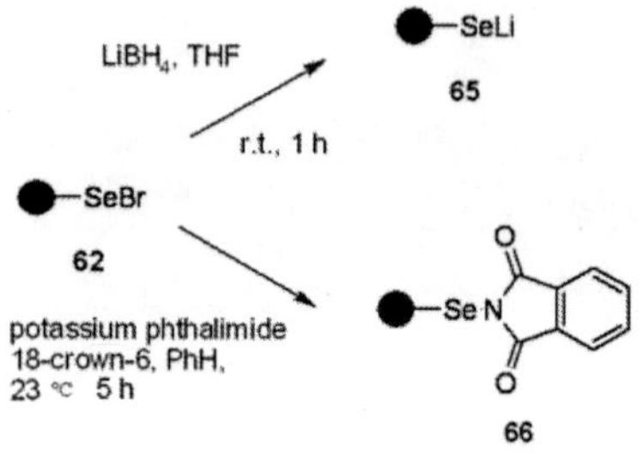

Scheme 22. Nicolaou's use of resin **65**.

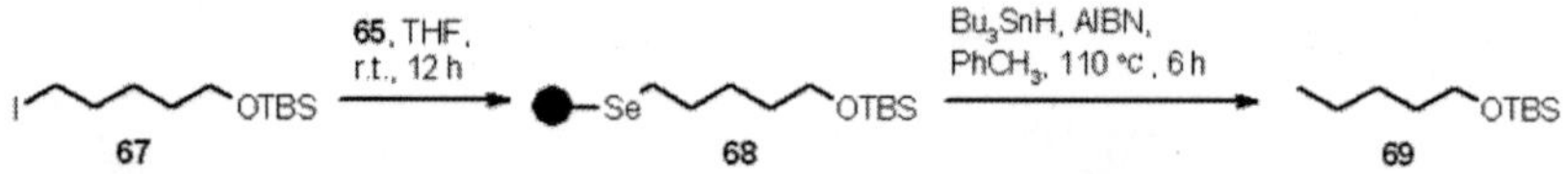

Scheme 21. Nicolaou's modification of PS-selenyl bromide.

In addition, the same resin 62 can give the phtalimido-derivatives 66, used for insertion of benzyl alcohol (product 71) onto the alkene moiety (70), after radical cleavage (Scheme 23).

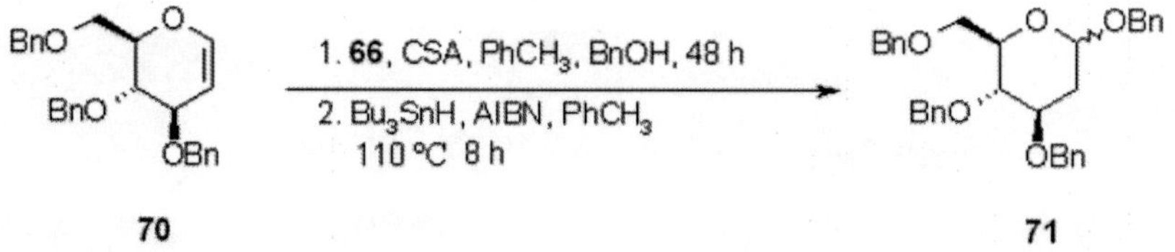

Scheme 23. Nicolaou's use of resin 66

In another interesting work, the same group [29,30] used selenium-based linkers, followed by reductive radical cleavage from the resin, for the synthesis of a small library of 2-deoxy glycosides (Scheme 24).

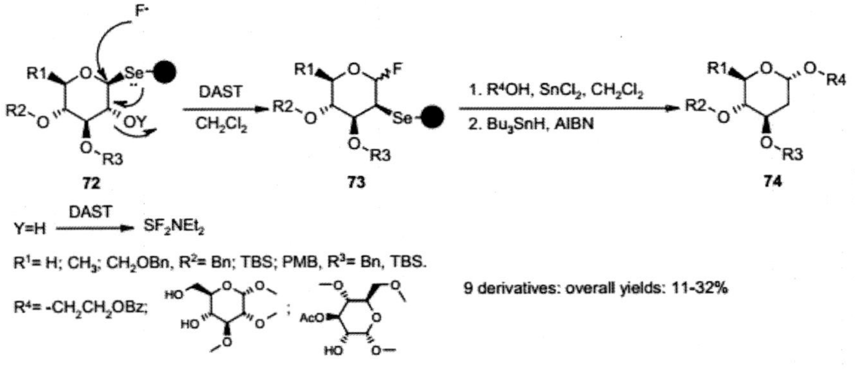

Scheme 24. Nicolaou's synthesis of 2-deoxy glycoside library.

A small library of 3 × 3 was generated (3 "donors"-the sugars described by R[1], R[2] and R[3]- and 3 "acceptors", described by R[4]), with overall yields between 11-32%. The key reaction was the 1,2-seleno migration promoted by diethylaminosulfur trifluoride (DAST), while radical cleavage was typical of selenium-based linkers. Selenium-functionalized resins were exploited by Nicolaou's group [31,32] also in the studies towards the synthesis of heterocycles, namely indolines, indoline-based policyclic systems and indoles, which were obtained *via* different mechanisms of radical cleavage (Scheme 25).

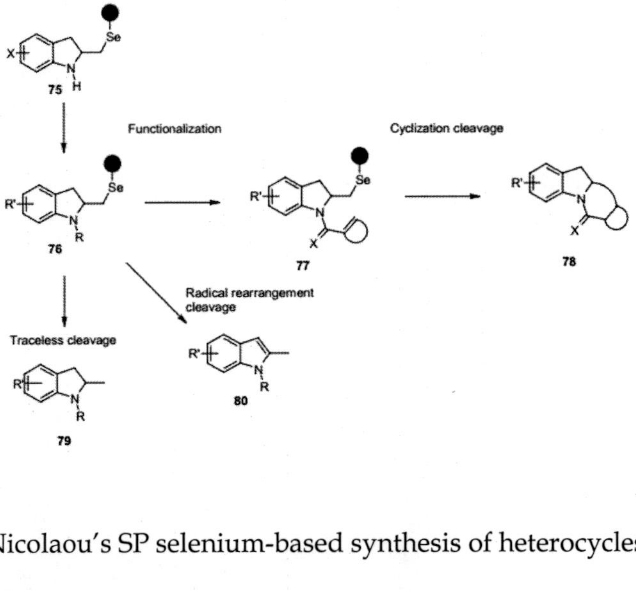

Scheme 25. Nicolaou's SP selenium-based synthesis of heterocycles.

Starting from solid-supported selenyl bromide, indolines 75 were prepared by Lewis-acid mediated cycloloaddition using the appropriate o-allyl anilines. Subsequent functionalization at the nitrogen atom and/or at the substituent on the benzene ring allowed the introduction of additional diversity elements.

Use of different cleavage strategies enabled then the preparation of small libraries of drug-like molecules. This represents an interesting extension of this methodology to combinatorial chemistry for the synthesis of bioactive compounds. In Scheme 26 a classical traceless cleavage is shown. The mechanism entails the homolytic cleavage of the Se-C bond, therefore generating a carbon-centred radical, quenched by the hydrogen-donor Bu3SnH.

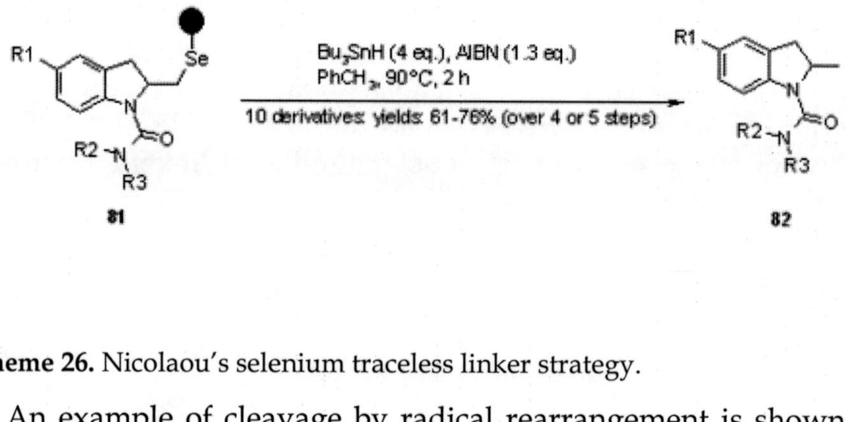

Scheme 26. Nicolaou's selenium traceless linker strategy.

An example of cleavage by radical rearrangement is shown in Scheme 28. In this case, the lack of H-donor makes the C-centred radical enough long-living to perform a shift. Although the nature of the shift is unknown, any of the supposed intermediates radical 88 and 89 can undergo oxidation to the indole ring by loss of a hydrogen atom.

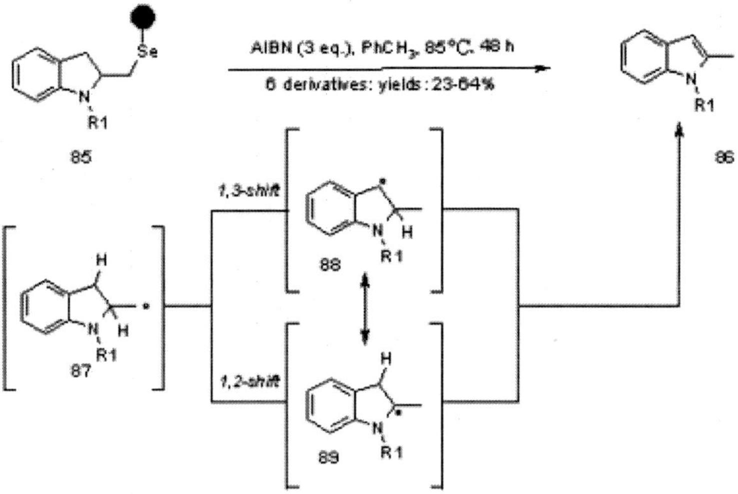

Scheme 28. Nicolaou's C-centred radical shift.

Very interesting are also other examples from the same group [33], which used the same cycloloadition/radical cleavage methodology to construct [3.3.1]bicycles from properly designed substrates (Schemes 29 and 30).

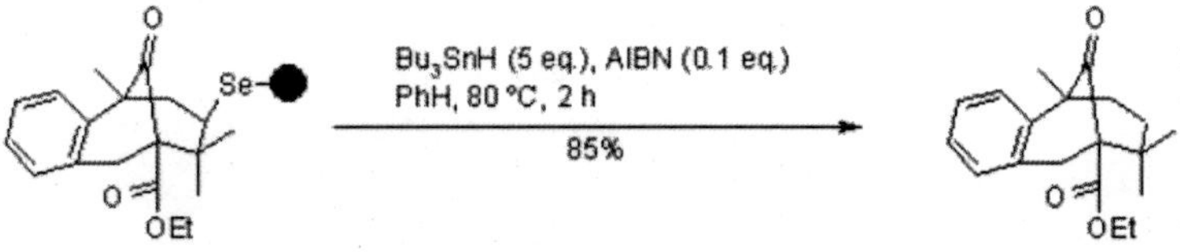

90 **91**

Scheme 30. Nicolaou's synthesis of bicycle **93**.

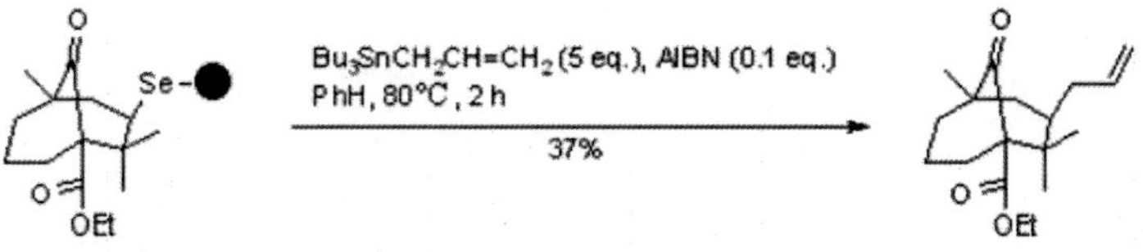

92 **93**

Scheme 29. Nicolaou's synthesis of bicycle 91.

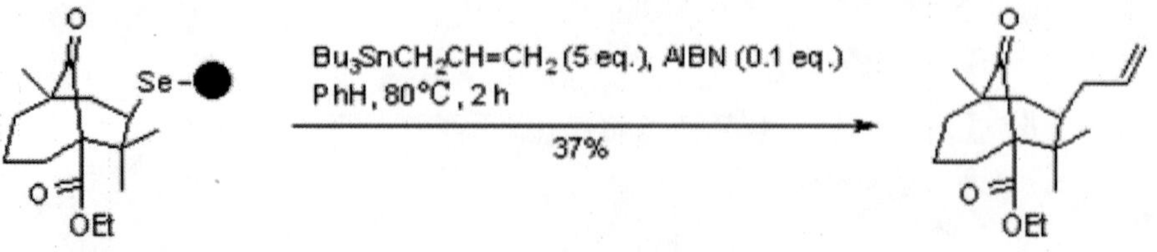

Scheme 30. Nicolaou's synthesis of bicycle 93.

As before, bicycles 90 (Scheme 29) and 92 (Scheme 30) were prepared by electrophilic cyclization with a solid-supported-SeBr resin, while cleavage was performed in a traceless manner (Scheme 29) or by trapping the resulting C-centred radical with allystannane (Scheme 30).

Although the methodologies shown above proved efficient for the synthesis of small compound libraries, it is important to highlight a major drawback that is implicit in the cleavage mechanism. In fact it was necessary to separate the desired products from the residues of the radical reaction, an option often tedious and difficult that biases the use of toxic tin derivatives for in-solution radical chemistry.

Another interesting cleavage methodology was reported by Engman and coworkers [34], who studied the radical carbonylation/ cyclization to make tetrahydrofuran-3-ones. The method was first developed in solution, exploiting alkyl phenyl selenides and then applied to a selenium-based resin 94, prepared in a few steps from cross-linked (1%) polystyrene (Scheme 31).

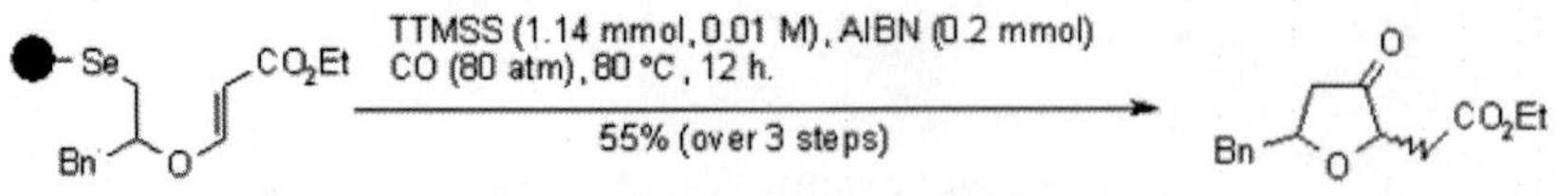

Scheme 31. Engman's use of a selenium linker.

TTMSS was chosen instead of the more common TBTH, which gave a large amount of reduced (non-cyclized) product, whereas, increasing the concentration of the H-donor resulted in poorer yields of the desired vs. reduction product. TBGH) gave worse results compared to TTMSS. Lower carbon monoxide pressure gave also more product of reduction. The final product 95 was isolated as a 9:1 mixture of cis/trans isomers (55% yield calculated over 3 steps, 2 of them for the synthesis of 94).

Fujita et al. [35] used PS-arylselenides for the selenolactonization on 3-butenoic acids with successive cleavage via radical reduction (Scheme 32). The yield of the process to give 98 was rather low (20%), a result explained by Ph3SnH-promoted side-reactions with the amide linker.

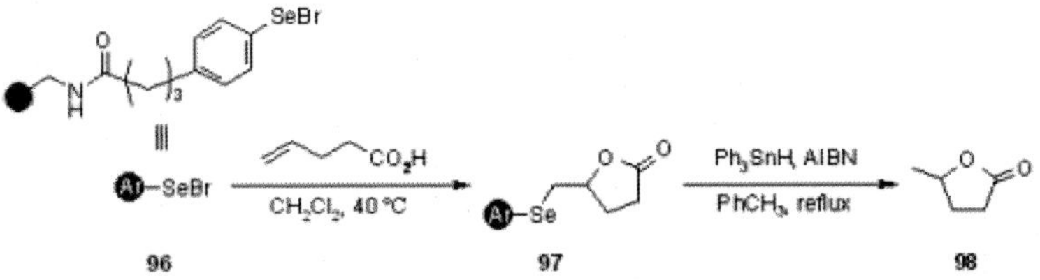

SeBr
O
N
H
3
Ar—SeBr
CO_2H
CH_2Cl_2, 40 °C
Ar—Se
O
O
Ph_3SnH, AIBN
$PhCH_3$, reflux
O
O
96
97
98

Scheme 32. Fujita's SP selenolactonization of 3-butenoic acids

Polystyrene-supported selenosulfonates can be conveniently prepared from the easily accessible selenobromide by reaction with arylsulfinates. Qian and Huang [36] studied the radical addition of such PS-selenosulfonates to alkenes, with subsequent oxidation-elimination of selenium to obtain vinyl sulfones (Scheme 33).

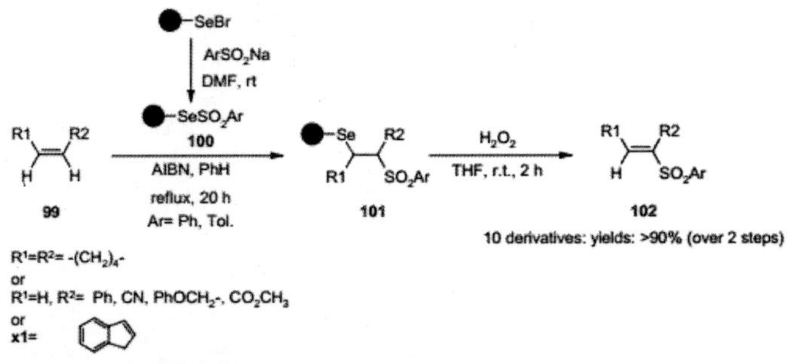

—SeBr
$ArSO_2Na$
DMF, rt
—$SeSO_2Ar$
100
R1 R2
H H
99
AIBN, PhH
reflux, 20 h
Ar= Ph, Tol.
—Se R2
R1 SO_2Ar
101
H_2O_2
THF, r.t., 2 h
R1 R2
H SO_2Ar
102
10 derivatives: yields: >90% (over 2 steps)
R^1=R^2= -$(CH_2)_4$-
or
R^1=H, R^2= Ph, CN, $PhOCH_2$-, CO_2CH_3
or
x1=

Scheme 33. Huang's synthesis of selenosulfonates.

Radical addition was performed under classical radical conditions and the overall yields after cleavage were rather high. The same group also performed the radical addition of PS-arylselenosulfonates to alkynes, followed by oxidation to selenoxide and syn-elimination to give acetylenic sulfones 105 [37] or by acidic hydrolytic cleavage from the SP, for the preparation of β-keto sulfones 106 [38] (Scheme 34).

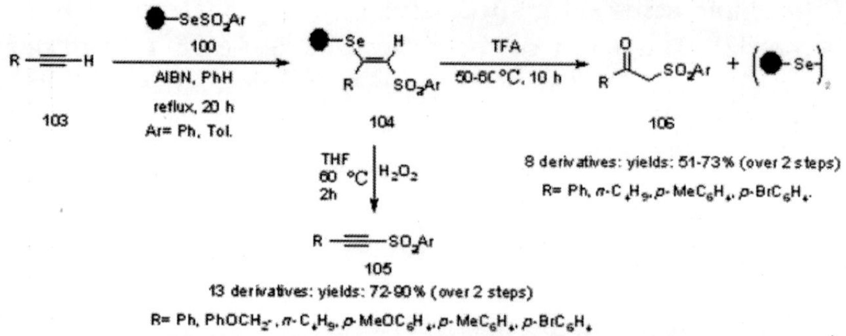

8 derivatives: yields: 51-73% (over 2 steps)
R= Ph, n-C_4H_9, p-MeC_6H_4, p-BrC_6H_4.

13 derivatives: yields: 72-90% (over 2 steps)
R= Ph, $PhOCH_2$-, n-C_4H_9, p-$MeOC_6H_4$, p-MeC_6H_4, p-BrC_6H_4

Scheme 34. Huang's synthesis of acetylenic sulfones and β-keto sulfones.

Cleavage under oxidation-elimination conditions (to give 105) gave higher yields compared to the acidic cleavage (to give 106), as shown. In any case, there was practically no need for further purification after cleavage from the resin and the PS-diselenide obtained after acidic cleavage could be regenerated.

In another interesting work, a solid-supported selenosulfone as radical transfer agent [39]. In this case however, the reaction led to a cyclization product that, therefore, remained anchored to the SP. Only successively this was cleaved form the resin by the classical oxidation-elimination protocol (Scheme 35).

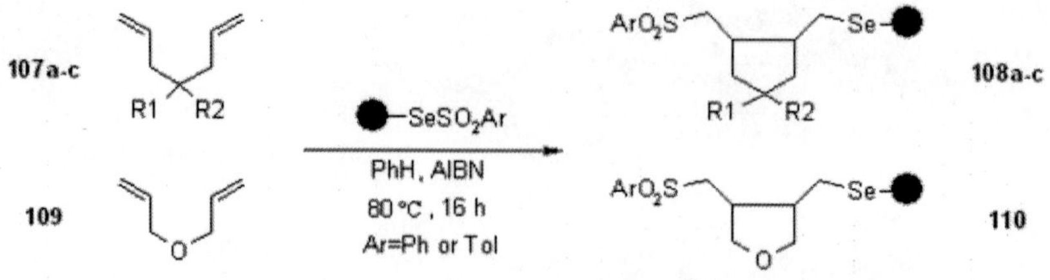

Scheme 35. Huang's SP-atom transfer cyclization.

The isolated yields were calculated after cleavage (44-58%, 12 examples), using two different selenylarylsulfonyl resins.

Oxygen-based Linkers

Samarium (II) iodide is a useful reagent, exploited for several scopes, having the characteristic of being a powerful single-electron reducing agent. All the examples below, where fission of either a carbon-oxygen or a nitrogen-oxygen bond has been exploited to release organic molecules from a solid support, make use of samarium iodide.

Cleavage of Hydroxylamine-based Linkers

Abell and colleagues [40] used SmI2 for the traceless cleavage of N-O bond on substrates appropriately linked to SP. With this methodology it was generated a small library of amides and ureas (Scheme 36).

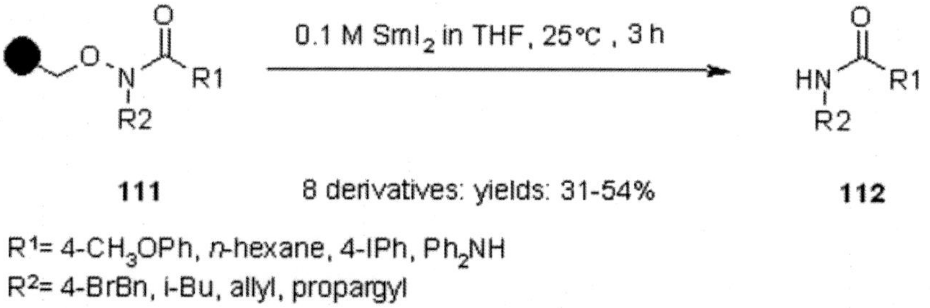

0.1 M SmI_2 in THF, 25°C , 3 h

111 8 derivatives: yields: 31-54% **112**

R1= 4-CH_3OPh, *n*-hexane, 4-IPh, Ph_2NH
R2= 4-BrBn, i-Bu, allyl, propargyl

Scheme 36. Abell's amides and ureas library.

Although yields were modest, the purities of the cleaved products 112 were rather high. The same cleavage technique was exploited by Andersson [41] that started from a polystyrene-based resin to synthesize tertiary amines (Scheme 37).

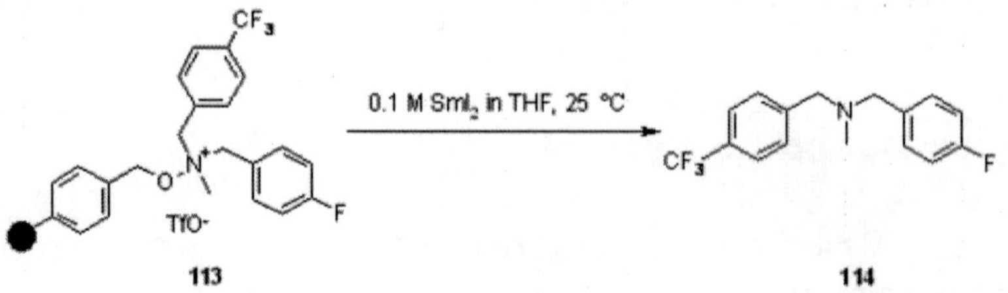

Scheme 37. Andersson's reductive cleavage.

Although SmI2 gave clean cleaved products, the cleavage method used for the library production exploited the alternative LiI, which worked under neutral conditions.

Another example of the reductive cleavage promoted by SmI2 was shown by Meloni and Taddei [42], who applied it to the synthesis of β-lactams (Scheme 38). Polystryerene-bound hydroxylamine was once again used as the starting material.

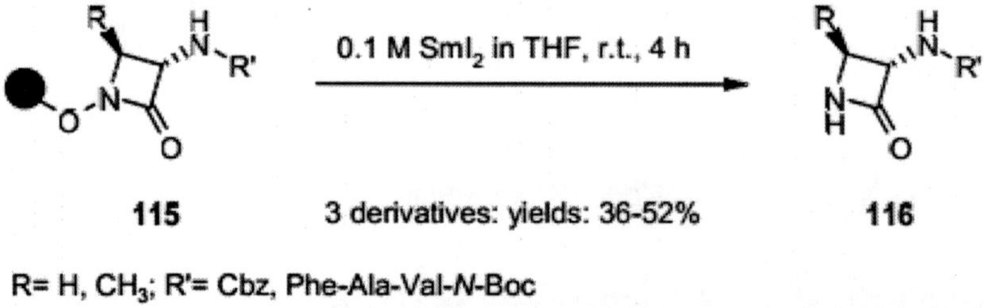

Scheme 38. Taddei's synthesis of β-lactams.

Addition of Ketyl Radicals to α-β Unsaturated Esters

Procter's group [43,44] used a Sm(II) based radical chemistry for the enantioselective synthesis of γ-butyrolactones on SP exploiting (1R, 2S)-ephedrine linker as the chiral auxiliary. Bromo-Wang resin was first functionalized with (1R, 2S)-ephedrine and successively esterified at the alcohol group with the appropriate α,β-unsaturated acyl chloride to give 117. This was, in turn, used in a reaction involving a ketyl radical addition to the anchored α,β-unsaturated ester. The subsequent cyclative cleavage led to the formation of lactone 119 (Scheme 39).

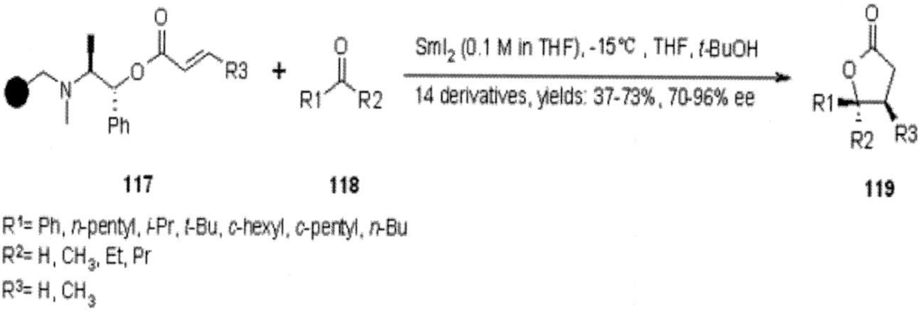

SmI_2 (0.1 M in THF), -15°C, THF, t-BuOH
14 derivatives, yields: 37-73%, 70-96% ee

117 118 119

R^1= Ph, n-pentyl, i-Pr, t-Bu, c-hexyl, c-pentyl, n-Bu
R^2= H, CH_3, Et, Pr
R^3= H, CH_3

Scheme 39. Procter's asymmetric lactones synthesis.

Under optimized conditions, the reaction gave rather high isolated yields (apart for a few exceptions) and diasteroselectivities. The same conditions were applied to the synthesis of lactone 121 (Scheme 40), a moderate DNA-binding metabolite isolated from Streptomyces GT61115.

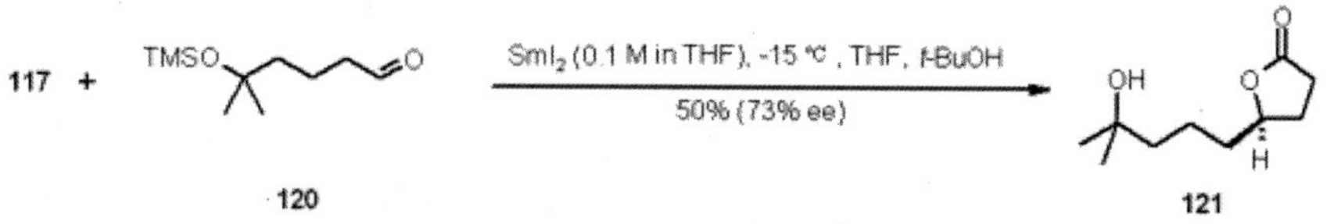

117 + 120

SmI_2 (0.1 M in THF), -15 °C, THF, t-BuOH
50% (73% ee)

121

Scheme 40. Synthesis of lactone 121.

Cyclative cleavage gives back, beside the desired product, the resin functionalized with the asymmetric linker, which can be reused. Following Procter's lead on asymmetric induction for samarium-mediated radical synthesis of γ-butyrolactones, Dai and coworkers [45] linked axially chiral crotonates on Rink amide resin for the same purpose (Scheme 41).

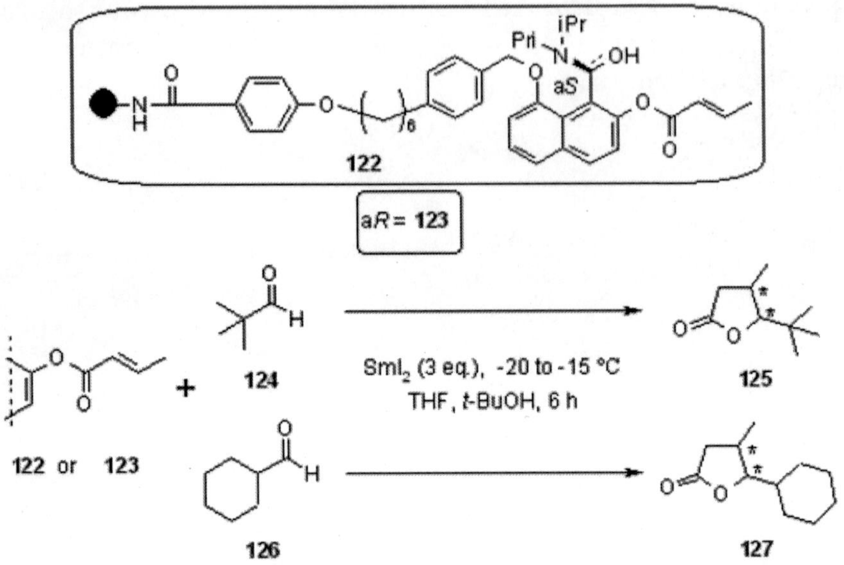

122

aR = 123

122 or 123 + 124 → 125

122 or 123 + 126 → 127

SmI_2 (3 eq.), -20 to -15 °C
THF, t-BuOH, 6 h

Scheme 41. Dai's chiral crotonates on Rink amide resin.

Reaction of 122 or 123 with aldehydes 124 and 126 gave reasonable isolated yields (125 = 58%, 127 = 66%), with a good cis/trans selectivity (125 = 71: 29; 127 = 89: 11) and good ee (125-cis: 94%, 125-trans: 89%; 127-cis = 127-trans: 88%). The mechanism of chiral transfer was supposed to pass through an intermediate were the samarium atom coordinates the oxygens from 122 (or 123) and the oxygen of the ketyl radical which performs the 1,4-addition to the crotonates.

HASC Traceless Linker

The HASC traceless linker developed by Procter and mentioned above (see Scheme 8) has been exploited also using oxygen-based linkers. Applications of the ether linkage involve the synthesis of functionalized carbonyl compounds [46,47] (Scheme 42). Resin 128 was synthesized starting from bromo Wang resin, derivatized with the appropriate phenol and functionalised with α-bromo-γ-butyrolactone. Lactone ring-opening was performed with some amines (pyrrolidine, morpholine, tetrahydroisoquinoline and methoxymethylamine) and the resulting alcohol protected accordingly.

When Weinreib hydroxylamine was used, the obtained tertiary amide was successively reacted with a Grignard reagent (BnMgCl, i-PrMgCl, c-PrMgCl) to yield the corresponding ketone. Cleavage of 129 from the resin gave products 130 in good yields after 4-5 steps.

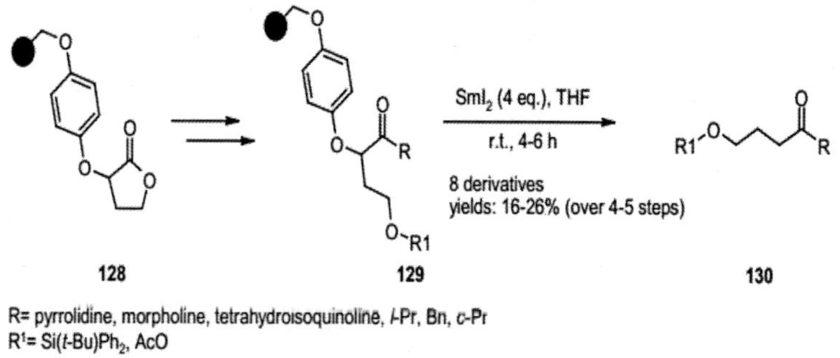

SmI_2 (4 eq.), THF
r.t., 4-6 h
8 derivatives
yields: 16-26% (over 4-5 steps)

128 129 130

R= pyrrolidine, morpholine, tetrahydroisoquinoline, *i*-Pr, Bn, *c*-Pr
R^1= Si(*t*-Bu)Ph_2, AcO

Scheme 42. Procter's application of HASC strategy.

"Carbon–Based" Linkers

Floreancig's group [48,49] used an oxidative release (often cyclorelease) on the appropriate polymer-supported precursors. The mechanism involved in the oxidative cleavage is presented in Scheme 43.

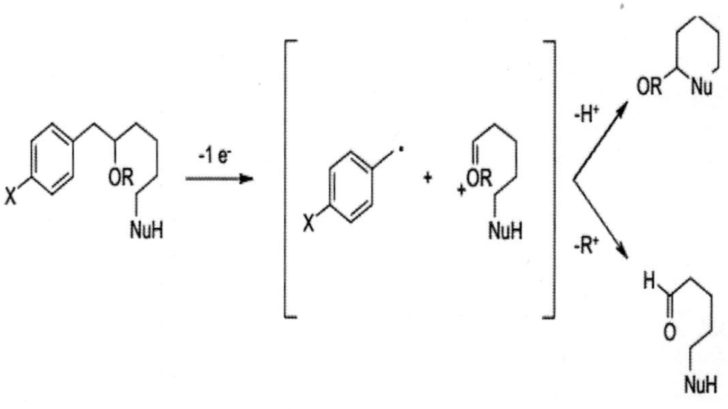

Scheme 43. Floreancig's proposed mechanism.

The process, (Electron Transfer Initiated Cyclization, ETIC) starts with single electron oxidation and mesolytic cleavage of the C-C bond of the homobenzylic ether, forming a benzylic radical and an oxocarbenium ion. The resulting cation is successively transformed in a neutral species, giving a cyclised product if an appropriate nucleophile is present in the molecule, or an aldehyde (or ketone) if there is loss of the oxygen-bound "R" group (when present). To test this methodology on support, appropriate soluble ROMP polymers were prepared (Scheme 44).

The five equations prove the feasibility of the method, either by intramolecular trapping (equations 1-3), or by oxidative traceless release (equations 4-5). In the former case either formation of a C-O bond (equation 1) or a C-C bond (equation 2 and 3), which occurred with high diasteroselectivity, is shown. In the latter case formation of the aldehyde 138 of equation 4 requires loss of a THP group.

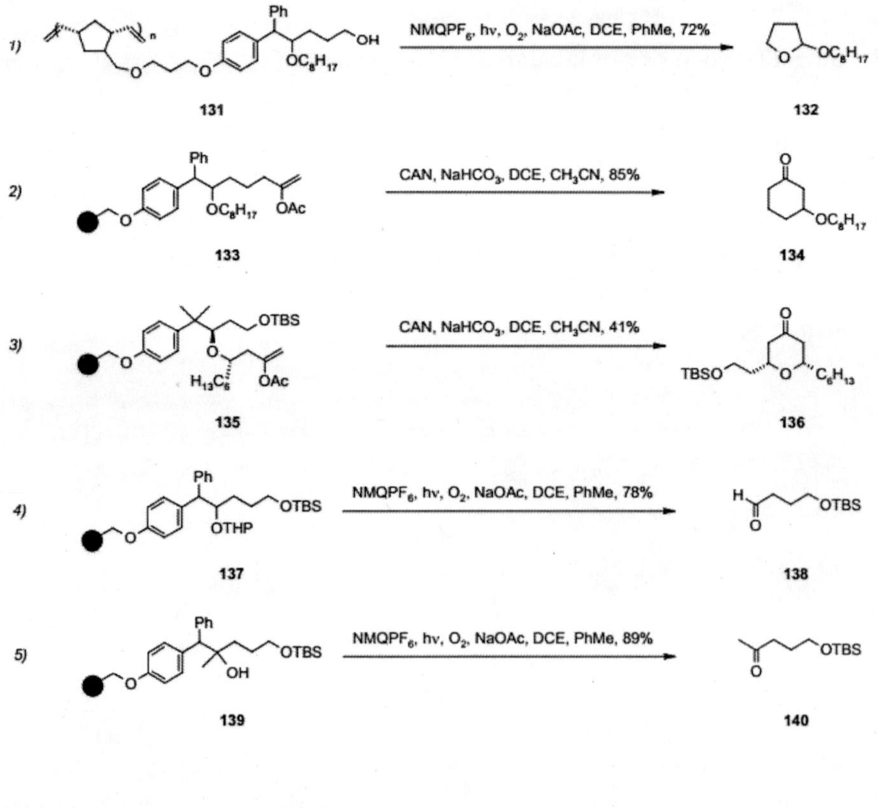

Scheme 44. Application of lROMP polymers.

The purity of the products after this particular cleavage was rather high, due to the characteristic mechanism of the reaction, which ensured cyclocleavage or oxidative release only when the appropriate nucleophile was correctly installed in the molecule. This means that any impurity will be retained on the support, in a kind of purification step. Therefore, it is possible to perform several steps on the support, without the fear of accumulating side-products that have to be separated at the end of the process.

An interesting application of thiohydroxamate (THA) linker strategy was proposed by Routledge [50]. In previous reports [51,52] the same group presented a traceless THA linker, cleavable under photolytic conditions, but in the need of novel SP-linkers, a second-generation THA linker, stable under diverse chemical and photolytic conditions, was studied (Scheme 45).

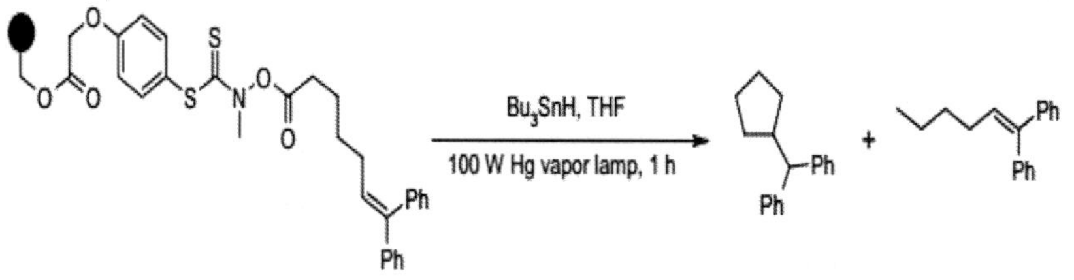

Bu_3SnH, THF

100 W Hg vapor lamp, 1 h

141

142 (37%; 2 steps)

143

Scheme 45. Routledge's THA linkeras polymer-supported radical source.

Although 141 proved to be stable to heat and light, it was sensitive to nucleophilic cleavage, which might make it of limited application for SP preparation of libraries. On the other hand, 141 was found useful as polymer-supported radical source. The nature of the linker allows the generation of a carbon-centred radical ultimately arising from a carbon-carbon bond fission, that can be appropriately exploited for successive reaction such as the one depicted above. The only product found was 142, even if isolated in moderate yield.

Solid-Supported Reagents

Another important aspect of radicals exploiting SP deals with the use of supported reagents for radical reactions. This combination allows toxic reagents, often difficult to separate from the desired products in solution, to be used with ease of purification and little contamination. Since the most important reagents in radical chemistry are represented by organostannanes, it was obvious that the initial efforts to translate in-solution radical chemistry to SP, would have involved tin-base reagents.

From this point of view, several works have tried to bind organostannanes on solid supports, in order to reproduce common radical reactions and make work-up easier with little or no contamination from the toxic metal. Within this class, organostannylhydrides occupy a dominant role. Neumann's group successfully immobilized dibutyltinhydride on SP with a two-carbon spacer between the metal and the aromatic ring of polystyrene to obtain resin 145, whose applications were highlighted in successive publications [53,54]. When employed for radical cyclization (Scheme 46), the results obtained were comparable with the same reaction run in solution with Bu3SnH (36%:0%:64%).

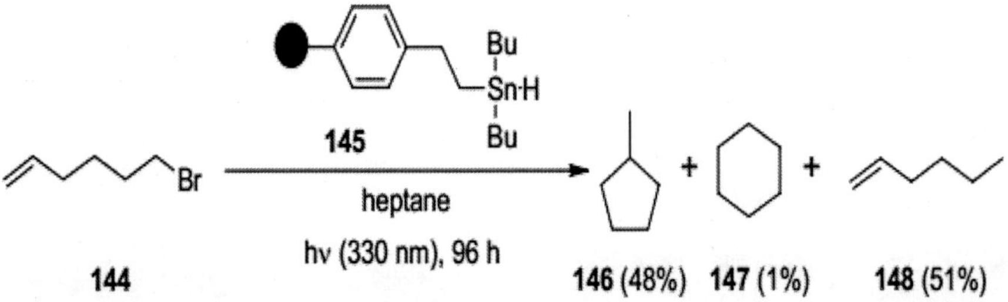

Scheme 46. Use of resin 145 for radical cyclization.

An application to the Barton dehydroxylation reaction [55], which has been exploited extensively later by the same group in a series of derivatized alcohols [56], is shown in Scheme 47. An example of deamination following preparation of the correspondent isocyanides is instead shown in Scheme 48.

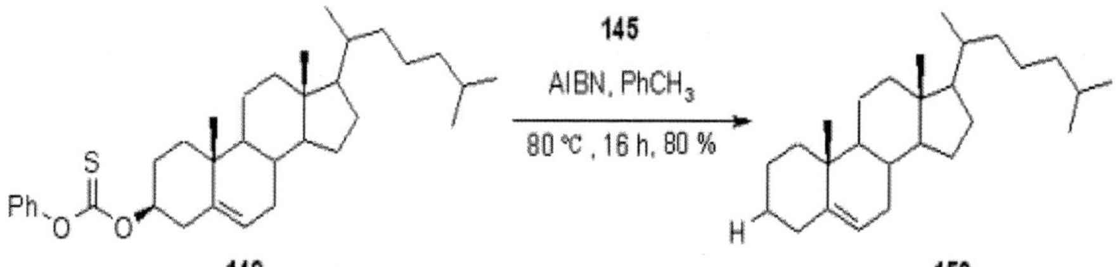

Scheme 47. Barton dehydroxylation using resin 145.

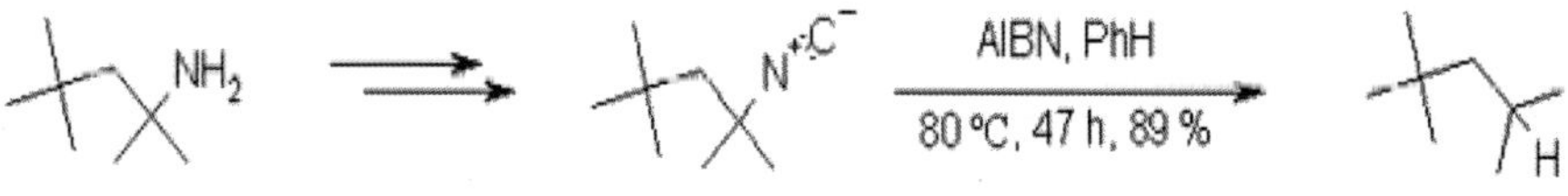

Scheme 48. Deamination reaction using resin 145.

The supported tin hydride was recycled by reduction of the filtered resin, maintaining a good efficiency. Following a similar path, Dumartin and coworkers [57,58] synthesized supported organostannanes with spacer of 2, 3 and 4 carbon atoms, in order to avoid β-elimination of the stannane as by-product, therefore producing recyclable and more stable reagents for radical reactions. In a successive paper [59], the same group used the anchored -Bu2SnX (X = halogen) with a C-4 spacer for catalytic reductions of haloalkanes in the presence of NaBH4 and AIBN. The length of the spacer proved to be important for reaction yields (due to more accessibility of the metal centre) and residual pollution by the metal.

Deleuze [60] prepared a macroporous organotin chloride by inclusion of a suitable organostannyl-styrene precursor in the polymerization reaction. The so prepared supported reagent (in combination with NaBH4) proved to be comparable to Bu3SnH in reducing alkylhalides in solution. Dumartin's group [61] investigated the possibility to use tin-based reducing agent in catalytic amount, as it had been done in solution [62]. In particular, their attention focused on the use of siloxanes, particularly polymethylhydrosiloxane (PMHS), for the regeneration of catalytic amounts of TBTH. However, PMHS proved ineffective using polymer-bound organostannane 154, evidently because it was incapable of penetrating the PS beads.

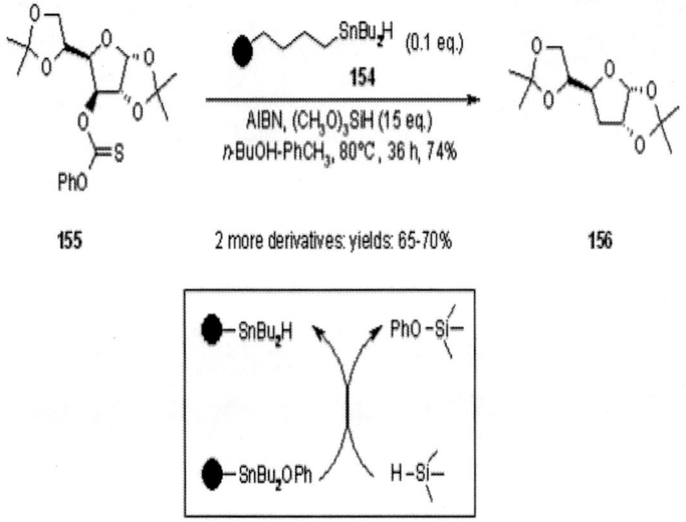

Scheme 49. Dumartin's catalytic approach.

An alternative was chosen and found in trimethoxysilane, giving good results. The reactivation of the tin by-product is highlighted in the scheme (Scheme 49). An alternative to organostannyl hydrides in radical chemistry is represented by the combination organo distannanes/sensitizer/H-donor via, for example hυ initiation. Neumann also studied anchored distannanes on SP [63,64], which can easily be prepared starting from the commercial polystyrene. The tin chloride derivative 157 was synthesized by conventional chemistry, and subsequently treated with Bogdanivic's magnesium (Mg-anthracene-3THF) to form the distannane 158.

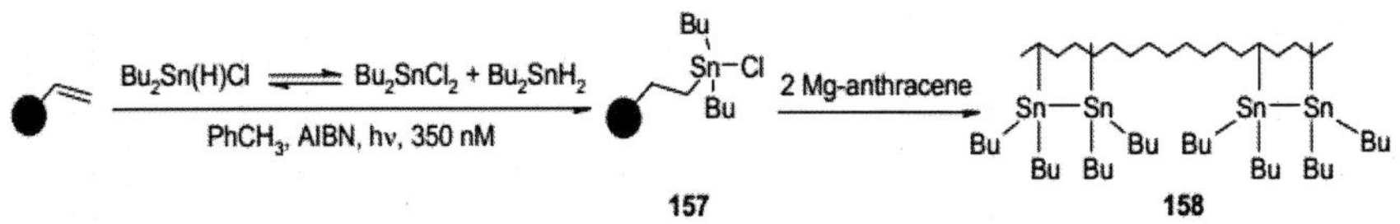

Scheme 50. Neumann's organodistannane.

This could be used as the equivalent non-supported reagent in the addition of tert-butyl iodide to acetylenes, yielding similar results. However, the advantage is that after the reaction the polymer containing SnI groups can be recovered by filtration, and completely regenerated in a straightforward manner by reaction with Bogdanivic's magnesium.

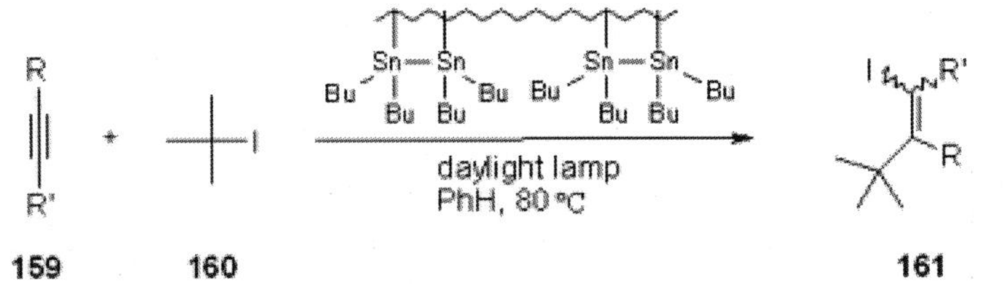

Scheme 51. Addition of tert-butyl iodide to acetylenes.

Interestingly, when the polymer was treated with the bromide 162 in the presence of acetone as long-lived triplet sensitizer, and 2-propanol as weak H-donor, the corresponding radical 163 thus generated evolved in a somewhat unexpected manner. In fact, instead of the anticipated reduction, a cyclization product was obtained.

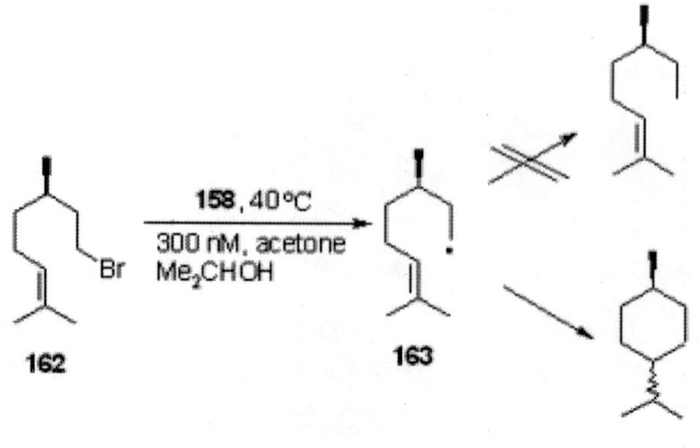

Scheme 52. Resin 158-mediated radical cyclization

Generally speaking, the possibility to chose the H-donor strength (and, therefore, to direct the radical reaction in one of the competitive pathways), the ease of separation of the tin-residues and the little contamination, make this option reagent a feasible suitable choice. Enholm and colleagues [65] studied the use of allystannane anchored to SP. This toxic reagent is used in the allylation of organohalides (or other homolytic functions) under radical conditions and suffers of the common problems encountered with tin-containing reagents. The use of supported tin reagent leads to ease of purification and minimal contamination of the products. In order to enhance the rates of the intermolecular reaction at test, it was chosen a soluble non-cross-linked polystyrene support (Scheme 53), which induces a 100-fold increase in the rate of intermolecular radical reaction compared to the heterogeneous classical SP reactions.

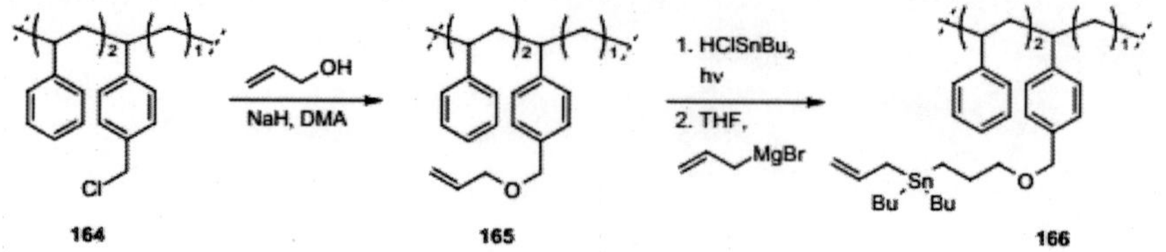

Scheme 53. Enholm's allylstannane.

It was found that the best combination <number of reactive sites-polymer properties> led to a 33% of loading capacity. Test reaction is shown in Scheme 54.

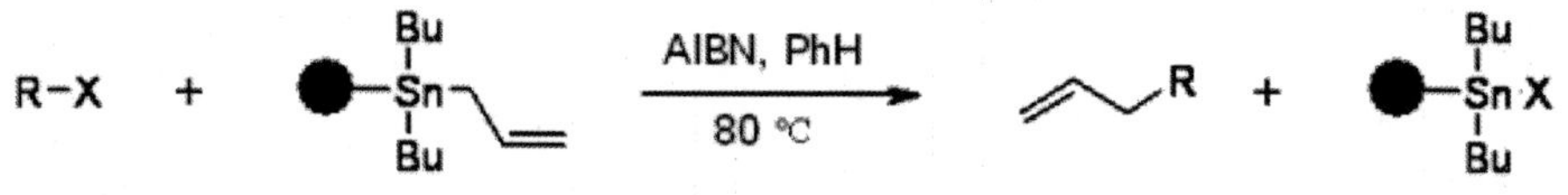

R–X + ●–Sn(Bu)₂–allyl → (AIBN, PhH, 80 °C) → allyl–R + ●–Sn(Bu)₂X

167 166 168 169

Scheme 54. Application of allylstannane 166.

Without going through the details, it is worth underlining some features of the methodology: electron-deficient substrates worked best, while electron-rich did not work at all. This behavior was attributed to the "philicity" of the generated tin radical, which is rather electron-rich (nucleophilic), generating a mismatch with the radical substrate. Interestingly, tertiary radicals, that gave no reaction in solution, gave a 50% yield. Finally, this procedure displayed a strong regioselectivity towards dihalides substrates, yielding a product of single allylation, strongly preferring electron-deficient radicals. In an extension of the precedent methodology, the same group [66] performed the preparation of another organotin-supported reagent for the reduction of organohalides (Scheme 55).

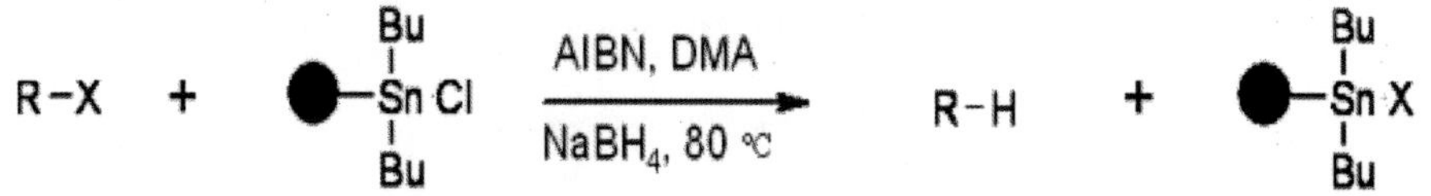

R–X + ●–Sn(Bu)₂Cl → (AIBN, DMA, $NaBH_4$, 80 °C) → R–H + ●–Sn(Bu)₂X

167 170 (cat.) 171 169

Scheme 55. Enholm's reduction of organohalides

The preparation of 170 was performed following the synthesis depicted in Scheme 53, where 165 was only treated with HClSnBu2. The catalytic methodology used by Enholm is well established [67].

It is widely used for radical reduction of organohalides and represents a useful way to avoid the excess of tin-derivatives, toxic and difficult to separate. The application of the catalytic strategy to the SP allows a low loading of tin-supported reagent and, in addition, the use of a soluble support speeds up the reaction times compared to heterogeneous analogue reactions. Primary alkyl halides reacted quickly with 0.1 eq. of 170 and when the amount of supported reagent was lowered, some reactions took longer. It is worth noting that use of a dihalides with only 1 equivalent of 170 gave only one product of reduction.

Contamination with tin was evaluated via degradation of the products followed by inductively coupled plasma mass spectrometry (ICP-MS), showing the presence of very low residual amount of the metal (around 10 ppm).

In a very early example of solid-supported organotin reagents, Ueno and coworkers [68] compared in-solution radical cyclization's under classical conditions (Bu3SnH/AIBN) with the use of polymer-bound radical carrier (Bu3SnCl/NaBH4) for the synthesis of 2-alkoxytetrahydrofurans (Scheme 56).

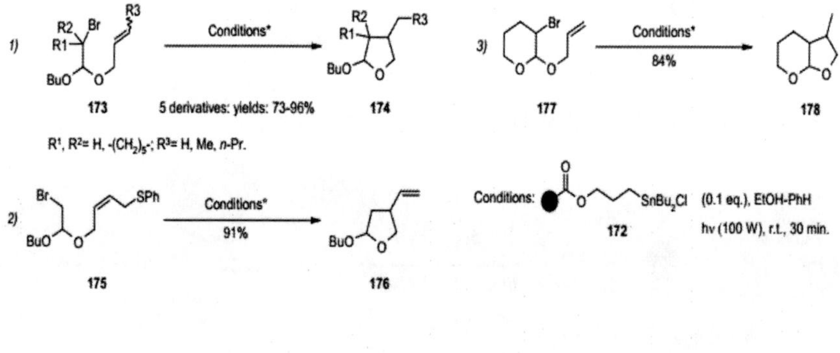

1) 173 —Conditions*→ 174; 5 derivatives: yields: 73-96%

R^1, R^2= H, -$(CH_2)_5$-; R^3= H, Me, *n*-Pr.

2) 175 —Conditions*, 91%→ 176

3) 177 —Conditions*, 84%→ 178

Conditions: 172 ($SnBu_2Cl$) (0.1 eq.), EtOH-PhH hv (100 W), r.t., 30 min.

Scheme 56. Ueno's synthesis of 2-alkoxytetrahydrofurans.

It was found that the reactions were higher-yielding when polystyrene-bound organotin chloride 172 was used as radical carrier. The supported reagent could be used several times after appropriate washing. Trialkylgermanium hydrides have shown a high potential in radical chemistry often hindered by their high costs, due to their lower toxicity in comparison to that of the corresponding organostannane. In the need of new reagents, Mochida's group [69] studied the synthesis of polymer-supported germanium hydrides, either via functionalization of polystyrene or by addition to the polymerization mixture of a suitable germanium-derivatized styrene. The supported reagents were tested for the reduction of organic halides, finding comparable results to known in-solution analogous reactions. In addition, it was demonstrated the opportunity of regeneration of the insoluble reagent by reduction with LiAlH4.

Bowman's group [70] also performed the synthesis of resin-bound germanium hydrides. Examples of their applications, which always gave reasonable results, are shown below (Scheme 57).

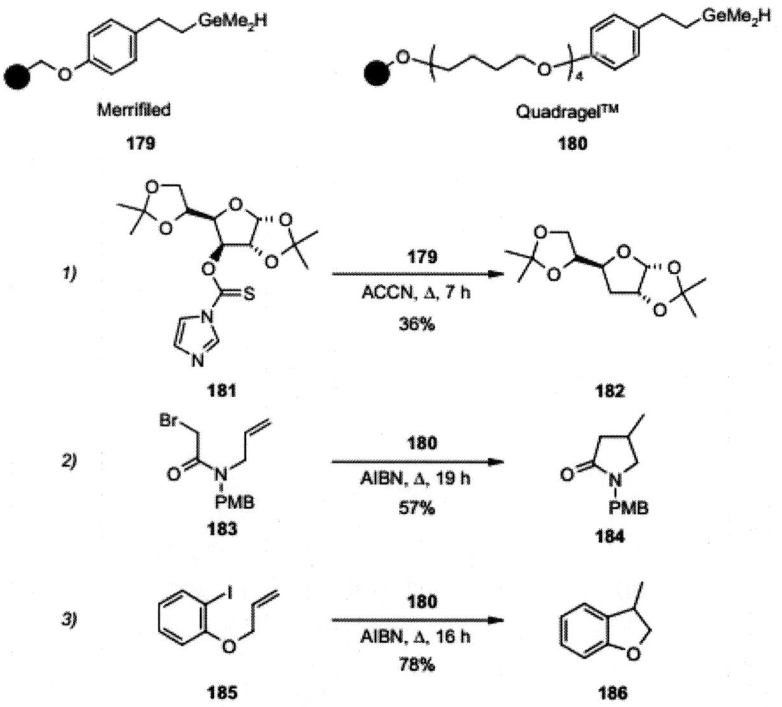

GeMe$_2$H
Merrifiled
179
GeMe$_2$H
Quadragel™
180

1) 181 → 179, ACCN, Δ, 7 h, 36% → 182

2) 183 → 180, AIBN, Δ, 19 h, 57% → 184

3) 185 → 180, AIBN, Δ, 16 h, 78% → 186

Scheme 57. Bowman's use of SP- germanium hydrides.

In the quest of radical sources on SP, the work presented by Murphy's group [71] represents an alternative to the immobilized organostannanes previously described. Tetrathiofulvalene (TTF) is the promoter for the radical-polar crossover reaction [72-74], a methodology used to generate aryl radicals from aryldiazonium salts. The reaction run in solution suffers from the need for separation of TTF by-products and Murphy's goal was immobilization of the promoter in order to simplify work-up (Scheme 58).

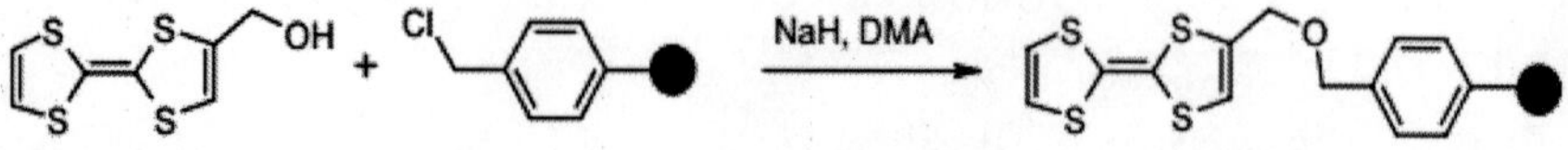

Scheme 58. Murphy's synthesis of PS-TTF.

Hydroxymethyltetrathiofulvalene 187 was prepared and loaded to a gel-type resin and used for test in radical cyclization reactions, as the example shown below (Scheme 59).

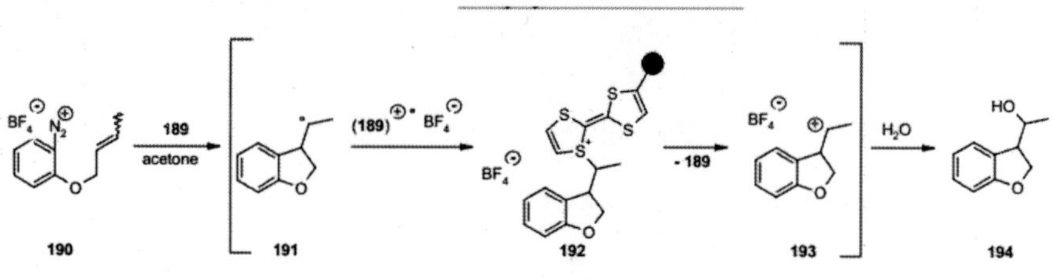

Scheme 59. Murphy's use of PS-TTF.

The reactions tested were successful, with comparable results with the homologous run in solution, even if slightly lower-yielding. A major concern was the presence of side-products such as the deoxygenated analogue of 194, possibly arising from H-abstraction by the radical intermediate 191 from the polymer benzylic hydrogens, or products of reaction of cation 193 with the aryl backbone of the resin. In both cases, it was proved that none of these concerns was factual and the reactions resulted rather clean. In addition, contamination of the products from TTF by-products was absent and it was also proven that the SP-TTF could be recycled after regeneration by reduction with NaBH4 in methanol.

Giacomelli and colleagues [75] successfully attempted a useful extrapolation of the classic Barton-Crich thiohydroxamates chemistry [76] in a work where a new polymer-supported radical source was prepared and tested. N-hydroxythiazole 2(3)-thione 195 was synthesized and loaded on a Gly-Wang resin to obtain radical source 196 (Scheme 60).

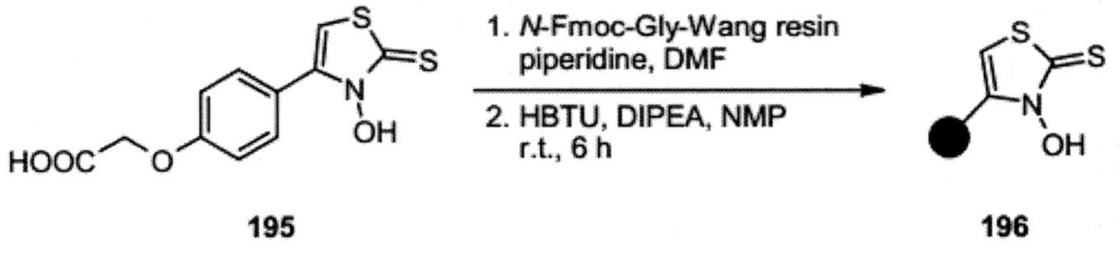

1. *N*-Fmoc-Gly-Wang resin piperidine, DMF
2. HBTU, DIPEA, NMP r.t., 6 h

195 196

Scheme 60. Giacomelli's interpretation of Barton-Crich thiohydroxamates chemistry.

The polymer-supported reagent was then tested for the generation of alkyl radicals (Scheme 61), where R = $C(CH_3)_3$; CH_2Ph; CH(Me)CH_2Ph; $CH_2CH_2CH(NHBoc)COOt$-Bu) and alkoxyl radical (Scheme 62).

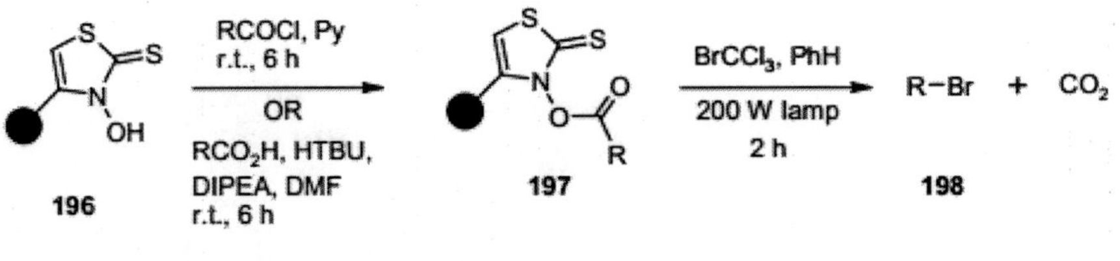

RCOCl, Py r.t., 6 h OR RCO_2H, HTBU, DIPEA, DMF r.t., 6 h

$BrCCl_3$, PhH 200 W lamp 2 h

R−Br + CO_2

196 197 198

Scheme 61. Use of resin 196 for the generation of alkyl radicals.

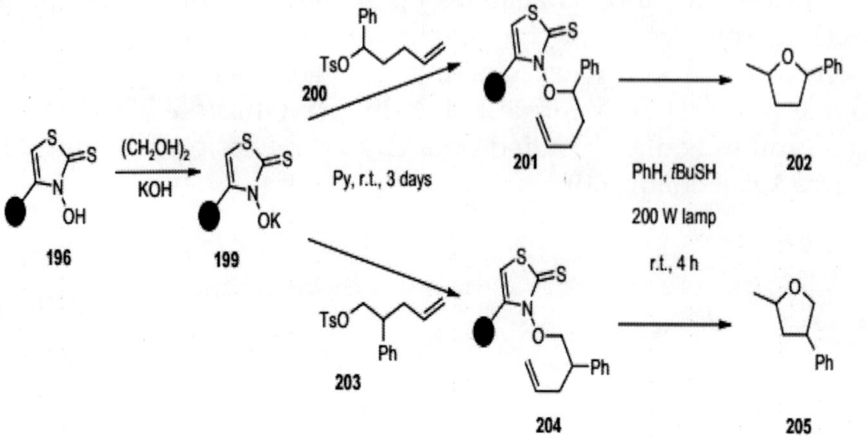

Scheme 62. Use of resin 196 for the generation of alkoxyl radicals.

In both cases the reactions were successful, proving this methodology valid for the generation of radicals and having the additional advantage of ease of purification compared to the often tedious experience of purifying the in-solution analogues.

Transition-metal-catalyzed atom transfer radical cyclization (ATRC) represents an alternative, tin-free, non-reductive, catalytic, radical cyclization methodology. Clark's group [77,78] studied the efficiency of this method for the construction of small molecule by anchoring the ligand to SP. In a first attempt [77], functionalized silica was used for the support of the N-alkyl-2-pyridylmethanimine ligands (Scheme 63).

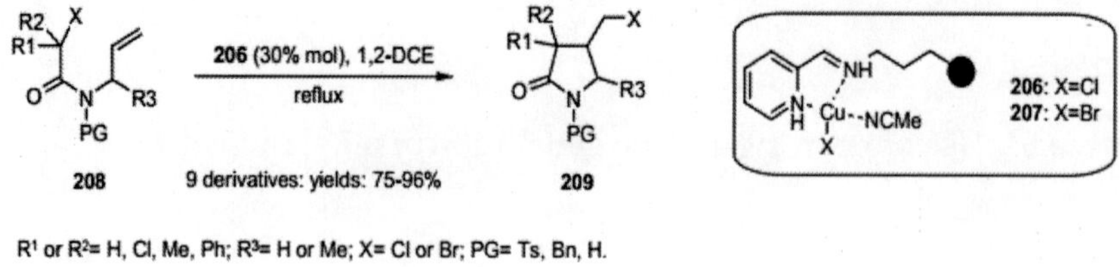

Scheme 63. Clark's ATRC approach.

All the reactions were high-yielding, the major difference among them being the time necessary to get to completion (from 1 to 36 h). The stereochemistry is not shown, although the trans isomer resulted to generally be the major one. The catalysts 206 and 207 were prepared by treatment of the immobilized ligand with CuX and 206 was also reused at least 3 times, giving still good yields (92→90→86%), but longer reaction times (3→24→36 h), a deactivation probably caused by the formation of an inactive CuCl2 complex. In a successive paper, Clark [78] used different supports and several ligands, in order to determine the best combination to improve the efficiency of the reaction (Scheme 64).

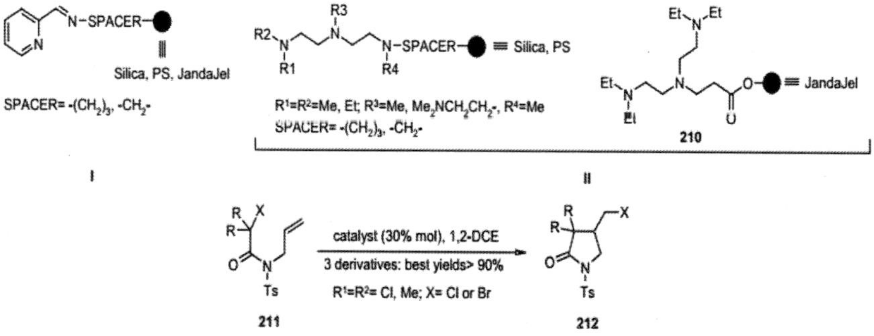

Scheme 64. Clark's catalyst improvements.

Running the above reaction at room temperature, some differences among the catalysts (obtained after treatment of supported ligands of type I or II with CuX) arose, since there was a clear superiority of those of type I vs. type II, in terms of reaction times and conversions (yields were, however, rather high). Raising the temperature to reflux, the reaction times were less than 30 min, independently of the catalyst used. Some more substrates were investigated, giving interesting results. For instance, in (Scheme 65) the catalyzed radical addition to a terminal alkyne is described. The absolute yields were high (88-98%), independently of the ligand used, even if type I catalysts were slightly better. E:Z ratio was almost the same for the ligand screening, however, the presence of reduced product 216 varied between 11-6%, being 0% when silica type I was used.

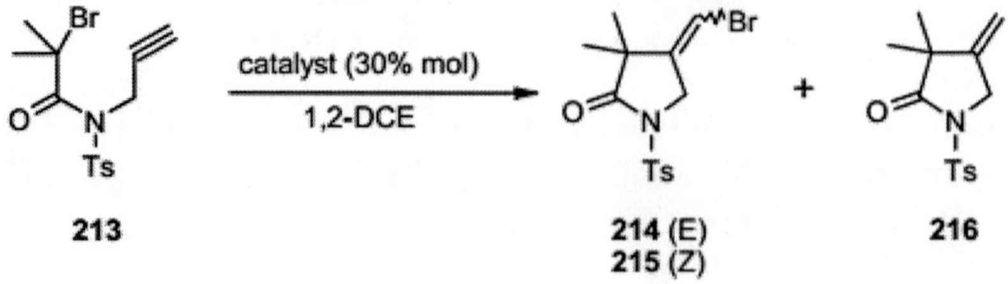

Scheme 65. Example of radical addition to a terminal alkyne.

In Scheme 66 is depicted a 5-endo-trig radical polar crossover addition, a normally disfavored process following Baldwin's rules [79,80], but feasible when polyamine-copper catalysts such as type II (Scheme 64) are used [81]. In this case a mixture of 218 and 219 was obtained. The best yield (100%) was found when 210 was used as ligand, while 220 was isolated in low yield (15%) as the sole product when type I Janda Jel ligand was used.

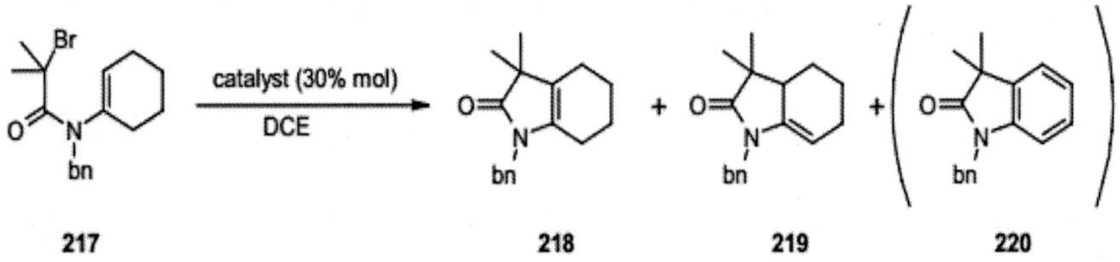

Scheme 66. Clark's 5-endo-trig radical polar crossover addition

Finally, in Scheme 67 a 4-exo-trig cyclization is represented. This also is a disfavored process according to Baldwin's rules, which was successful only when 210 was used as ligand.

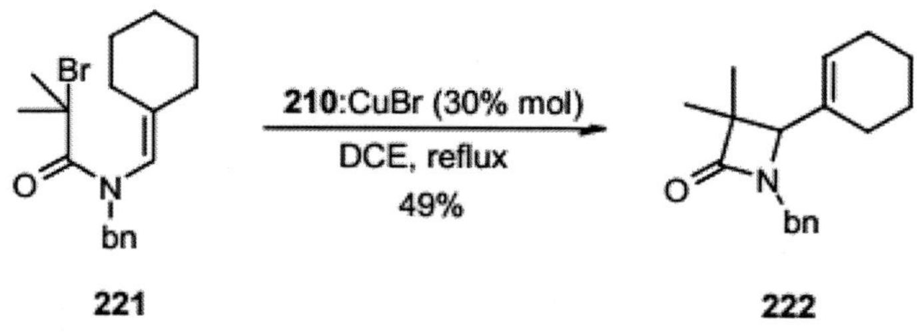

Scheme 67. Clark's 4-exo-trig radical cyclization.

Nagashima and coworkers [82] studied the same reaction, using different copper complex as the catalyst. In this case, in fact, the metal-ligand complex used for the catalysis was immobilized on a very peculiar SP the (PMHS). The siloxane gel [bipy]@Si contained, embedded, a bypyridine ligand and it was synthesized from PHMS and 4,4′-bis(allyloxymethyl)-2,2′-bipyridine (DAbipy) by the platinum-catalyzed hydrosilylation of alkenes. The catalyst was then tested on several radical precursors (Scheme 68).

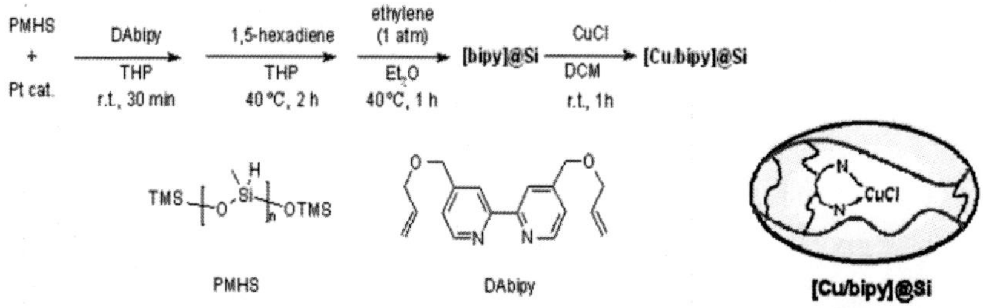

Scheme 68. Nagashima's Cu-catalytic complex.

The synthesis of the actual catalyst [Cu/bipy]@Si was completed after treatment of [bipy]@Si with CuCl. The metal was easily soaked into the gel, forming the active complex, which resulted more stable to air than the not-embedded version and that can be filtrated from the organic solvents used for the reaction and re-used with small loss of TON. Metal leakage inferior to the detection limit was observed.

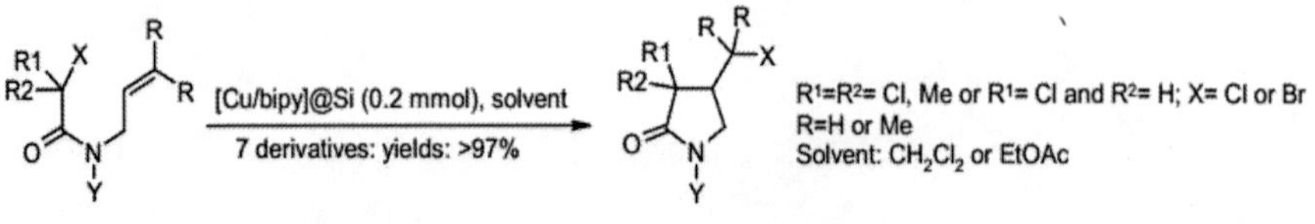

R1 X R2 O N Y R R [Cu/bipy]@Si (0.2 mmol), solvent 7 derivatives: yields: >97% R1 R2 R R X O N Y

R^1=R^2= Cl, Me or R^1= Cl and R^2= H; X= Cl or Br
R=H or Me
Solvent: CH_2Cl_2 or EtOAc

223 224

Scheme 69. Application of [Cu/bipy]@Si to 5-exo-trig radical cyclization.

CCl_3 O N bn [Cu/bipy]@Si EtOAC, 80 °c, 2 h 99% Cl Cl Cl O N bn

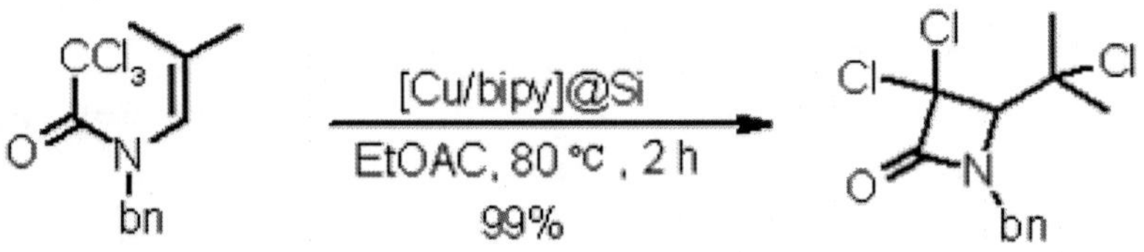

225 226

Scheme 70. Application of [Cu/bipy]@Si to 4-exo-trig radical cyclization.

The reactions shown in Schemes 69 and 70 were all pretty successful (97-99%) and when the reaction gave low yields at room temperature, simple heating gave almost complete conversions. Hypervalent iodine reagents can be used for the generation of alkoxy radicals under irradiation conditions. Teduka and Togo [83] used polystyrene-supported (diacetoxyiodo)benzene for the synthesis of 2-substituted 1,3-dioxolanes (Scheme 71), 2-substituted tetrahydrofurans (Scheme 72) and 2-substituted 1,3-dioxanes (Scheme 73).

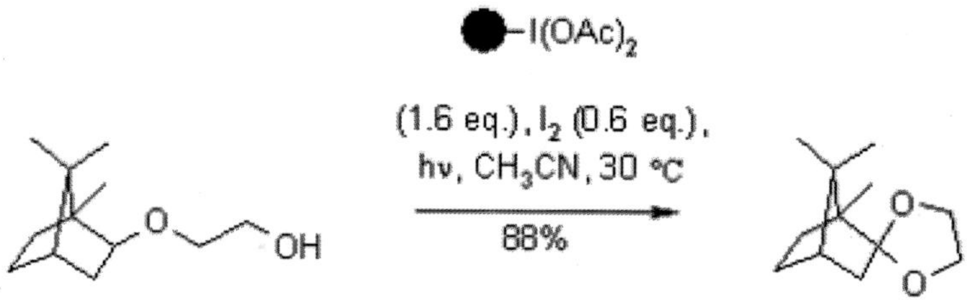

●–I(OAc)$_2$

(1.6 eq.), I_2 (0.6 eq.),
hν, CH_3CN, 30 °C

88%

227 228

Scheme 71. Togo's synthesis of 1,3-dioxolanes.

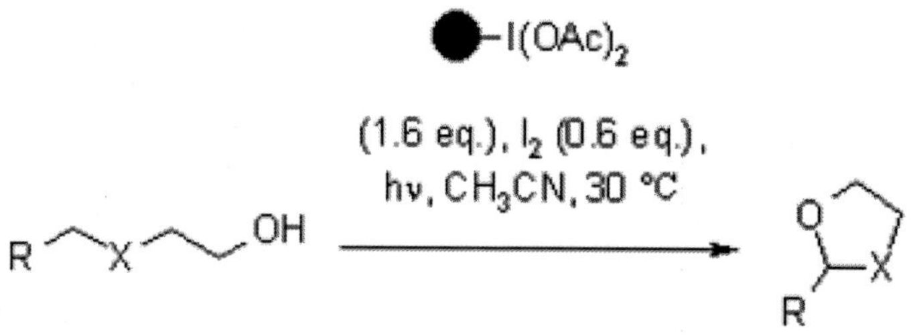

●–I(OAc)$_2$

(1.6 eq.), I_2 (0.6 eq.),
hν, CH_3CN, 30 °C

229 9 derivatives: yields: 47-76% 230

Scheme 72. Togo's synthesis of 2-substituted tetrahydrofurans and dioxolanes.

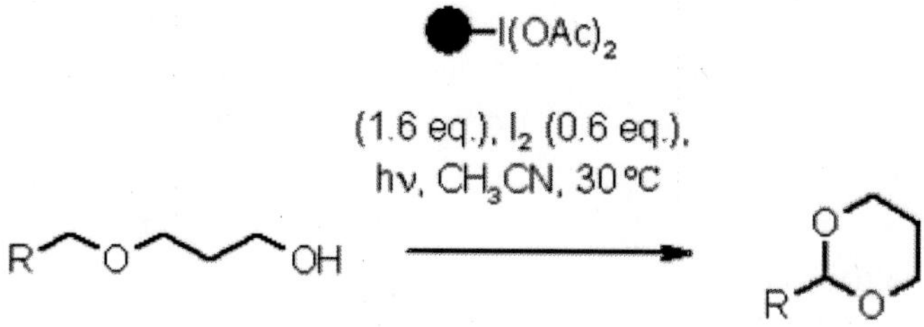

●–I(OAc)$_2$

(1.6 eq.), I_2 (0.6 eq.),
hν, CH_3CN, 30 °C

229a 2 derivatives: yields: 62-70% 231

R= Ph, 4-NO_2-C_6H_4

Scheme 73. Togo's synthesis of 2-substituted-1,3-dioxanes.

Oxidation of alcohols to the corresponding aldehyde or ketone represents a very common reaction in daily organic synthesis; however, there is always a need for new and milder methods for sensitive molecules, especially in SP organic chemistry. Kashiwagi's group [84] studied a mild and efficient methodology, using a catalytic two-phase system, under basic conditions. (Scheme 74)

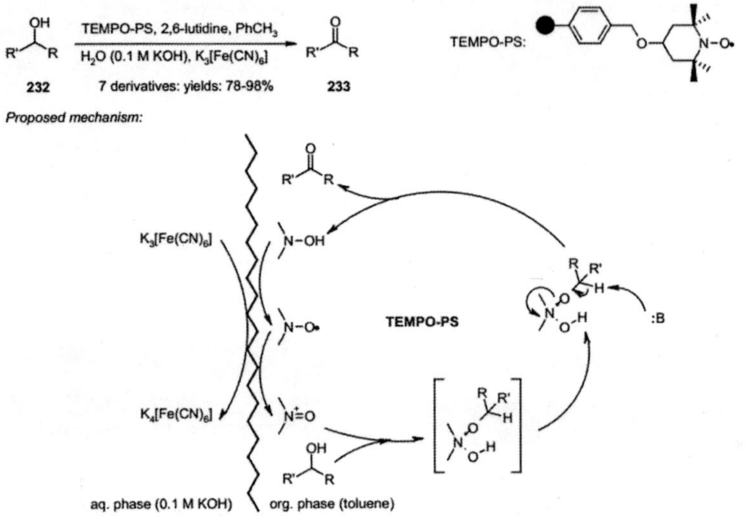

Scheme 74. Kashiwagi's catalytic two-phase system.

The oxidation mechanism is displayed above and consists of an interface interaction between the oxidant (K3[Fe(CN)6]) in aqueous, basic solution and the hydroxylamine polymer-bound (TEMPO-PS) in organic phase (toluene), formed after oxidation of the alcohol 232. The base (2,6-lutidine) was necessary to close the catalytic cycle, yielding the desired carbonyl 233. Interestingly, in was found that this method has a rather high chemoselectivity, preferring primary alcohols oxidation to secondary alcohols.

EXPLOITATION OF RADICAL CHEMISTRY IN SP SYNTHESIS

The importance of radicals in organic chemistry has grown considerably over the past three decades and several methodologies have been studied and set up to face the need of new routes for the synthesis of target molecules or classes of compounds. In this section, the studies and applications of those methodologies to SP will be discussed. Following our goal to provide a reference for the day-to-day use of solid-phase radical reactions, we have not based our classification on mechanistic grounds, but more simply on the formal way to generate a new bond.

Carbon-Carbon Bond Formation through Intermolecular Reactions

Substitutions Alpha to Carbonyl Groups

The first examples of intermolecular free radical allylation reaction on solid support was reported by Sibi and Chandramouli [85], for the synthesis of simple α-allyl acetic acids (Scheme 75).

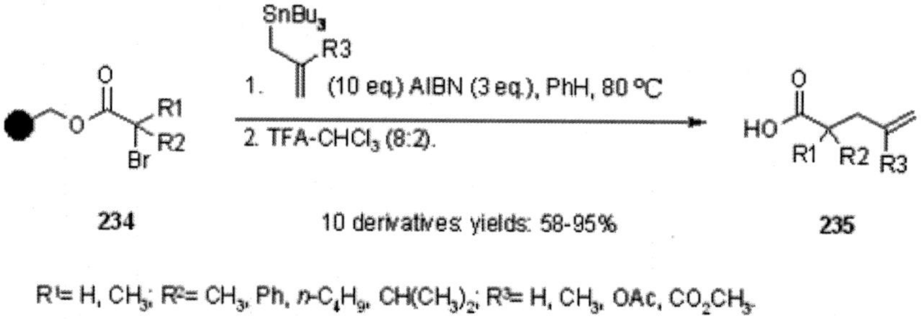

1. $SnBu_3$ / R3 (10 eq.) AIBN (3 eq.), PhH, 80 °C

2. TFA-$CHCl_3$ (8:2).

234 → 235

10 derivatives yields: 58-95%

R1= H, CH_3; R2= CH_3, Ph, n-C_4H_9, $CH(CH_3)_2$; R3= H, CH_3, OAc, CO_2CH_3.

Scheme 75. Sibi's allylation reaction on supported α-bromo.

Wang resin was used as the solid support and the reactions were carried out using allyltin derivatives. As shown in the scheme, the reactions were rather successful and allylation was more effective (95% yield) when R^3 was either and electron-donating (OAc) or electron-withdrawing (CO2CH3) group, while substitution around the C-centered radical gave lower yield (58%) for R^1 = R^2 = CH_3. Allyl transfer reaction was studied also by Enholm's group [86] who worked on a soluble support (Scheme 76).

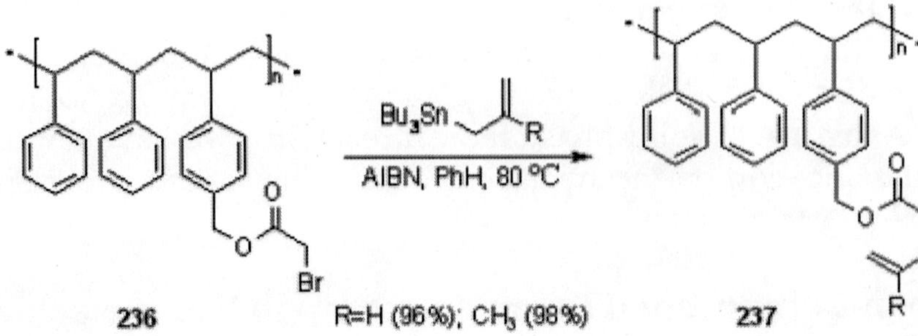

Scheme 76. Enholm's allyl transfer reaction.

The polymer (non-cross-linked polystyrene) was functionalized to obtain the radical precursors. As shown in Scheme 76, 236 could be reacted with allyltributyltin derivatives under radical conditions, allowing allylation of the carbonyl α-carbon.

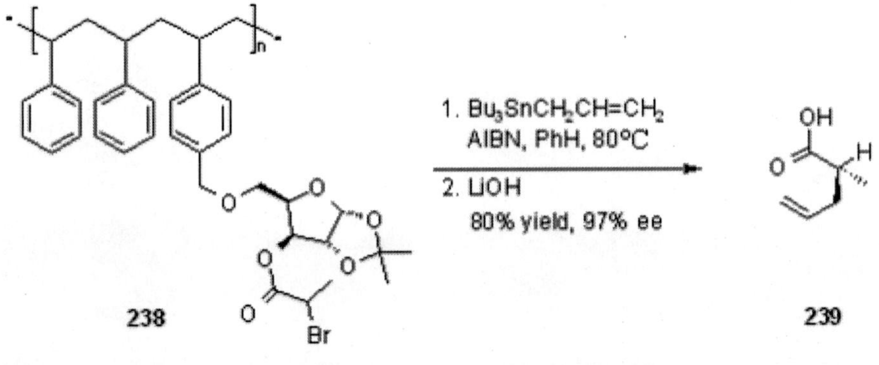

Scheme 77. Enholm's allyl transfer reaction.

Allylation of the same radical precursor (α-bromo carboxylate) linked this time to the polymer by means of a sugar moiety is shown in Scheme 77. When the reaction was performed in solution, use of a Lewis acid increased stereoselectivity. However, in the case shown in Scheme 77 any Lewis acid (LA) used gave premature cleavage. Fortunately, in this case a good stereoselectivity (97% ee) was found without any LA addition: this can be explained invoking the Lewis acidity of Bu3SnBr (formed during the reaction), which coordinates preferentially to one face of the sugar molecule directing the allyl attack. Enholm studied a series of intermolecular radical reactions also exploiting ROMP polymers [87] (see below).

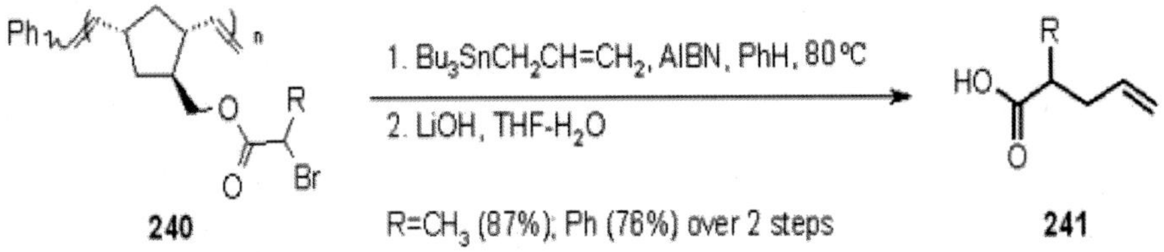
Ph
n
O
R
O
Br
240
1. Bu3SnCH2CH=CH2, AIBN, PhH, 80 °C
2. LiOH, THF-H2O
R=CH3 (87%); Ph (76%) over 2 steps
R
HO
O
241

Scheme 78. Enholm's use of ROMP polymers for the synthesis of 241.

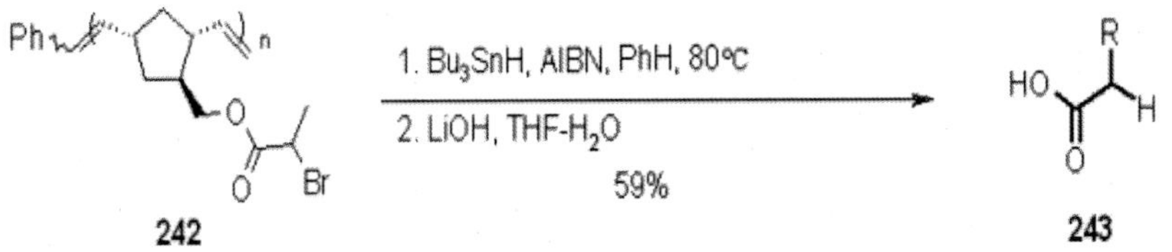
Ph
n
O
O
Br
242
1. Bu3SnH, AIBN, PhH, 80 °C
2. LiOH, THF-H2O
59%
R
HO
H
O
243

Scheme 79. Enholm's use of ROMP polymers for the synthesis of 243.

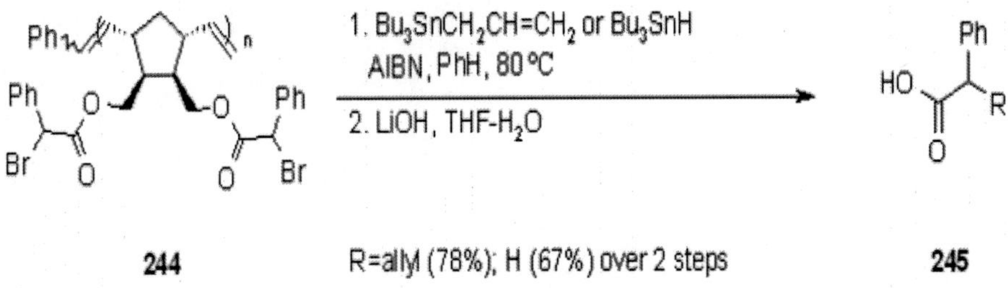
Ph
n
Ph
O
O
Ph
Br
O
O
Br
244
1. Bu3SnCH2CH=CH2 or Bu3SnH
AIBN, PhH, 80 °C
2. LiOH, THF-H2O
R=allyl (78%); H (67%) over 2 steps
Ph
HO
R
O
245

Scheme 80. Enholm's use of ROMP polymers for the synthesis of 245.

The polymers 240, 242 and 244 (Schemes 78-80) were prepared by building the appropriate norbornene precursors that were subsequently treated with Grubbs' catalyst (see below). Treatment of the so obtained soluble polymers under radical conditions gave the desired products of allylation (Scheme 78) or reduction (Scheme 79) in good yields. A tremendous advantage of this type of polymers is represented by the presence of at least one molecule of substrate in each monomer, giving a 100% loading capacity to the polymer. In the case of Scheme 80, where each monomer contains two reactive sites, a 200% theoretical loading of the resin was obtained. Linhardt and coworkers [88] used SmI2 as a SET reagent for the C-glycosilation of Neu5Ac on SP.

Since carboxyl TentaGel resin failed to give any desired product, the reaction was run on a less hydrophilic material, namely a functionalized controlled pore glass (CPG). The glass was derivatized in order to present an amine group handle (Scheme 81). Methyl ester of neuraminic acid was first esterified with a spacer that was subsequently attached to the support. As shown in the scheme below, the results of the C-glycosilation reaction were rather satisfactory.

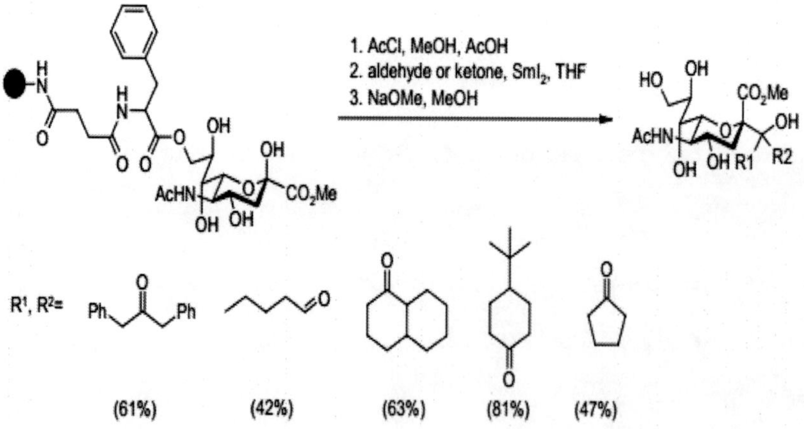

1. AcCl, MeOH, AcOH
2. aldehyde or ketone, SmI_2, THF
3. NaOMe, MeOH

R^1, R^2= (61%) (42%) (63%) (81%) (47%)

Scheme 81. Linhardt's C-glycosilation.

Radical Addition to Alkenes

Radical addition to unsaturated C-C bonds represents one of the most common reactions in radical chemistry. The efficacy of the intermolecular versions of such reactions is heavily dependent on the nature of reaction itself. An early example of this chemistry on SP was presented by Zhu and Ganesan [89], that proved the efficiency of alkyl radical addition to acrylates attached to a chosen resin (Scheme 82).

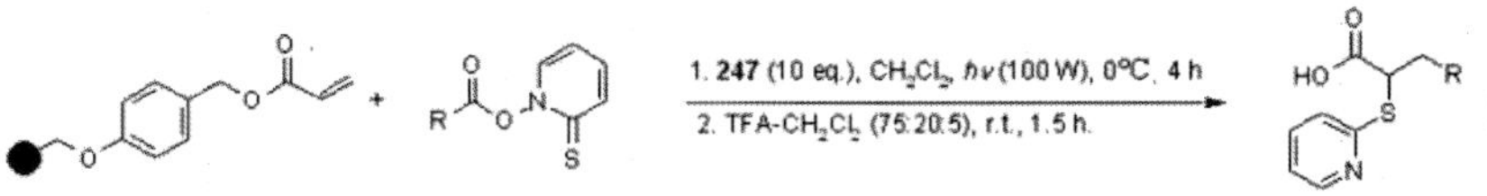

1. 247 (10 eq.), CH_2Cl_2, $h\nu$ (100 W), 0°C, 4 h
2. TFA-CH_2Cl_2 (75:20:5), r.t., 1.5 h.

246 247 248

R= 1-adamantyl (63%); *t*-Bu (91%); *c*-Hex (91%); Bn (32%); $PhCH_2CH_2$- (94%)

Scheme 82. Ganesan's radical addition to supported alkenes.

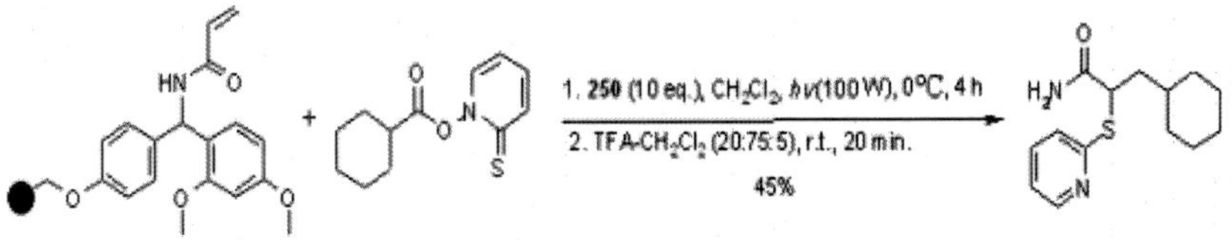

1. 250 (10 eq.), CH_2Cl_2, $h\nu$(100 W), 0°C, 4 h
2. TFA-CH_2Cl_2 (20:75:5), r.t., 20 min.
45%

249 250 251

Scheme 83. Ganesan's radical addition using Rink resin.

The alkyl radicals were generated by photolysis of the corresponding Barton esters 247 and addition was generally good, with the exception of the stabilized benzyl radical that gave poor conversion (yields refer to the use of PS-Wang resin, as Tentagel-Wang gave lower yields). The reaction was performed using a Wang (Scheme 82) or a Rink (Scheme 83) resin. In the latter case yield was lower, a result in accord to the in-solution reaction, probably due to the differences between acrylate ester 246 and acryl amide 249. 1,2-Radical initiated addition of haloalkanes to C=C bonds is employed in commercial routes for the preparation of dihaloethenylcyclopropane carboxylic acids, and the development of cleaner and easier methods has brought Kumar and colleagues [90] to attempt the reaction anchoring the olefin to SP (Scheme 84).

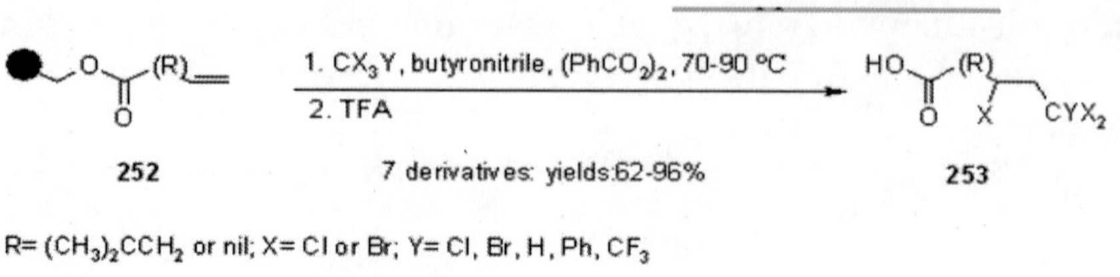

1. CX_3Y, butyronitrile, $(PhCO_2)_2$, 70-90 °C
2. TFA

252 7 derivatives: yields 62-96% 253

R= $(CH_3)_2CCH_2$ or nil; X= Cl or Br; Y= Cl, Br, H, Ph, CF_3

Scheme 84. Kumar's radical addition to supported alkenes.

The olefins were linked to Wang or Merrifiled resins by their carboxylic unit and the reactions were run in the presence of butyronitrile as a cosolvent (except when CCl4 was used as reagent). Yields varied from moderate to high, for reaction times between 24 and 48 h, making this method effective, but rather slow. The synthesis of racemic α-amino acids was described by Yim [91] that applied the mercury method for the generation of alkyl radicals in the presence of polymer-supported dehydroalanine (Scheme 85).

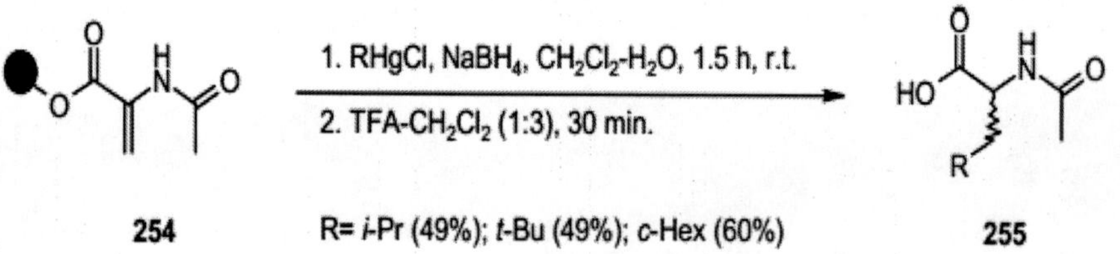

1. RHgCl, $NaBH_4$, CH_2Cl_2-H_2O, 1.5 h, r.t.
2. TFA-CH_2Cl_2 (1:3), 30 min.

254 R= *i*-Pr (49%); *t*-Bu (49%); *c*-Hex (60%) 255

Scheme 85. Yim's radical addition to supported alkene.

Addition of the alkyl radical on the double bond was effective and after cleavage reasonable yields of the desired products were isolated. When the same reaction was performed under more usual radical conditions (Bu3SnH, AIBN, t-BuI, PhH, under reflux), only trace amounts of the product (8%) were isolated, while prolonged heating led only to decomposition. The reason for this behavior was attributed to the poor swelling of the Wang resin in the solvent used in the attempt (benzene). Caddick's group [92] prepared a small array of amide derivatives by intermolecular radical addition of C-centered radicals to an activated acrylate anchored to a resin. Aminomethyl-polystyrene resin was chosen as the SP where to anchor the acrylate ester of 2,3,5,6-tetrafluoro-4-hydroxybenzoic acid (256).

The choice of such electron-poor acrylate was made in order to facilitate the radical reaction and obtain an amide after cleavage with an appropriate amine. The general synthetic scheme of this high-yielding methodology is represented in Scheme 86.

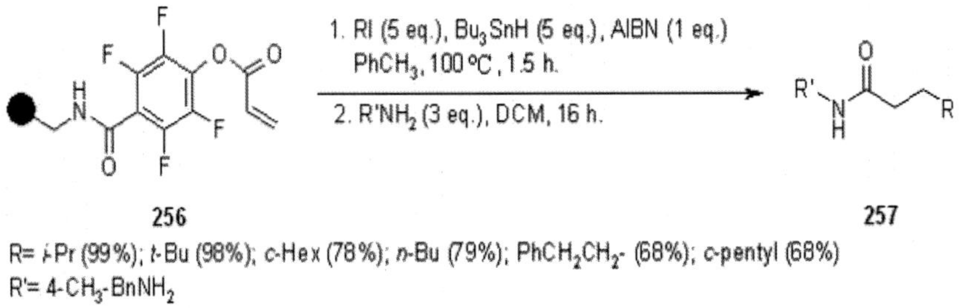

1. RI (5 eq.), Bu_3SnH (5 eq.), AIBN (1 eq.) $PhCH_3$, 100 °C, 1.5 h.
2. R'NH_2 (3 eq.), DCM, 16 h.

256 257

R= *i*-Pr (99%); *t*-Bu (98%); *c*-Hex (78%); *n*-Bu (79%); $PhCH_2CH_2$- (68%); *c*-pentyl (68%)
R'= 4-CH_3-BnNH_2

Scheme 86. Caddick's radical addition to supported alkene.

In Schemes 87 and 88 the use of carbohydrate derivatives as radical precursors (258 and 260 respectively) is shown. In these cases, α-amino acids were used as amines, although the yields were lower compared to the previous case.

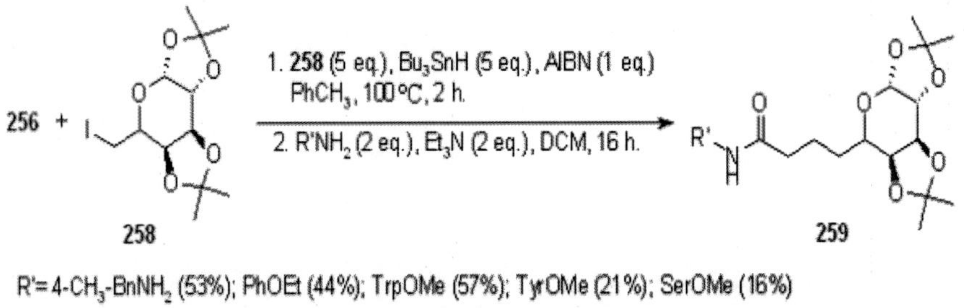

256 +

1. **258** (5 eq.), Bu_3SnH (5 eq.), AIBN (1 eq.) $PhCH_3$, 100 °C, 2 h.
2. R'NH_2 (2 eq.), Et_3N (2 eq.), DCM, 16 h.

258 259

R'= 4-CH_3-BnNH_2 (53%); PhOEt (44%); TrpOMe (57%); TyrOMe (21%); SerOMe (16%)

Scheme 87. Synthesis of compound 259.

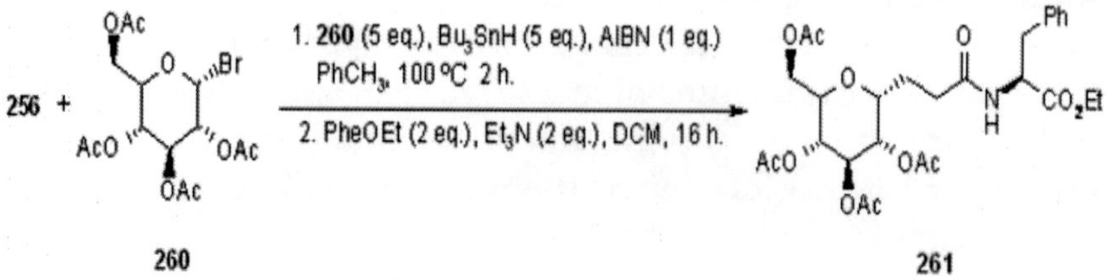

256 + 260 → 261

1. 260 (5 eq.), Bu_3SnH (5 eq.), AIBN (1 eq.) $PhCH_3$, 100 °C 2 h.
2. PheOEt (2 eq.), Et_3N (2 eq.), DCM, 16 h.

Scheme 88. Synthesis of compound 261.

Radical Addition to Oxime

Among the literature references regarding the use of radical chemistry on SP, Naito's group contributed by studying the reactivity of SP-supported oxime ethers. In their works [93-97], they studied the addition of appropriate alkyl iodides to polymer-supported glyoxylic oxime ether, giving the first example of intermolecular radical reaction exploiting triethyl borane or diethyl zinc as radical initiators. The methodology was used for the synthesis of unnatural aminoacids (Scheme 89).

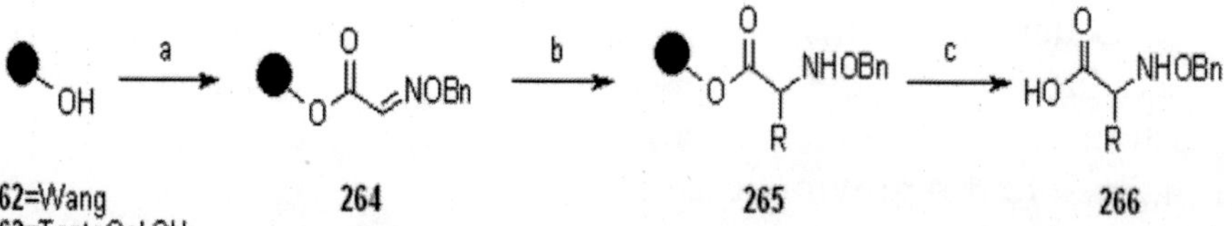

262=Wang
263=TentaGel OH

(a) HO_2CC=NOBn, DCC, DMAP, CH_2Cl_2, 20 °C, 12 h (for **262**);
HO_2C=CNOBn, 2,6-dibenzoyl chloride, Py, DMF, 20 °C, 12 h (for **263**)
(b) Et_3B, 1 h or: Et_2Zn, 15 min.; or: RI, Bu_3SnH, Et_3B, 1 h or: RI, Et_3B or Et_2Zn;
(c) TFA/$CHCl_3$ (1:3, v/v), 20 °C, 30 min.

Scheme 89. Naito's radical addition to supported oxime.

After loading the oxime to the solid phase (264), a first attempt was made using only Et3B in DCM, which worked as initiator and terminator to afford, after cleavage, aminoacid 266 where R = Et. Running the reaction with an alkyl iodide R-I in the presence of

Bu3SnH, a range of products were obtained, thus, when R = i-Pr, c-hexyl, t-Bu, s-Bu, the corresponding unnatural amino acids were recovered in good yields, while the unstable primary radical i-Bu and the bulky adamantyl gave lower yields. In all the reactions, a small amount (5-30%) of 266 (R = Et) was found as by-product from the competitive addition of ethyl radical arising from Et3B. The reaction was also run in the presence of Et2Zn, which gave comparable results to the Et3B; in both cases the radical chain was effective also at low temperatures (-78 °C). Successively, they tried to run the reaction under iodine-atom transfer mechanism, using Et3B and i-PrI, but avoiding Bu3SnH. In this case a large amount of product where R = Et was found. The increased competition of the side-reaction was not explained, however, a change in solvent and temperature (toluene, 80 °C) as well as the use of a larger amount or alkyl iodide (60 eq.) gave a better selectivity towards the desired product (5.7:1 vs. 2:1). Such selectivity, however, remained lower compared to the reaction run in solution.

In general, when the reaction was run in the presence of an organic solvent, Wang resin proved to be slightly better compared to TentaGel OH. However, when the reaction was tried in aqueous media, which is normally effective in solution, Wang resin did not work at all because of its poor swelling properties. Success was achieved with TentaGel OH resin, where it was necessary to use a solution of Et3B in THF or MeOH (compared to the hexane solution used in the previous attempts) to obtain a monophasic solution. The reaction was run in a mixture H2O-MeOH (2:1, v/v, at 70 °C), under iodine atom transfer conditions and the overall yields were acceptable though lower compared to the ones run in organic solvents; a comparable amount of ethylated product was obtained.

Although the methodology described by Naito was biased by the presence of the undesired product of ethyl radical addition, the use of aqueous solvents and the absence of the toxic tin hydride make this procedure quite appealing. The same group also studied the stereoselective version [96-99] of the above-described reaction, using a chiral auxiliary (Scheme 90).

1. a
2. b

267

268

(a) Et_3B or Et_2Zn, -78 °C or: RI, Et_3B or Et_2Zn, RI-$PhCH_3$ (4:1, v/v) 0°C.
(b) TFA/CH_2Cl_2 (1:5, v/v), 20°C.

Scheme 90. Naito's stereoselective radical addition to supported oxime.

Oppolzer's camphorsultam was chosen as an auxiliary, while the reactions were tested in SP (using Wang resin) and in solution. The oxime was attached to a carboxylic acid derivative which, in the SP version, served as the spacer between the reactive moiety and the polymeric support (compound 267).

In a first attempt, it was decided to try ethylation of 267 following the method described in Scheme 88, using Et3B or Et2Zn at low temperatures (-78 °C) in order to facilitate stereoselectivity. In both cases the results were very promising, high yields (59-74%), with little variation between solvents (DCM slightly better than toluene) and radical source (Et3B slightly better than Et2Zn), with diastereoselectivity higher than 95%. In order to prove the general validity of the method, radical addition from a different radical source was tried. At first Bu3SnH was used together with an alkyl iodide, but results were disappointing: low yield (around 40%) and a mixture of products were obtained (addition of ethyl radical prevailed). When the reaction was run without the tin hydride, yield improved (78%), but the great majority was still the undesired product (5:1), which was in contrast with the analogous solution-phase reaction. In order to overcome these problems, the conditions were changed: the reaction was performed using a mixture of R-I (30 eq.) and toluene (4:1) under iodine atom transfer conditions in the presence of Et3B or Et2Zn (5 eq.), and at higher temperatures (0 °C).

This led to higher yields (R = i-Pr: 69% Et3B, 53% Et2Zn; R = c-hexyl: 58% Et3B, 41% Et2Zn), and higher stereoselectivities (>90%)

compared to the in-solution reactions. The reason for the presence of such large amount of by-product in SP was not explained, however, this tendency increased at lower temperatures, suggesting that triethylborane concentrate on solid-support as Lewis acid, releasing a large amount of ethyl radical around the surface of the polymer.

In an approach similar to the one described by Naito in Scheme 89, Kim and colleagues [100] performed the synthesis of α-aminoesters by addition of alkyl radical to an anchored phenylsulfonyl oxime ether (Scheme 91).

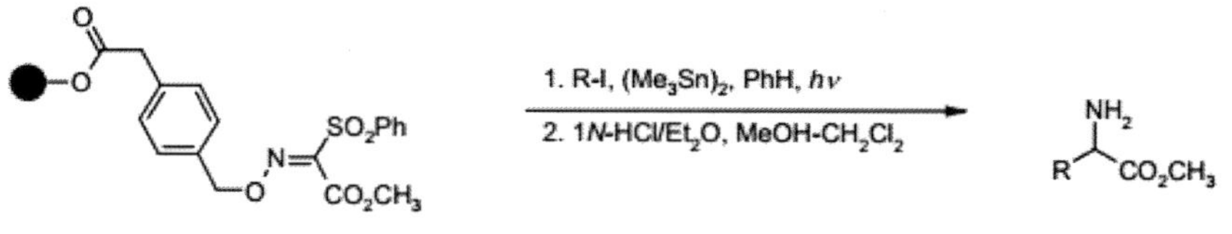

1. R-I, $(Me_3Sn)_2$, PhH, *hν*
2. 1*N*-HCl/Et_2O, MeOH-CH_2Cl_2

269 **270**

R=PhO$(CH_2)_4$- (50%); $CH_3OCO(CH_2)_2$- (49%); *c*-Hex (44%); 1-adamantyl (39%)
4-*t*-Bu-PhCH_2- (22%); Bn (25%); R'OCOCH_2- (no reaction). Yields calculated on 3 steps.

Scheme 91. Kim's radical addition to supported oxime ether.

The reaction was successful for unactivated alkyl radicals, lower-yielding in the case of addition of benzyl radicals (either from benzyl iodide or bromide), while no reaction was observed for α-carbonyl radicals, which suggests that the process is governed by radical philicity.

Carbon-Carbon Bond Formation through Intramolecular Reactions

Radical Addition to Oximes

In successive papers [96,97,101-103], Naito's group continued to study the reactivity of oxime ethers on solid-support by testing a 5-exo-trig radical cyclization for the synthesis of pyrrolidines (Schemes 92 and 93).

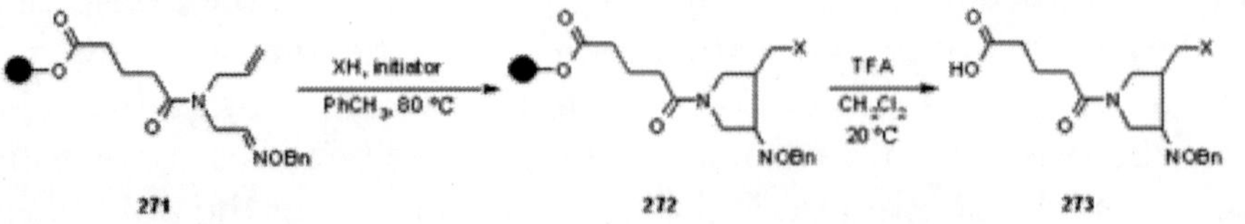

Scheme 92. Naito's SP-atom transfer radical cyclization into oxime-alkene.

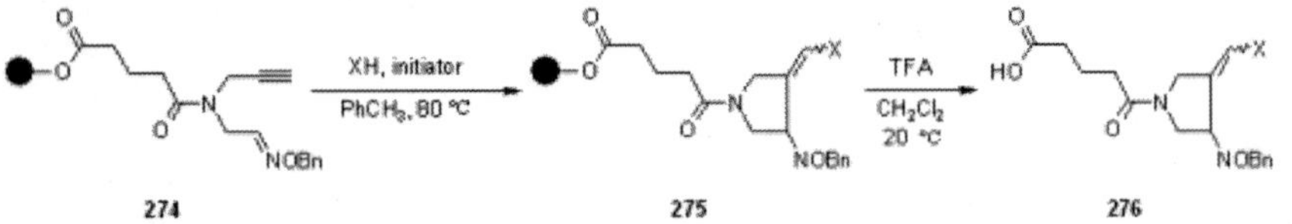

Scheme 93. Naito's SP-atom transfer radical cyclization into oxime-alkyne.

Resins 271 (Scheme 92) and 274 (Scheme 93) were prepared and reacted to test their cyclization potentiality. The major difference between the reactions depicted in Schemes 89-91 was failure of cyclization at low temperature. For this reason, it was necessary to use toluene as solvent, heated up to 80 °C. The use of Bu3SnH as XH gave reasonable yield when AIBN was used as radical initiator (Scheme 92: 47%). Better results were obtained using Et3B (Scheme 92: 64%; Scheme 93: 77%), while further variation of the radical initiator did not improve the process: in fact, 9-BBN gave only trace of product and Et2Zn gave successful cyclization, but low yield (Scheme 92: 26%). Use of TTMSS in place of Bu3SnH gave lower yield (Scheme 92: 50%). As observed previously (see Scheme 89), the desired reaction is always in competition with ethyl radical addition-cyclization and when a less reactive radical precursor such as Et3SiH was used, the sole product was X = Et. The same reaction was successfully tried on an appropriate substrate containing a solid-supported α,β-unsaturated amide (Scheme 94) and extended to check the possibility of stereoselection using an asymmetric substrate (Scheme 95).

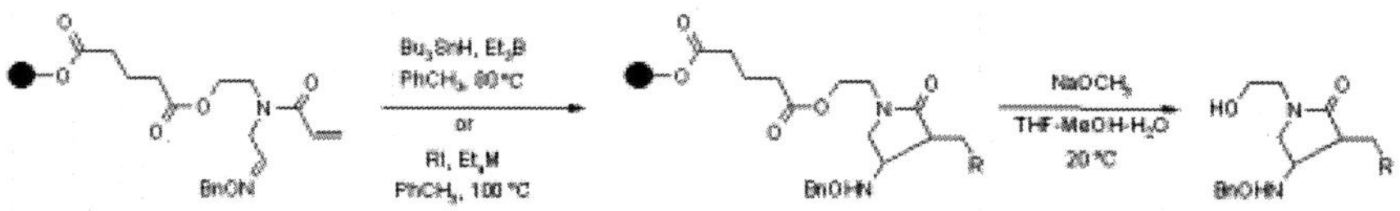

Scheme 94. Radical addition to an electron-poor alkene.

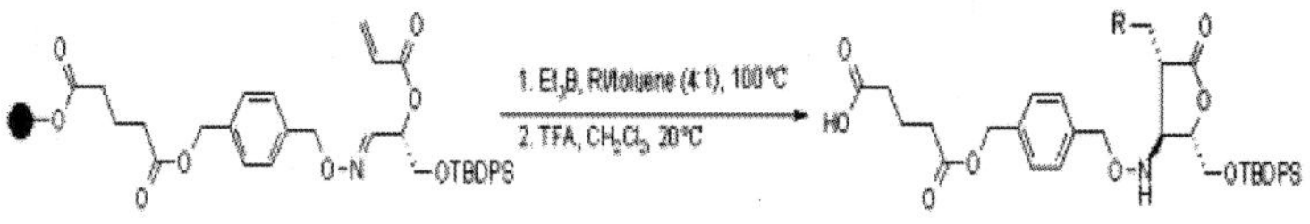

Scheme 95. Example of stereoselective addition/cyclization reaction

The radical addition to an electron-poor alkene is shown in Scheme 94. In a first attempt the addition of a stannyl radical was tested and product 279 was recovered after cleavage in good yield (64%). The same reaction was then run under iodine-atom transfer conditions, using different R-I. The results were satisfying, although a large amount of alkyl iodide had to be used (30 eq.). The only reaction that failed was the one with the bulky t-Bu radical (R = i-Pr: 69%; c-hexyl: 54%; c-pentyl: 59%; s-Bu: 55%). The results obtained using EtnM without R-I were rather interesting, thus, when Et3B was used, a 72% of 279 product (R = Et) was isolated, while the use of Et2Zn gave only a 12% of the same product.

Scheme 95 displays the stereoselective addition/cyclization reaction promoted by alkyl radicals, using asymmetric substrate 280. The reaction was run in a strong excess of R-I (4:1 v/v with toluene), because, with smaller amounts, the product of ethyl radical addition was predominant. The products 281 were recovered in reasonable yields (R = i-Pr: 57%; c-hexyl: 50%; Et: 92%, when the reaction was run with only Et3B), but with a constant diasterometic ratio of 8:1, independently of the nature of the adding radical.

Radical Addition to Alkenes, Alkynes and Arenes

A radical cyclization for the synthesis of 2-oxindoles was studied by Fukase's group [104]. In this case, N-(2-bromophenyl)acrylamides were anchored to the SP and treated under classical radical conditions (Bu3SnH, AIBN) (Scheme 96).

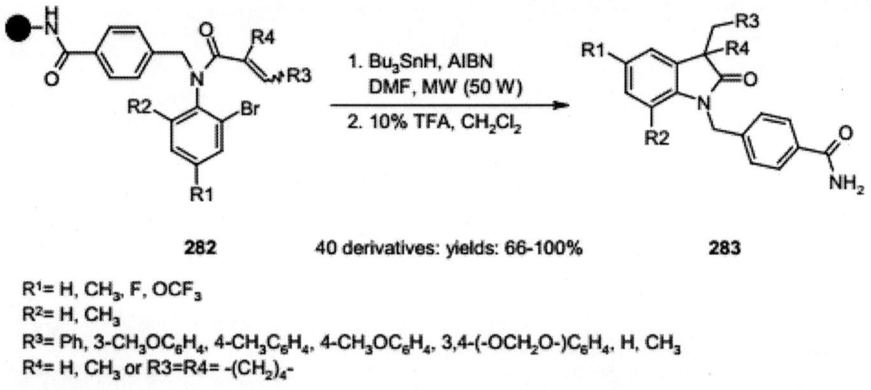

1. Bu_3SnH, AIBN DMF, MW (50 W)
2. 10% TFA, CH_2Cl_2

282 40 derivatives: yields: 66-100% 283

R^1= H, CH_3, F, OCF_3
R^2= H, CH_3
R^3= Ph, 3-$CH_3OC_6H_4$, 4-$CH_3C_6H_4$, 4-$CH_3OC_6H_4$, 3,4-(-OCH_2O-)C_6H_4, H, CH_3
R^4= H, CH_3 or R3=R4= -$(CH_2)_4$-

Scheme 96. Fukase's SP-radical cyclization into alkenes.

Oxindoles such as 283 were obtained smoothly under MW irradiation (<1 h), compared to conventional heating (24 h). Interestingly, the reaction was run in DMF, a solvent where Bu3SnH is rather insoluble. This characteristic forced an increase in concentration of the H-donor on the surface of the resin. In addition, the same reaction run in solution gave no product at all, sustaining the thesis of concentration effect [105].

Bowman and coworkers [106] studied the addition of reactive aryl radicals to azoles and extended the methodology to SP using less toxic and troublesome TBGH and TTMSS (Scheme 97).

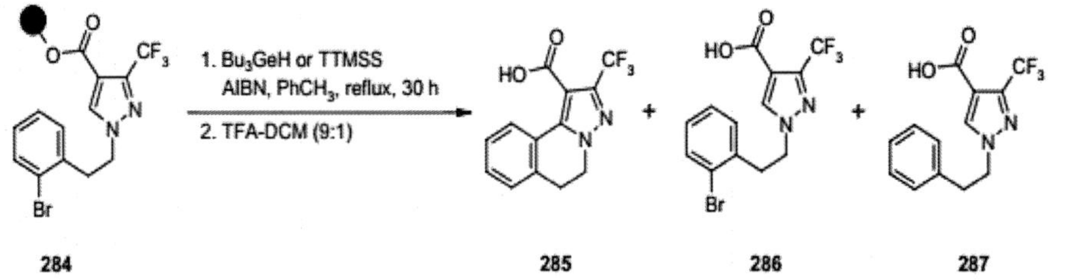

1. Bu_3GeH or TTMSS AIBN, $PhCH_3$, reflux, 30 h
2. TFA-DCM (9:1)

284 285 286 287

Scheme 97. Bowman's SP-radical addition into pyrazole.

The probe scaffold was anchored to a Wang resin. The results were rather disappointing compared to the relative reactions carried out in solution: in fact, when TBGH was used only 20% of 285 was isolated after cleavage, together with 70% of unreacted 286. However, switching to TTMSS improved the yield of 285 (53%), but this time, the side product was the reduced, uncyclized 287 (27%).

DeMaesmaekerandWendeborn[107]studiedthefunctionalization of cyclohexendiol derivatives via radical cyclization on solid phase. Cyclization was performed using an aryl radical (Scheme 98) and a vinyl radical (Scheme 99), giving in both cases a mixture of products.

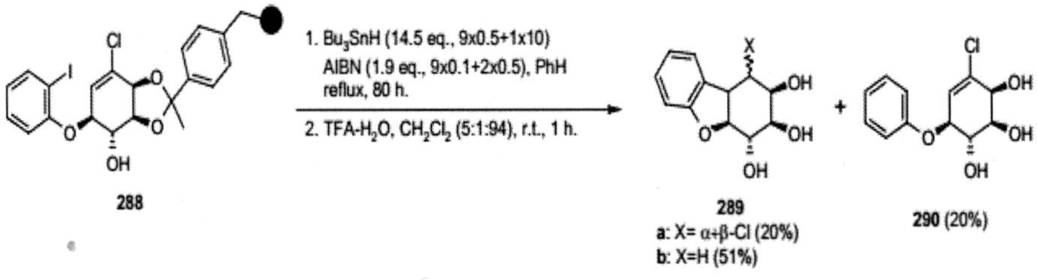

1. Bu_3SnH (14.5 eq., 9x0.5+1x10) AIBN (1.9 eq., 9x0.1+2x0.5), PhH reflux, 80 h.
2. TFA-H_2O, CH_2Cl_2 (5:1:94), r.t., 1 h.

288

289
a: X= α+β-Cl (20%)
b: X=H (51%)

290 (20%)

Scheme 98. De Maesmaeker's SP-aryl radical cyclization into alkene.

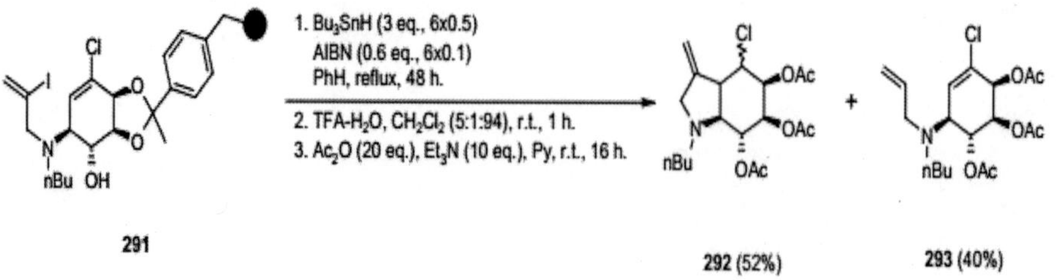

Scheme 99. De Maesmaeker's SP-vinyl radical cyclization into alkene.

In the first case (Scheme 97), product 289a was obtained as the minor component together with the prematurely reduced starting material 290, and the product of over-reduction of 289a (289b, X = H).

Unfortunately, it resulted impossible to achieve complete removal of the chlorine atom without increasing the amount of 290. In the second reaction (Scheme 98), a large amount of reduced product 293 was found, while chlorine atom resisted in the desired product 292. Du and Armstrong [108] provided a milder approach to the synthesis of solid-supported benzofurans via radical cyclization, which is based on the use of SmI2 (Scheme 100).

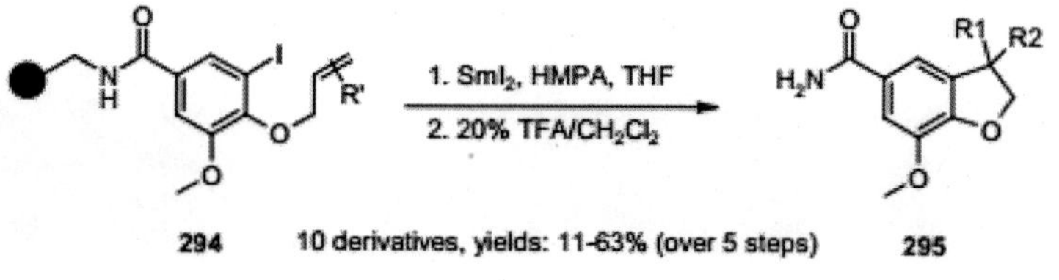

Scheme 100. Armstrong's SP-aryl radical cyclization into alkenes.

Yields were measured over 5 steps (4 of them necessary to synthesize 294). The reaction was run at room temperature and a range of diverse substituted alkenes was tested. The cyclization reaction took place in almost all cases, being the α,β-unsaturated ester (294 with R′ = COOEt) derivative an exception. However, in some cases, the yields were lower due to a variable amount of disproportionation product. This side-reaction proved to be substrate-dependent more than SP-dependent, since parallel test reactions in solution were carried out, revealing similar problematic.

A major drawback of this approach could be the use of toxic reagents in excess (HMPA was necessary for the reaction to take place) and the difficulty in washing out the samarium ions when ungrafted polystyrene resin (Rink) was used. The latter problem was resolved using polyethylene glycol grafted resin, which swells well in aqueous solvents, making wash-up with NaHCO3 saturated solution more efficient for the elimination of Sm3+ impurities from the beads. In a subsequent paper [109], the same group brought forward the reaction depicted in the previous scheme by capture of the anionic metal derivative, formed after cyclization, with an electrophile (Scheme 101).

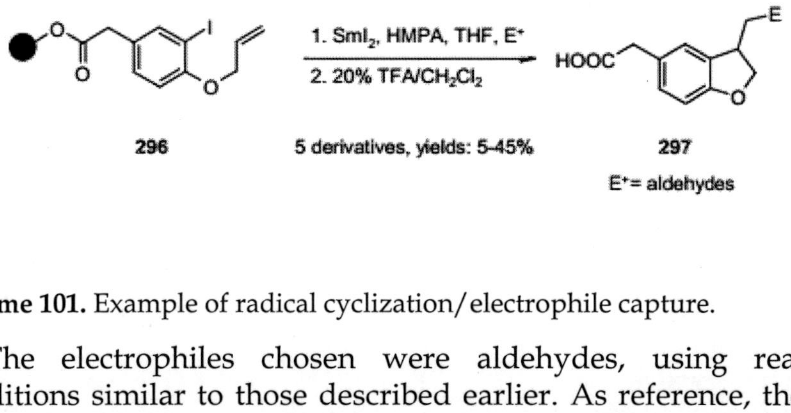

1. SmI_2, HMPA, THF, E^+

2. 20% TFA/CH_2Cl_2

296 5 derivatives, yields: 5-45% 297

E^+= aldehydes

Scheme 101. Example of radical cyclization/electrophile capture.

The electrophiles chosen were aldehydes, using reaction conditions similar to those described earlier. As reference, the test reactions were also run in solution, where it was acknowledged the short half-life of the anion formed.

Compared to the previous example, attachment to the SP was performed via an appropriate handle on the benzene ring. In a first attempt Rink resin was used, giving no product of electrophilic quenching of the anion. Since the amide nitrogen in the linking group of Rink was supposed to be responsible for early interception of the anion generated after radical cyclization, it was decided to use the ester linkage of TentaGel S BHP resin (a resin also swelling better in the solvent used for the reaction). In this case, interception was effective, even if in a mixture with simple cyclized product.

Since the yields were strongly dependent on the electrophile nature, this method did not present the quality necessary to be used in combinatorial chemistry. Following the same idea, Berteina and De Maesmaeker [110] tested the 5-exo-trig cyclization to form 2,3-dihydro-benzofurans. Either the aryl iodide or the unsaturated counterpart, tethered by means of an oxygen atom, was directly anchored to aminomethylphenyl functionalized polystyrene beads. Usually, a spacer was inserted between the resin and the reactive molecule. As first attempt, the radical cyclization was performed under usual radical conditions (Scheme 102).

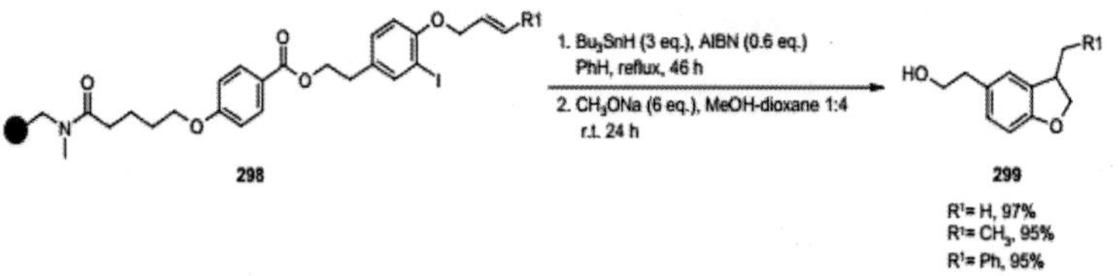

Scheme 102. De Maesmaeker's SP-aryl radical cyclization into alkenes.

The radical cyclization depicted in Scheme 102 was effective, high yielding and no product of early reduction or starting material was found. Interception of the alkyl radical formed after cyclization with allytributyltin was also attempted (Scheme 103). For R^1 = H, the reaction gave a good yield of desired product 300 (beside 22% of simple cyclized product 299), while for R^1 = CH_3 or Ph, yields were pretty low and after cleavage the main product was found to be the unreacted starting material (beside 10% of 299), demonstrating a slower reaction that can be explained on the basis of steric hindrance.

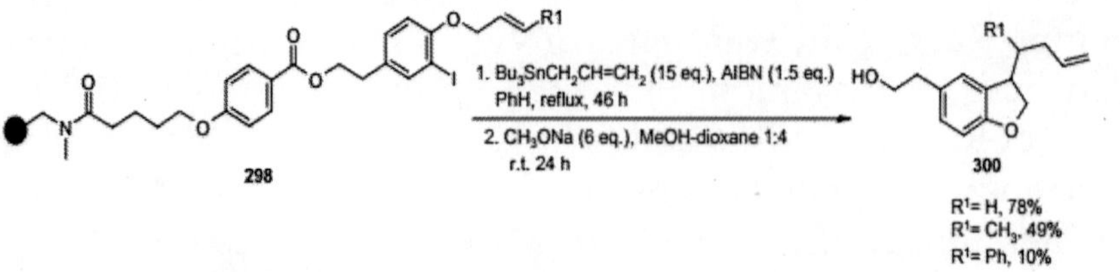

Scheme 103. Example of radical cyclization/allyl interception.

In a second attempt, it was chosen to have the unsaturated moiety, in this case an alkyne, directly connected to the SP. The reaction led to a mixture of two products 302 and 303 (Scheme 104).

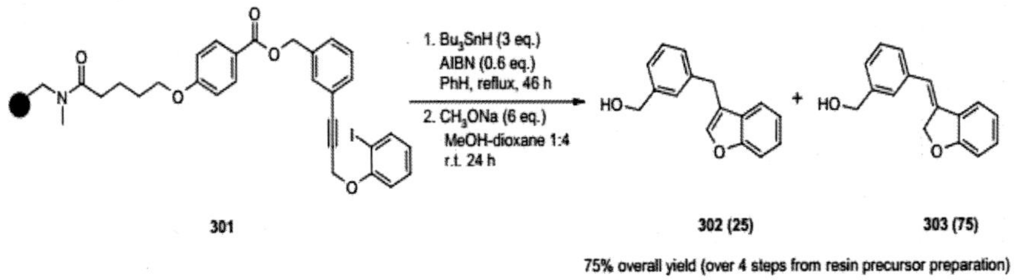

Scheme 104. De Maesmaeker's SP-aryl radical cyclizatino into alkyne.

In another publication [111], the same group used a similar 5-exo-trig cyclization to form 1,3-dihydro-isobenzofurans (Scheme 105).

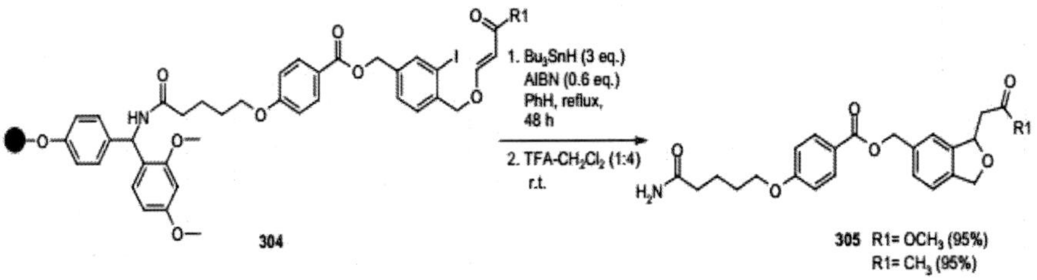

Scheme 105. De Maesmaeker's synthesis of benzofurans 305.

The acid-labile Rink linker was necessary since, following the strategy adopted in Schemes 103 and 104, low yields of desired products (due to partial decomposition under the basic cleavage conditions) were obtained. This change of plan gave almost quantitative yields after cleavage.

In another experiment, it was evaluated the radical cyclization onto an aromatic ring, which yielded a benzoquinoline (Scheme 106).

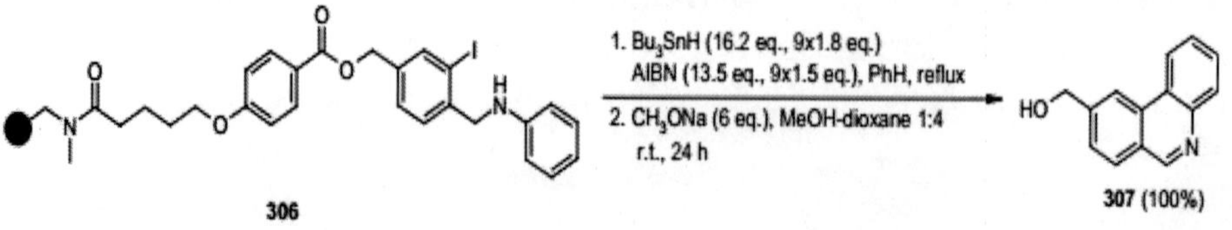

Scheme 106. De Maesmaeker's SP-aryl radical cyclization into a benzene ring.

In this case, it was necessary to use large amounts of H-donor and, most importantly, of radical initiator (AIBN), added in aliquots during the reaction. Lower amounts of the reagents gave uncompleted conversions. The need of a large excess of AIBN is probably due to its role played in the aromatization step after cyclization.

Also Routledge's group [112] used the intramolecular radical cyclization to form tetrahydrofurans derivatives (such as the one depicted in Scheme 100) testing the use of hypophosphite salts as radical carriers in alternative to the toxic TBTH on SP (Scheme 107).

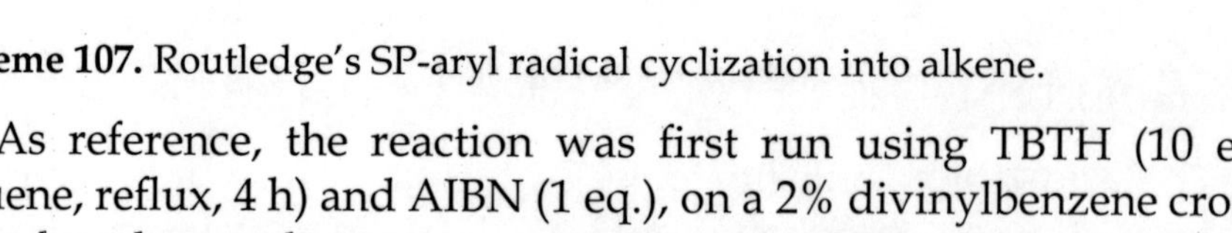

Scheme 107. Routledge's SP-aryl radical cyclization into alkene.

As reference, the reaction was first run using TBTH (10 eq., toluene, reflux, 4 h) and AIBN (1 eq.), on a 2% divinylbenzene cross-linked carboxypolystyrene resin, giving a 95% yield of cyclized product 311a (beside 4% of 309 and 1% of 310). The reaction was then run using N-ethylpiperidine hypophosphite (EPHP, 20 eq.) and AIBN (3 eq.), but at best, only a 57% yield of 311a was isolated (besides 38% of 309 and 5% of 310). The reason for this behavior was initially supposed to lay in the nature of the solvent used (toluene), since EPHP is a salt and its concentration on the surface of the resin could be low in apolar solvents. However, when the reaction was run in polar mixtures (THF-EtOH 4:1), no real change of product distribution was observed (63% 311a, 31% 309, 6% 310). The attention was then shifted to the nature of the solid phase and a range of cross-linked and PEG grafted solid-supports was tested. The idea was that resins more hydrophilic than polystyrene would allow the radical carrier to approach the solid surface. For the purpose the following resins were used: NovaSyn® TG carboxy resin, a macroporous resin (ArgoPore®), PEG grafted resins, HypoGel® and ArgoGel® and the polytetrahydrofuran cross-linked resin, JandaJel®. The best results were obtained using the latter, which gave a 98% of 311a (beside a 2% of 310), under optimized conditions (20 eq. EPHP, initiator 2 eq., THF-EtOH 4:1, 48 h, under reflux). To confirm the validity of the choice made, a different reaction was carried out, using a primary alkyl radical precursor 312 (Figure 108).

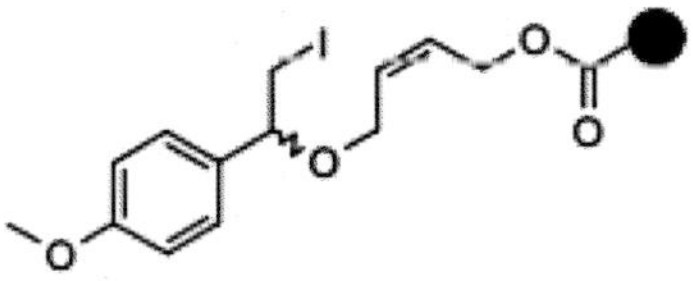

312

Figure 108. Structure of alkyl radical precursor 312.

The reaction was successful, with a 92% yield of desired cyclized product. The use of alternatives such as EPHP on SP could allow radical chemistry to become a more useful method for SP synthesis, with potential extension to combinatorial and medicinal chemistry.

In a recent paper [113], the same group studied the influence of microenvironment effects in the reaction depicted in Scheme 107 by determination of deuterium incorporation in the cyclized product. In order to do so, cyclization was carried out using deuterated EPHP or Bu3SnD in deuterated solvents (d8-THF: d6-EtOH 4:1 or 1:1; d4-MeOH; d6-benzene). The ratio of deuterated product 311b (R=D) versus 311a (R=H) was measured by GC-MS analysis.

The objective was the rationalization of the role played by the polymer backbone/linker in the radical H-abstraction, a side-reaction encountered in radical chemistry on SP. A first conclusion was that, as already observed by Curran [8], all resins can compete with relatively slow H-transfer reactions (such as H-abstraction from

EPHP by an alkyl radical). However, the overall picture indicated that there are multiple factors influencing radical chemistry on SP: nature of the resin and linker (flexibility), solvent (swelling properties) and nature of the radical carrier (bond strength andsolubility). Therefore, a general prediction of the effects of a specific resin on kinetics and mechanism of a radical reaction is rather difficult and the combination of the above-mentioned factors need to be considered case-by-case.

A more traditional approach was used by Balasubramanian and coworkers [114], who performed aryl radicals cyclization into alkenes to prepare benzofuran derivatives (Scheme 109) and alkyl radicals cyclization into alkynes to prepare functionalized furans (Scheme 110). The reactions were run using a very large excess of the H-donor, Bu3SnH. When the same excess of organostannyl hydride was used in solution, large amount of early reduction products were normally obtained. Therefore, somehow, use of the solid phase seemed to neutralize this inconvenience.

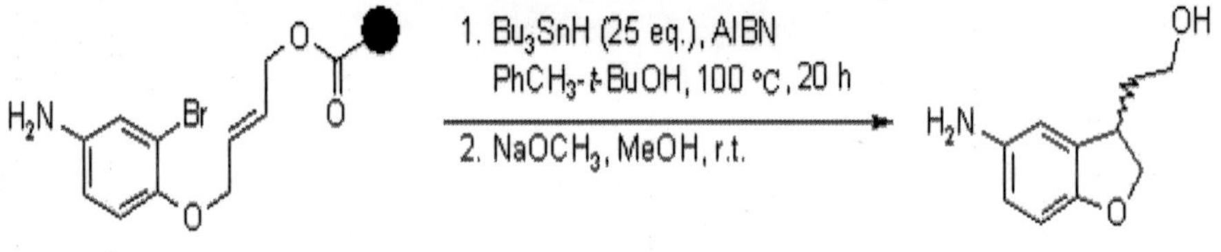

313 314

Scheme 109. Balasubramanian's SP-aryl radical cyclization into alkene.

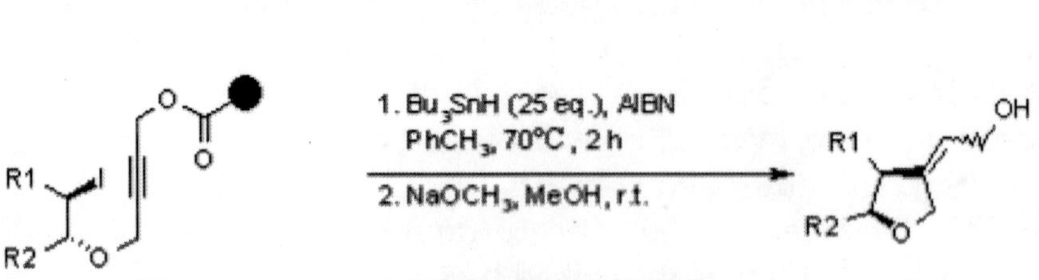

315 4 derivatives: yield: 63-80% 316

R^1= Ph, t-Bu; R^2= H
or
R^1=R^2= -$(CH_2)_3$-, -$(CH_2)_4$-

Scheme 110. Balasubramanian's SP-alkyl radical cyclization into alkene.

Both carboxylated PS and TentaGel-COOH were used as resins. The main difference among them was the amount of radical initiator needed; in fact, in the former case, an excess of AIBN had to be used, while in the latter, only 6% mol was sufficient to perform complete conversion. The reason for this different behavior was attributed to chain-terminating processes involving the benzylic hydrogen on the polystyrene backbone, absent in the TentaGel resin. The lack of reduced side-products was tested on a slower (three orders of magnitude) reaction (shown in Scheme 110) usually more sensitive to early reduction, but even in this case it was not observed.

In an application to the synthesis of molecule of biological interest, a 5-exo-trig radical cyclization was performed by Lown's group [115] for the synthesis of 1-chloromethyl-1,2-dihydro-3H-benz[e]indole (seco-CBI) and a polyamide conjugate (Scheme 111).

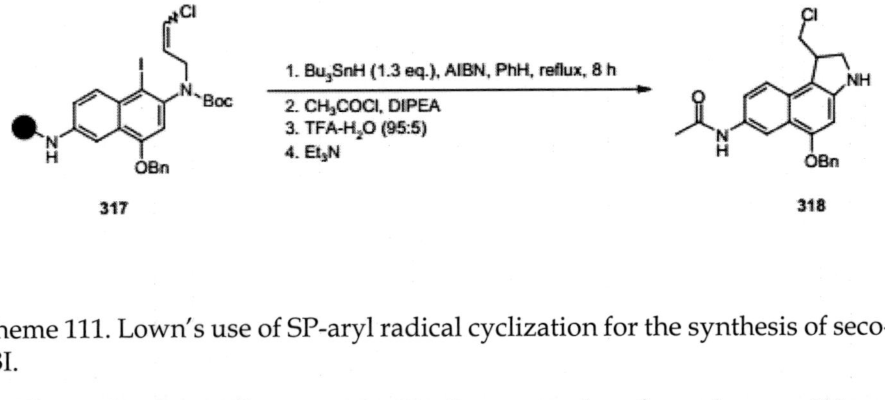

1. Bu_3SnH (1.3 eq.), AIBN, PhH, reflux, 8 h
2. CH_3COCl, DIPEA
3. TFA-H_2O (95:5)
4. Et_3N

317 318

Scheme 111. Lown's use of SP-aryl radical cyclization for the synthesis of seco-CBI.

The radical reaction was carried out starting from bromo-Wang resin according to the procedures applied in solution, commonly used for the synthesis of CC-1065 derivatives. To measure the efficacy of the reaction on SP, the product of cyclization was first acetylated and then cleaved, to yield a total 74% (48% plus a 26% of the product with the Boc protecting group still in place). The synthesis of a bis-polyamide-seco-CBI conjugate was completed using 318 and performing additional amide couplings both on solid- and solution-phase. Toru and colleagues [116] performed the intramolecular radical addition of an alkyl radical to unsaturated C-C bonds, in order to form γ-butyrolactones (Scheme 112).

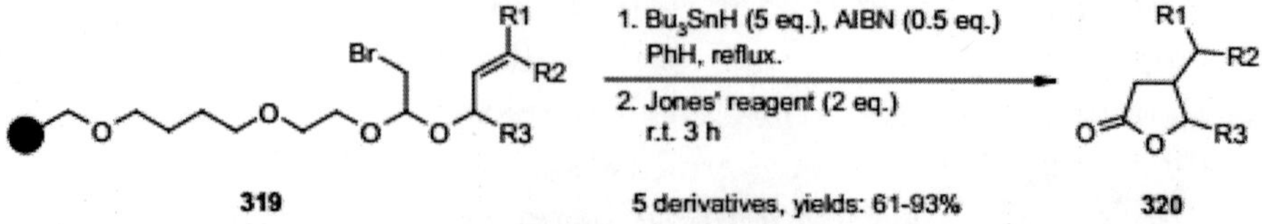

1. Bu_3SnH (5 eq.), AIBN (0.5 eq.) PhH, reflux.
2. Jones' reagent (2 eq.) r.t. 3 h

319 — 5 derivatives, yields: 61-93% — 320

Scheme 112. Toru's SP-alkyl radical cyclization into alkenes.

Merrifield resin was functionalized with an appropriate linear spacer and radical precursors 319 were prepared on the SP by addition of allyl (or propargyl) alcohols to the loaded vinyl ether, in the presence of NBS. The radical reaction was performed under common conditions and the excess of H-donor did not give uncyclized, reduced side-products. The toxic organotin by-products were separated by filtration and washing, while cleavage of the desired products from the resin was achieved using Jones' oxidation. Yields were rather satisfactory, even when a 5-exo-dig cyclization on a triple bond was tested (47%, E:Z = 1:9).

During his studies towards the use of supported radical chemistry, Enholm [117] proposed the first example of stereoselective radical cyclization on a soluble support. The use of soluble supports (precipitating at –78 °C in MeOH) represent a great advantage for SP radical reactions since, according to the authors, the rates of the reactions involved are similar to those run in solution. Thus, a custom polymer incorporating a chiral auxiliary unit was synthesized by ring opening metathesis polymerization (ROMP). The radical reaction studied was a 6-exo-trig cyclization (Scheme 113).

After building the monomer, the polymer was prepared employing a Grubbs-type ruthenium-based catalyst. Each monomer contained a succinimide anchor tethered to an (+)-isosorbide unit (the chiral auxiliary) esterificated with the radical precursor. After the ROMP, the polymer obtained reacted under radical conditions in the presence of a Lewis acid. Subsequent cleavage with lithium hydroxide afforded the desired acid 323. Constructing the polymer in this way, it was possible to achieve a 100% loading, a result not viable using commercial resins. The results were encouraging since, beside an 80% isolated yield, there was a >99% ee (being ZnCl2 the best solution).

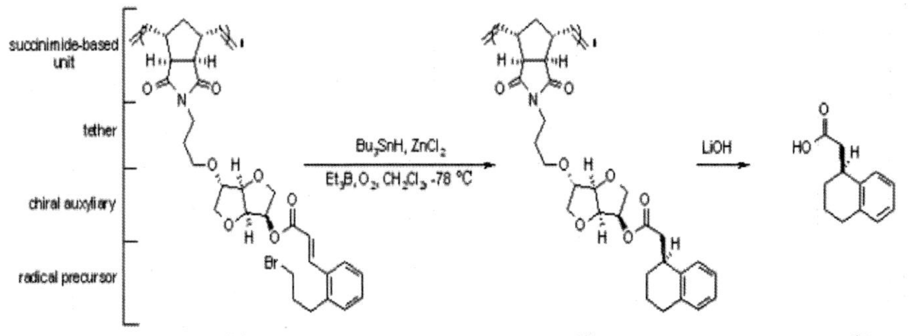

Scheme 113. Enholm's stereoselective alkyl radical cyclization on ROMP polymer.

The use of a combination of low temperatures and Lewis acid for stereoselective radical cyclization is known [118-123], however, in this case, the use of zinc chloride proved to be far more better than other LA (such as MgBr2 or Yb(OTf)3), probably owing to the presence of the polymer-embedded sugar which creates a favorable environment for zinc-mediated stereoselection. Kilburn's group [124] exploited the reactivity of isocyanides to perform thiol-mediated radical cyclizations to obtain pyrroline or pyrrolidinone derivatives. In Scheme 114 a Wang-type resin was used. Treatment under radical cyclization conditions in the presence of 2-mercaptoethanol gave resin 325 that after trifluoroacetic acid-mediated cleavage gave pyrrolidinone 326 in reasonable yields.

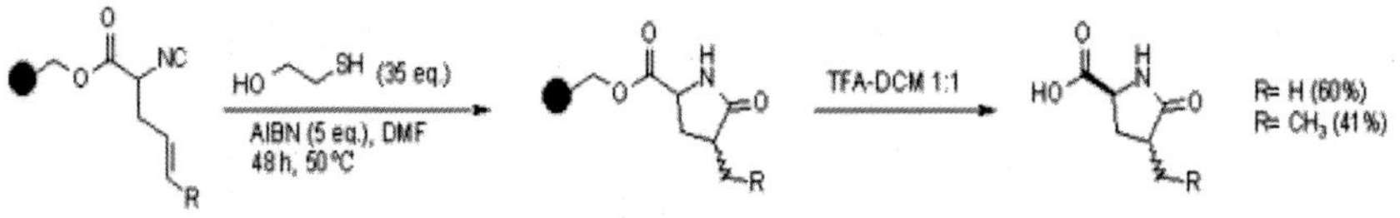

Scheme 114. Kilburn's thiol-mediated radical cyclization into SP-isocyanides.

When HMBA-AM resin was used (Scheme 115), the use of ethanethiol during the cyclization step gave rise to resin 328, which could be cleaved under nucleophilic conditions preserving the pyrroline structure to yield amide derivative 329. In both cases the starting isonitrile was built on resin, which has the advantage of avoiding the handling of the unpleasant substance.

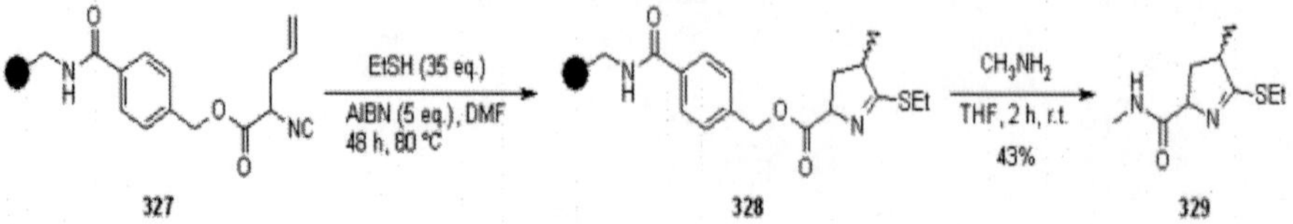

Scheme 115. Kilburn's pyrroline synthesis.

Harrowven and colleagues [125] reported two new methods for radical cyclization on SP using supported dienes and, respectively, thiyl and tosyl radical. In this case heating of the loaded resin in the presence of AIBN and the radical precursor (PhSH or TsSePh) in an appropriate solvent was sufficient (Scheme 116).

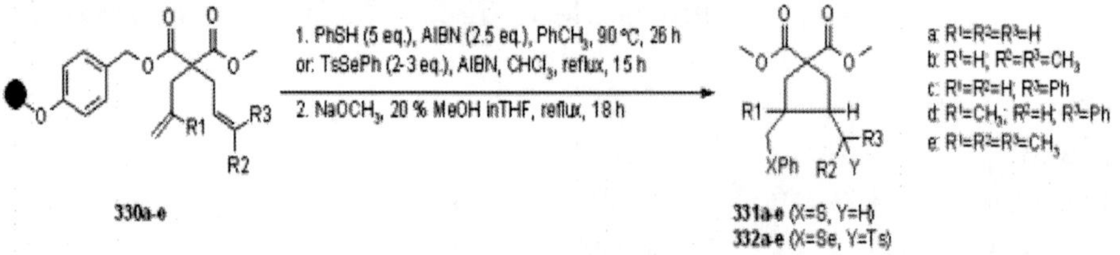

Scheme 116. Harrowven's SP-atom transfer radical cyclization into alkenes-alkenes

Radical precursors 330a-e were loaded on a PS-Wang resin. The results obtained when using phenylthiol (X = S, Y = H) as radical source/cyclization promoter were rather satisfactory, with yields of 331 ranging between 69-78%; when using p-tolyl benzeneselenosulfonate (X = Se, Y = Ts) on the same substrates, the desired products 332 were recovered in comparable yields (66-74%). The isomeric product distribution followed Beckwith's guidelines for radical cyclizations [126,127].

Carbon-Heteroatom Formation

An interesting radical cyclization initiated by IBX was discovered [128-130] and thoroughly studied by Nicolaou. Mechanism and limits of the reaction as well as an application in solid-phase were subsequently reported [131] (Scheme 117). The interesting mechanism comprises a radical reaction initiated by IBX (A), where THF actively participate (B) and explains the reason for the need of an aryl amide for the success of the reaction. While it was tested and performed with good results in solution, the extension to the SP resulted in poor yield of the desired product (15%), probably due to oxidation of the resin by the IBX excess.

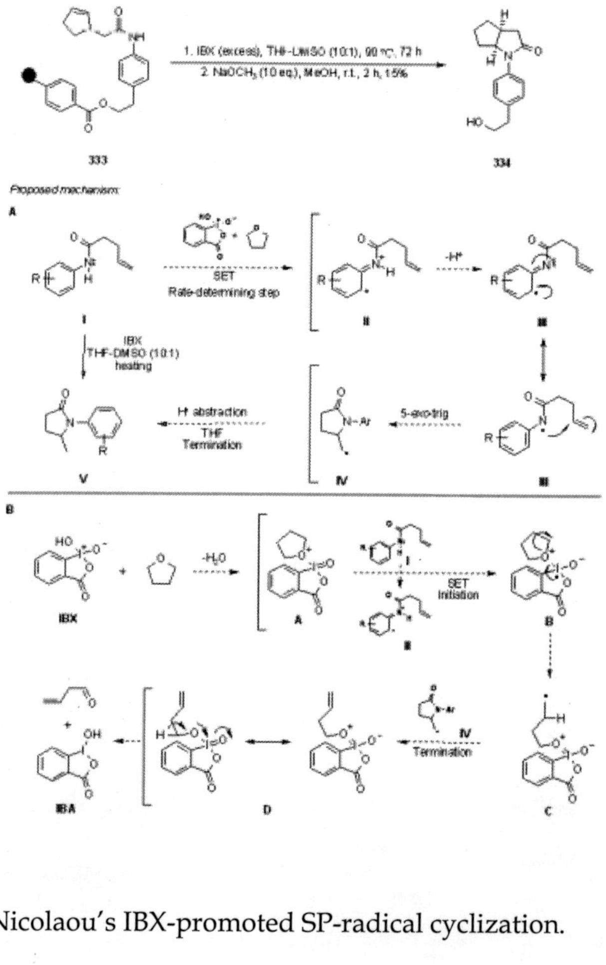

Scheme 117. Nicolaou's IBX-promoted SP-radical cyclization.

Caddick and coworkers performed the addition of sulfonyl radicals to isolated alkenes (Scheme 118) and alkynes (Scheme 119) [132] and the best conditions are displayed below. Interestingly, the nature of solvent, linker (eq. 1,2 of Scheme 118 vs. eq. 1,2 of Scheme 119) and resin did have great influence on the reaction outcome, spanning between combinations were there was no reaction at all, to conditions giving decomposition.

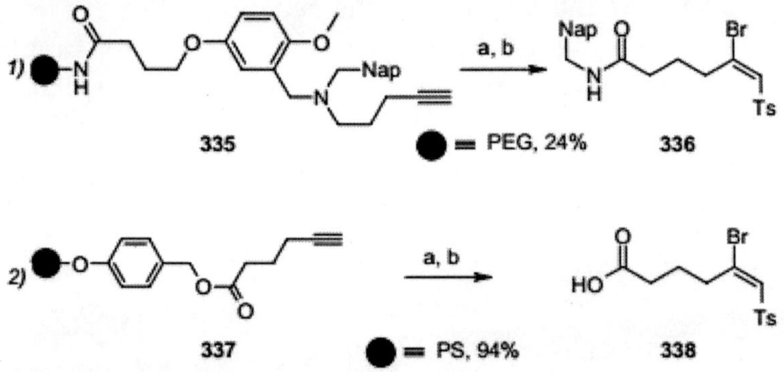

1) 335 → (a, b) 336; ● = PEG, 24%

2) 337 → (a, b) 338; ● = PS, 94%

a) TsBr, AIBN, $PhCH_3$, 65-70 °C; b) 95% TFA (aq.)

Scheme 118. Caddick's radical addition to alkynes.

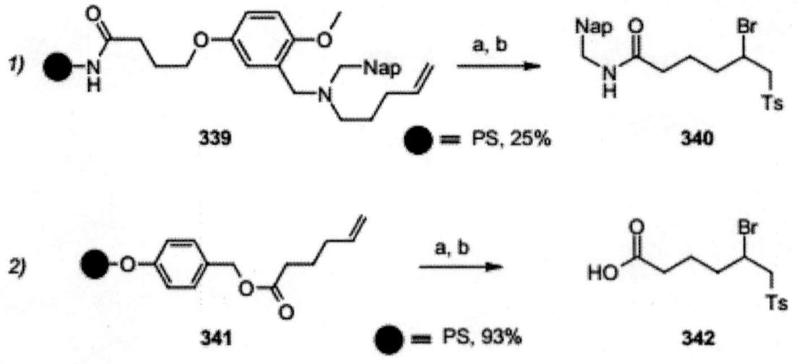

1) 339 → (a, b) 340; ● = PS, 25%

2) 341 → (a, b) 342; ● = PS, 93%

a) TsBr, AIBN, $PhCH_3$, 65-70 °C; b) 95% TFA (aq.)

Scheme 119. Caddick's radical addition to alkenes.

During the studies about the transformation of hydrophilic to hydrophobic groups, in order to change the pharmacological properties of molecules, Plourde Jr. [133] explored the intermolecular radical addition of thiols to supported alkenes (Schemes 120 and 121).

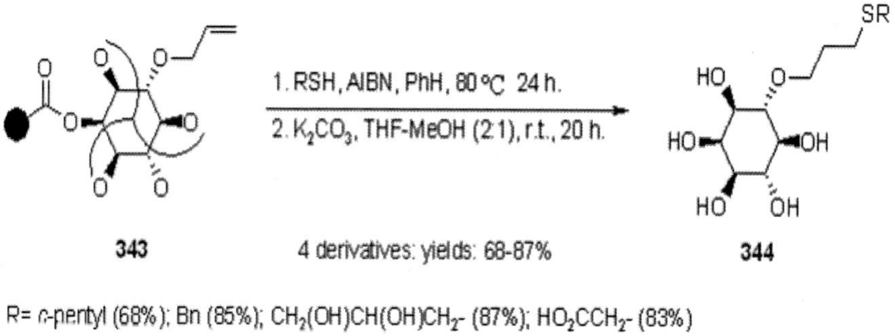

1. RSH, AIBN, PhH, 80 °C 24 h.
2. K_2CO_3, THF-MeOH (2:1), r.t., 20 h.

343 4 derivatives: yields: 68-87% 344

R= n-pentyl (68%); Bn (85%); $CH_2(OH)CH(OH)CH_2$- (87%); HO_2CCH_2- (83%)

Scheme 120. Plourde Jr.'s radical addition to supported alkenes.

The reactions were run under the usual radical conditions, even if when the carbamate linker was used (Scheme 121) an uncommon solvent for radical reactions such as DMF was chosen. Yields were generally good, however, when R = Ar or HetAr, no reactions was observed (not shown). Addition of a thiol-bearing resin to an alkene was employed also for the synthesis of a new linker useful for the synthesis of amines. In order to improve the scope of the 2-(thiobenzyl) ethyl carbamate linker normally used for such a purpose, Timár and Gallagher [134] proposed to start from Merrifield SH resin.

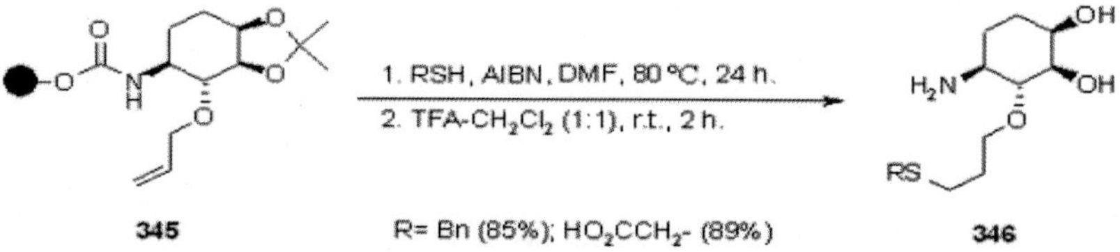

1. RSH, AIBN, DMF, 80 °C, 24 h.
2. TFA-CH_2Cl_2 (1:1), r.t., 2 h.

345 R= Bn (85%); HO_2CCH_2- (89%) 346

Scheme 121. Example of Plourde Jr.'s use of carbamate linker.

This (Scheme 122) was treated with the previously prepared N-vinyloxycarbonyl derivative 348, under radical conditions. Addition of the benzyl thiyl radical to the vinyloxy moiety loaded the desired starting material to the resin, giving 349, which was treated under oxidative/basic conditions for cleavage. This modification offers an entry for secondary amines, complementing the original procedure (exploiting the reaction between a supported alcohol and an isocyanate), only viable for primary amines.

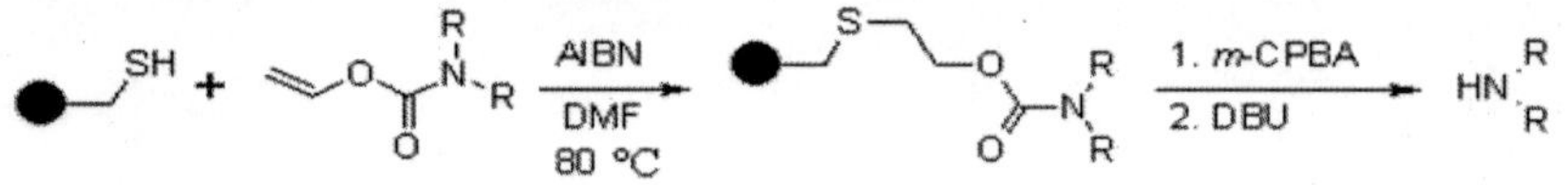
SH
AIBN
DMF
80 °C
1. m-CPBA
2. DBU
HN
347
348
349
350

Scheme 122. Gallagher's SP-radical synthesis of amines.

A possible problem in SP intermolecular radical reactions consists of the lower rate constants compared to the reactions run in solution, probably due to the heterogeneous nature of SP reactions. Zard's group [135] tried to address this problem by synthesizing a Wang-type soluble support. This support was then compared with the classical polystyrene Wang resin, via the radical addition of supported xanthates to olefins in solution and by the addition of immobilized olefins to free xanthates. Soluble support 357 was synthesized by xanthate polymerization and employed for intermolecular test radical addition. (Scheme 123)

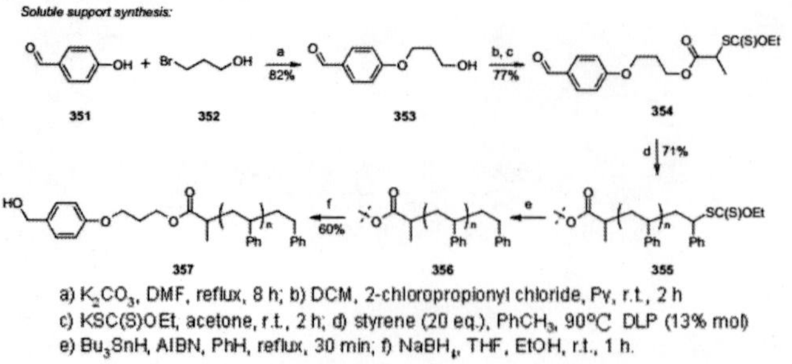
Soluble support synthesis:
351
352
353
354
355
356
357
SC(S)OEt
82%
77%
71%
60%
a) K_2CO_3, DMF, reflux, 8 h; b) DCM, 2-chloropropionyl chloride, Py, r.t., 2 h
c) KSC(S)OEt, acetone, r.t., 2 h; d) styrene (20 eq.), $PhCH_3$, 90°C DLP (13% mol)
e) Bu_3SnH, AIBN, PhH, reflux, 30 min; f) $NaBH_4$, THF, EtOH, r.t., 1 h.

Scheme 123. Zard's synthesis of soluble support for radical cyclization.

In Scheme 124 the supported xanthate was reacted with an excess olefin 360 under tin-free radical conditions. For both resins, the reaction was biased by the formation of by-products, compared to the one run without exploiting polymers. However, 359b (soluble support) gave cleaner reaction profile and higher yields (54% vs. 26%) compared to 359a (Wang). For Wang resin a large excess of olefin 360 and 30% mol of lauroyl peroxide (DLP) needed to be used, while for the soluble support, a 3-fold excess of 360 and 40% DLP were needed to complete the reaction.

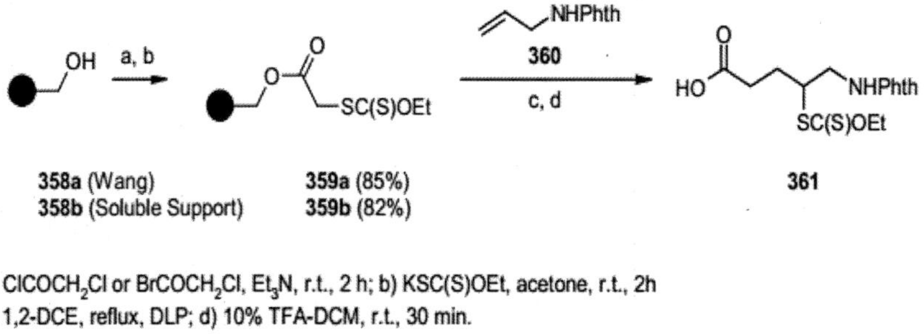

358a (Wang)
358b (Soluble Support)
359a (85%)
359b (82%)
361

a) $ClCOCH_2Cl$ or $BrCOCH_2Cl$, Et_3N, r.t., 2 h; b) KSC(S)OEt, acetone, r.t., 2h
c) 1,2-DCE, reflux, DLP; d) 10% TFA-DCM, r.t., 30 min.

Scheme 124. Zard's use of supported xanthates.

When the reagents were inverted (Scheme 125), in both cases, the reaction did not go to completion and side-product 365 was found in similar proportion to the desired product. Despite this, the reaction was substantially cleaner compared to the one described in Scheme 124, even if it was still needed a consistent amount of DLP (50% mol) compared to the polymer-free reaction. Final yields were reasonable (52% for Wang, 70% for soluble support). The presence of side product 365 says that at certain xanthate concentrations, the intermolecular reaction is in competition with the intramolecular process, preventing the reaction to go forward in the desired direction. These results suggest that intermolecular radical processes are viable on SP and that the use of soluble support has more benefits compared to the heterogeneous classical SP synthesis.

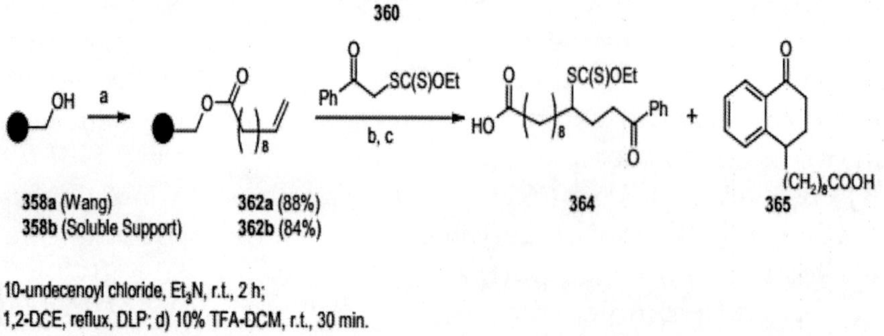

a) 10-undecenoyl chloride, Et_3N, r.t., 2 h;
c) 1,2-DCE, reflux, DLP; d) 10% TFA-DCM, r.t., 30 min.

Scheme 125. Zard's use of supported alkenes.

Back and Zhai [136] performed a radical addition of PS-arylselenosulfonates to alkynes (Scheme 126).

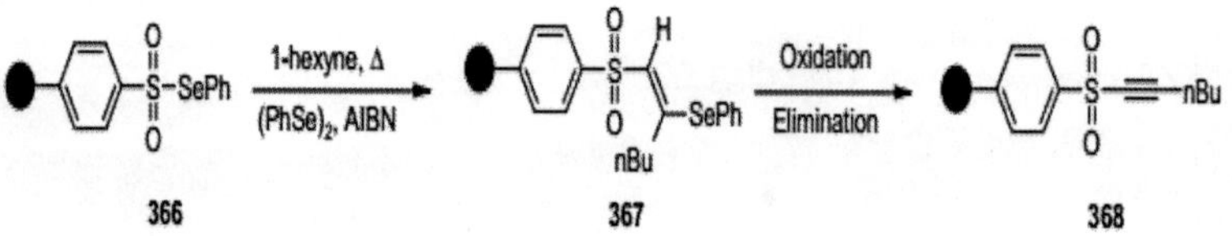

Scheme 126. Back's SP-radical addition of resin 366 to free alkynes.

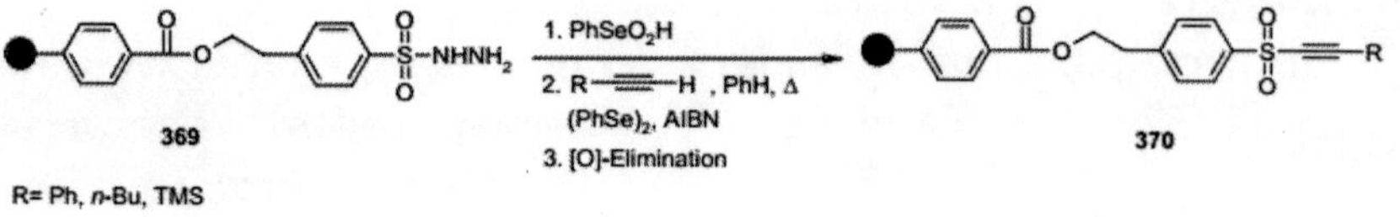

Scheme 127. Back's SP-radical addition of resin 369 to free alkynes.

Acetylenic sulfones of the kind of 368 (or 370) are useful synthones for a variety of reactions (such as cycloadditions). Their first attempt to perform their synthesis on SP through radical addition to 1-hexyne is displayed in Scheme 127. Radical reaction and subsequent oxidation-elimination to restore the triple bond were effective, however, when

368 was further reacted under cyclization conditions, it did not yield any product. For this reason, it was added a spacer to the SP, therefore, resin 369 was prepared and reacted to yield products 370. Finally, 370 were successfully reacted under cyclization conditions, giving the desired products in good yields.

In the quest for new methodologies viable for the synthesis of small molecules on SP, Attardi and Taddei [137] presented a radical approach for the construction of unnatural amino acids and their small oligopeptides (Scheme 128).

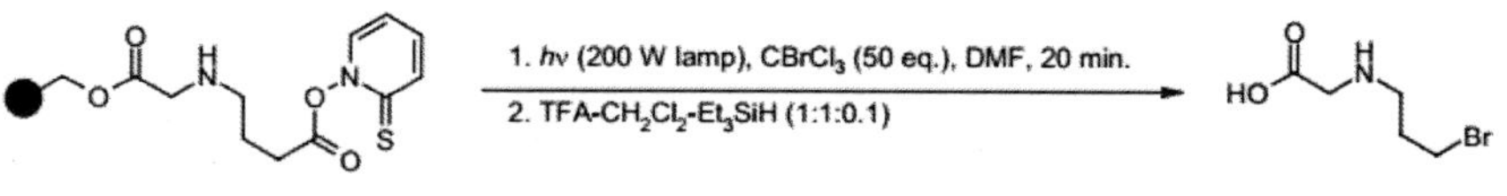

1. $h\nu$ (200 W lamp), $CBrCl_3$ (50 eq.), DMF, 20 min.
2. TFA-CH_2Cl_2-Et_3SiH (1:1:0.1)

371 372

Scheme 128. Taddei's SP-synthesis of unnatural AAs by radical approach.

In the example reported above the Wang resin anchored Barton ester 371 was obtained by coupling of the correspondent acid with 1-hydroxy-2-pyridinethione. The photolabile product was submitted to irradiation to give radical fragmentation and loss of CO2, followed by trapping with CBrCl3. To minimize side-products it was necessary the use of DMF as solvent and 50 equivalents of the halogenated chemical reagent. After cleavage, a respectable 72% yield was obtained. The same process was applied to di- and tri-peptides, while the resulting alkylbromides were further reacted with nucleophiles to obtain the desired unnatural aminoacid side-chain.

Other

Troc (2,2,2-trichloroethoxycarbonyl) group is a widely used protecting group in organic, especially oligosaccharide, chemistry. Since its removal is often performed under heterogeneous conditions, it is not ideal for SP chemistry. Fukase [105] studied a methodology for Troc removal on a substrate anchored to SP, performed using a radical methodology (Scheme 129). The conditions used were tested in solution before extension of the methodology to SP and it was found that DMF was the best solvent (the reaction did not proceed in toluene), probably due to a concentration effect of the organotin derivative to the polymer support; and that the reactions proceeded slowly under conventional heating, therefore, microwave irradiation seemed to be the best choice. Since the amount of $(Bu_3Sn)_2$ was substoichiometric, an involvement of DMF in the catalytic cycle was also suggested. Beside the high yields obtained, the method was also effective because of the absence of dichloroethoxycarbonylated byproducts.

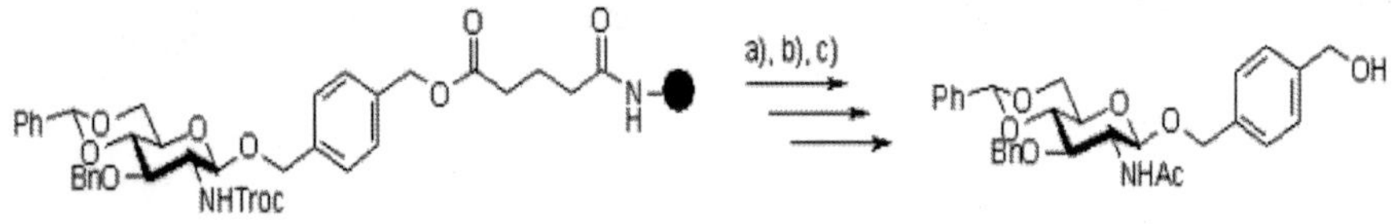

373

374

88% yield (5 steps)

a) $(Bu_3Sn)_2$ (0.1 eq.), DMF, MW (50 W, 20 psi, 200°C 30 min)
b) Ac_2O
c) NaOMe (1 M), THF-MeOH (3:1)

Scheme 129. Fukase's SP Troc radical deprotection.

The couple 2-aminobenzoic acid (Abz) and 3-nitrotyrosine (Tyr(3-NO2)) is used as fluorophore-quencher combination in FRET (Fluorescence Resonance Energy Transfer) methodology for monitoring structural, functional, or aggregation changes in evaluated biomolecules. Since the couple can be easily incorporated in peptides, it would be of great advantage having a method for selective nitration of tyrosine incorporated on a peptide anchored to SP. Savinov and colleagues [138] did find this very mild, SP-friendly methodology for selective nitration of phenols. After a careful optimization of the reaction conditions and a mechanistic study, the reaction was tested on resin-bound substrates (Scheme 130).

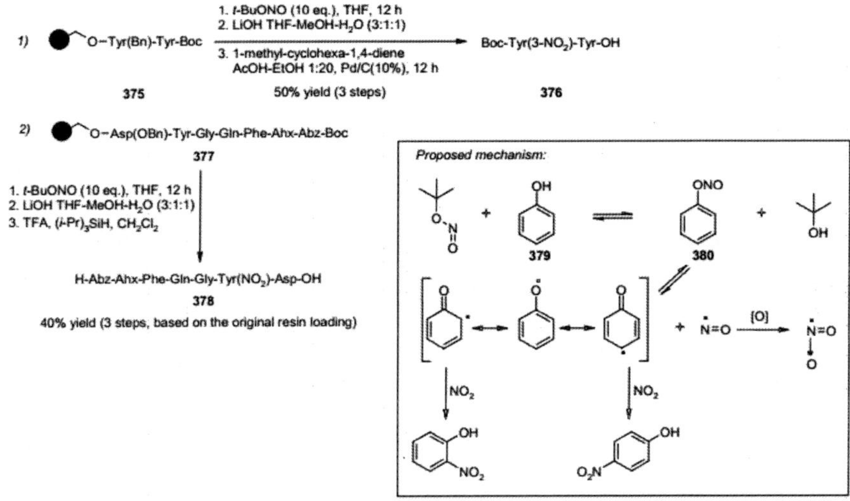

Scheme 130. Savinov's nitration of phenols

The right solvent for the reaction (THF) was first identified in solution, where some evidence of chemoselectivity was also found. Most importantly, the reaction worked only when a phenol hydroxyl group was present. Subsequently, the protocol was tested on dipeptide 375, confirming the need for a free phenolic hydroxyl group and, finally, on a peptide such as 376, giving the desired nitrotyrosine derivative 378 in good yield. The hypothesized mechanism was confirmed by the isolation of the O-nitroso derivative (such as 380) and the observation of the stabilized phenoxyl radical of 379, with its characteristic blue color. Decomposition of an O-nitroso derivative such as 380 led to nitric oxide, which underwent oxidation (probably from air oxygen) to nitrogen dioxide that subsequently coupled with the phenoxyl radical to give the nitration products. It is clear from the proposed mechanism, that the phenol hydroxyl group participates in the reaction, in fact, when anisole was subjected to the same experimental conditions, no nitrated products were observed.

CONCLUSIONS

Radical chemistry, despite having a long-standing tradition and deep relationships with polymers, has become one of the available means for solid-phase synthesis only in relatively recent years. Nevertheless, the wealth of the examples we have provided above witnesses the fruition of such a merging.

On the other hand, in the same period of time the concept of diversity-oriented synthesis (DOS) [12] has abruptly emerged in organic chemistry, with the goal of providing collections of small molecules. A number of tools are available to achieve such an end and in this context, as this review has illustrated, radical chemistry applied to SP represents an important and valid alternative

ACKNOWLEDGEMENTS

The authors would like to gratefully acknowledge Paolo Trifirò for his helpful assistance.

REFERENCES AND NOTES

1. Gomberg, M. An Instance of Trivalent Carbon: Triphenylmethyl. J. Am. Chem. Soc. 1900, 22, 757-771.
2. Geng, Y.; Discher, D.E.; Justynska, J.; Schlaad, H. Grafting Short Peptides onto Polybutadiene-block-poly(ethylene oxide): A Platform for Self-Assembling Hybrid Amphiphiles. Angew. Chem. Int. Ed. Engl. 2006, 45, 7578-7581.
3. Siegenthaler, K.O.; Schäfer, A.; Studer, A. Chemical Surface Modification via Radical C-C Bond-Forming Reactions. J. Am. Chem. Soc. 2007, 129, 5826-5827.
4. Hensarling, R.M.; Doughty, V.A.; Chan, J.W.; Patton, D.L. Clicking Polymer Brushes with Thiol-yne Chemistry: Indoors and Out. J. Am. Chem. Soc. 2009, 131, 14673-14675.
5. Walling, C. Some properties of radical reactions important in synthesis. Tetrahedron 1985, 41, 3887-3900.
6. Newcomb, M. Competition Methods and Scales for Alkyl Radical Reaction Kinetics. Tetrahedron 1993, 49, 1151-1176.
7. Gansäuer, A.; McGhee, A.; Procter, D. Radical Chemistry on Solid Support, In Radicals in Synthesis II, Springer: Berlin/Heidelberg, Germany, 2006; Vol. 264, pp. 93-134.
8. Curran, D.P.; Yang, F.; Cheong, J.H. Relative Rates and Approximate Rate Constants for Inter-and Intramolecular Hydrogen Transfer Reactions of Polymer-Bound Radicals. J. Am. Chem. Soc. 2002, 124,

14993-15000.

9. Lloyd-Williams, P.; Albericio, F.; Giralt, E. Convergent solid-phase peptide synthesis. Tetrahedron 1993, 49, 11065-11133.
10. Bochet, C.G. Photolabile protecting groups and linkers. J. Chem. Soc. Perkin Trans. 1 2002, 125-142.
11. Reitz, A.B. Recent advances in traceless linkers. Curr. Opin. Drug Discov. Dev. 1999, 2, 358-364.
12. Schreiber, S.L. Target-Oriented and Diversity-Oriented Organic Synthesis in Drug Discovery. Science 2000, 287, 1964-1969.
13. Scott, P.J.H.; Steel, P.G. Diversity Linker Units for Solid-Phase Organic Synthesis. Eur. J. Org. Chem. 2006, 2006, 2251-2268.
14. Jung, K.W.; Zhao, X.Y.; Janda, K.D. A linker that allows efficient formation of aliphatic C-H bonds on polymeric supports. Tetrahedron Lett. 1996, 37, 6491-6494.
15. Barluenga, S.; Dakas, P.Y.; Ferandin, Y.; Meijer, L.; Winssinger, N. Modular Asymmetric Synthesis of Aigialomycin D, a Kinase-Inhibitory Scaffold. Angew. Chem. Int. Ed. Engl. 2006, 45, 3951-3954.
16. Parsons, A.F.; Pettifer, R.M. A radical approach to N-desulfonylation. Tetrahedron Lett. 1996, 37, 1667-1670.
17. Luo, J.; Huang, W. A new strategy for solid phase synthesis of a secondary amide library using sulfonamide linker via radical traceless cleavage. Mol. Divers. 2003, 6, 33-41.
18. D'Herde, J.N.P.; J. de Clercq, P. Carbon-carbon bond formation on solid support. Application of the classical Julia-Lythgoe olefination. Tetrahedron Lett. 2003, 44, 6657-6659.
19. Allin, S.M.; Bowman, W.R.; Karim, R.; Rahman, S.S. Aromatic homolytic substitution using solid phase synthesis. Tetrahedron 2006, 62, 4306-4316.
20. Roberts, B.P. Polarity-reversal catalysis of hydrogen-atom abstraction reactions: concepts and applications in organic chemistry. Chem. Soc. Rev. 1999, 28, 25-35.
21. McAllister, L.A.; Brand, S.; de Gentile, R.; Procter, D.J. The first Pummerer cyclisations on solid phase. Convenient construction of oxindoles enabled by a sulfur-link to resin. Chem. Commun. (Camb.) 2003, 2380-2381.
22. Turner, K.L.; Baker, T.M.; Islam, S.; Procter, D.J.; Stefaniak, M. Solid-Phase Approach to Tetrahydroquinolones Using a Sulfur Linker Cleaved by SmI2. Org. Lett. 2006, 8, 329-332.
23. McAllister, L.A.; Turner, K.L.; Brand, S.; Stefaniak, M.; Procter, D.J.

Solid Phase Approaches to N-Heterocycles Using a Sulfur Linker Cleaved by SmI2. J. Org. Chem. 2006, 71, 6497-6507.

24. McAllister, L.A.; McCormick, R.A.; Procter, D.J. Sulfide- and selenide-based linkers in phase tag-assisted synthesis. Tetrahedron 2005, 61, 11527-11576.

25. Ruhland, T.; Andersen, K.; Pedersen, H. Selenium-Linking Strategy for Traceless Solid-Phase Synthesis: Direct Loading, Aliphatic C-H Bond Formation upon Cleavage and Reaction Monitoring by Gradient MAS NMR Spectroscopy. J. Org. Chem. 1998, 63, 9204-9211.

26. Ruhland, T.; Torang, J.; Pedersen, H.; Madsen, J.C.; Bang, K.S. Traceless Solid Phase Synthesis with Polystyrene-Bound Tellurium and in Comparison with Polystyrene-Bound Selenium. Synthesis 2004, 2323-2328.

27. Ruhland, T.; Torang, J.; Pedersen, H.; Madsen, J.C.; Bang, K.S. Green Traceless Cleavage from Resin-Bound Selenium and Tellurium and Analysis by 2D 29Si/1H HR-MAS NMR Spectroscopy. Synthesis 2005, 1635-1640.

28. Nicolaou, K.C.; Pastor, J.; Barluenga, S.; Winssinger, N. Polymer-supported selenium reagents for organic synthesis. Chem. Commun. (Camb.) 1998, 1947-1948.

29. Nicolaou, K.C.; Mitchell, H.J.; Fylaktakidou, K.C.; Suzuki, H.; Rodríguez, R.M. 1,2-Seleno Migrations in Carbohydrate Chemistry: Solution and Solid-Phase Synthesis of 2-Deoxy Glycosides, Orthoesters, and Allyl Orthoesters. Angew. Chem. Int. Ed. Engl. 2000, 39, 1089-1093.

30. Nicolaou, K.C.; Fylaktakidou, K.C.; Mitchell, H.J.; van Delft, F.L.; Rodríguez, R.M.; Conley, S.R.; Jin, Z. Total Synthesis of Everninomicin 13,384-1—Part 4: Explorations of Methodology; Stereocontrolled Synthesis of 1,1'-Disaccharides, 1,2-Seleno Migrations in Carbohydrates, and Solution- and Solid-Phase Synthesis of 2-Deoxy Glycosides and Orthoesters. Chem. Eur. J. 2000, 6, 3166-3185.

31. Nicolaou, K.C.; Roecker, A.J.; Pfefferkorn, J.A.; Cao, G.Q. A Novel Strategy for the Solid-Phase Synthesis of Substituted Indolines. J. Am. Chem. Soc. 2000, 122, 2966-2967.

32. Nicolaou, K.C.; Roecker, A.J.; Hughes, R.; van Summeren, R.; Pfefferkorn, J.A.; Winssinger, N. Novel strategies for the solid phase synthesis of substituted indolines and indoles. Bioorg. Med. Chem. 2003, 11, 465-476

33. Nicolaou, K.C.; Pfefferkorn, J.A.; Cao, G.Q.; Kim, S.; Kessabi, J. A Facile Method for the Solution and Solid-Phase Synthesis of Substituted [3.3.1] Bicycles. Org. Lett. 1999, 1, 807-810.

34. Berlin, S.; Ericsson, C.; Engman, L. Radical Carbonylation/Reductive Cyclization for the Construction of Tetrahydrofuran-3-ones and Pyrrolidin-3-ones. J. Org. Chem. 2003, 68, 8386-8396.
35. Fujita, K.I.; Watanabe, K.; Oishi, A.; Ikeda, Y.; Taguchi, Y. Preparation of Polymer-Supported Selenocyanates and their Application to Solid-Phase Oxyselenenylation-Deselenylation. Synlett 1999, 1999, 1760-1762.
36. Qian, H.; Huang, X. Polystyrene-Supported Selenosulfonates: Efficient Reagents for Regio- and Stereocontrolled Synthesis of Vinyl Sulfones. Synlett 2001, 1913-1916.
37. Qian, H.; Huang, X. Polystyrene-supported selenosulfonates: efficient reagents for the synthesis of acetylenic sulfones. Tetrahedron Lett. 2002, 43, 1059-1061.
38. Qian, H.; Huang, X. Solid-Phase Synthesis of alpha-Keto Sulfones. Synthesis 2006, 1934-1936.
39. Qian, H.; Huang, X. Radical Cyclization of 1,6-Diene Using Polystyrene-Supported Selenosulfones. J. Comb. Chem. 2003, 5, 569-576.
40. Myers, R.M.; Langston, S.P.; Conway, S.P.; Abell, C. Reductive Cleavage of N-O Bonds Using Samarium(II) Iodide in a Traceless Release Strategy for Solid-Phase Synthesis. Org. Lett. 2000, 2, 1349-1352.
41. Gustafsson, M.; Olsson, R.; Andersson, C.M. General combinatorial synthesis of tertiary amines on solid support. A novel conditional release strategy based on traceless linking at nitrogen. Tetrahedron Lett. 2001, 42, 133-136.
42. Meloni, M.M.; Taddei, M. Solid-Phase Synthesis of beta-Lactams via the Miller Hydroxamate Approach. Org. Lett. 2001, 3, 337-340.
43. Kerrigan, N.J.; Hutchison, P.C.; Heightman, T.D.; Procter, D.J. Application of an ephedrine chiral linker in a solid-phase, 'asymmetric catch-release' approach to gamma-butyrolactones. Chem. Commun. (Camb.) 2003, 21, 1402-1403.
44. Kerrigan, N.J.; Hutchison, P.C.; Heightman, T.D.; Procter, D.J. Development of a solid-phase 'asymmetric resin-capture-release' process: application of an ephedrine chiral resin in an approach to gamma-butyrolactones. Org. Biomol. Chem. 2004, 2, 2476-2482.
45. Zhang, Y.; Wang, Y.; Dai, W.M. Efficient Remote Axial-to-Central Chirality Transfer in Enantioselective SmI2-Mediated Reductive Coupling of Aldehydes with Crotonates of Atropisomeric 1-Naphthamides. J. Org. Chem. 2006, 71, 2445-2455.
46. McKerlie, F.; Procter, D.J.; Wynne, G. Reduction of alpha-aryloxy

carbonyl compounds with samarium(II) iodide. A new traceless linker for the solid phase synthesis of carbonyl compounds. Chem. Commun. (Camb.) 2002, 21, 584-585.

47. McKerlie, F.; Rudkin, I.M.; Wynne, G.; Procter, D.J. Evaluation of a new linker system cleaved using samarium(II) iodide. Application in the solid phase synthesis of carbonyl compounds. Org. Biomol. Chem. 2005, 3, 2805-2816.

48. Liu, H.; Wan, S.; Floreancig, P.E. Oxidative Cyclorelease from Soluble Polymeric Supports. J. Org. Chem. 2005, 70, 3814-3818.

49. Floreancig, P.E. Development and Applications of Electron-Transfer-Initiated Cyclization Reactions. Synlett 2007, 38, 191-203.

50. Whitehead, D.M.; Helliwell, P.A.; McKeown, S.C.; Routledge, A. Synthesis, chemical stability and application of a thiohydroxamic acid linker (THA) for solid-phase organic synthesis. React. Funct. Polym. 2009, 69, 884-890.

51. Horton, J.R.; Stamp, L.M.; Routledge, A. A photolabile 'traceless' linker for solid-phase organic synthesis. Tetrahedron Lett. 2000, 41, 9181-9184.

52. Whitehead, D.M.; Jackson, T.; McKeown, S.C.; Wilson, K.; Routledge, A. Application of X-ray Photoelectron Spectroscopy in Determining the Structure of Solid-Phase Bound Substrates. J. Comb. Chem. 2002, 4, 255-257.

53. Gerigk, U.; Gerlach, M.; Neumann, W.P.; Vieler, R.; Weintritt, V. Polymer-Supported Organotin Hydrides as Immobilized Reagents for Free Radical Synthesis. Synthesis 1990, 1990, 448-452.

54. Gerlach, M.; Joerdens, F.; Kuhn, H.; Neumann, W.P.; Peterseim, M. A polymer-supported organotin hydride and its multipurpose application in radical organic synthesis. J. Org. Chem. 1991, 56, 5971-5972.

55. Barton, D.H.R.; McCombie, S.W. A new method for the deoxygenation of secondary alcohols. J. Chem. Soc., Perkin Trans. 1 1975, 1574-1585.

56. Neumann, W.P.; Peterseim, M. Elegant Improvement of the Deoxygenation of Alcohols Using a Polystyrene-Supported Organotin Hydride. Synlett 1992, 1992, 801-802.

57. Ruel, G.; Ngo, K.T.; Dumartin, G.; Delmond, B.; Pereyre, M. Un nouvel hydrure organostannique greffé sur support insoluble. J. Organomet. Chem. 1993, 444, C18-C20.

58. Dumartin, G.; Ruel, G.; Kharboutli, J.; Delmond, B.; Connil, M.F.; Jousseaume, B.; Pereyre, M. Straightforward Synthesis and Reactivity of Polymer-supported Organotin Hydrides. Synlett 1994, 1994, 952-954.

59. Dumartin, G.; Pourcel, M.; Delmond, B.; Donard, O.; Pereyre, M. In situ generated, polymer-supported organotin hydrides as clean reducing agents. Tetrahedron Lett. 1998, 39, 4663-4666.

60. Chemin, A.; Deleuze, H.; Maillard, B. Preparation and reactivity of a macroporous polymer-supported organotin hydride catalyst. Eur. Polym. J. 1998, 34, 1395-1404.

61. Boussaguet, P.; Delmond, B.; Dumartin, G.; Pereyre, M. Catalytic and supported Barton-McCombie deoxygenation of secondary alcohols: a clean reaction. Tetrahedron Lett. 2000, 41, 3377-3380.

62. Lopez, R.M.; Hays, D.S.; Fu, G.C. Bu3SnH-Catalyzed Barton-McCombie Deoxygenation of Alcohols. J. Am. Chem. Soc. 1997, 119, 6949-6950.

63. Harendza, M.; Lessmann, K.; Neumann, W.P. A Polymer-Supported Distannane as Photochemical, Regenerable Source of Stannyl Radicals for Organic Synthesis. Synlett 1993, 1993, 283-285.

64. Junggebauer, J.; Neumann, W.P. An improved synthesis of a polymer-supported distannane and its application to radical formation. Tetrahedron 1997, 53, 1301-1310.

65. Enholm, E.J.; Gallagher, M.E.; Moran, K.M.; Lombardi, J.S.; Schulte, J.P. An Allylstannane Reagent on Non-Cross-Linked Polystyrene Support. Org. Lett. 1999, 1, 689-691.

66. Enholm, E.J.; Schulte, J.P. Convenient Catalytic Free Radical Reductions of Alkyl Halides Using an Organotin Reagent on Non-Cross-Linked Polystyrene Support. Org. Lett. 1999, 1, 1275-1277.

67. Corey, E.J.; Suggs, J.W. Method for catalytic dehalogenations via trialkyltin hydrides. J. Org. Chem. 1975, 40, 2554-2555.

68. Ueno, Y.; Chino, K.; Watanabe, M.; Moriya, O.; Okawara, M. Homolytic carbocyclization by use of a heterogeneous supported organotin catalyst. A new synthetic route to 2-alkoxytetrahydrofurans and gamma-butyrolactones. J. Am. Chem. Soc. 1982, 104, 5564-5566.

69. Mochida, K.; Sugimoto, H.; Yasuo, Y. First polymer-supported organogermanium hydrides and their reduction of organic halides. Polyhedron 1997, 16, 1767-1770.

70. Bowman, W.R.; Krintel, S.L.; Schilling, M.B. New Solid Phase Triorganogermanium Hydrides for Radical Synthesis. Synlett 2004, 2004, 1215-1218.

71. Patro, B.; Merrett, M.; Murphy, J.A.; Sherrington, D.C.; Morrison, M.G.J.T. Radical-polar crossover reactions with polymer-supported tetrathiafulvalene. Tetrahedron Lett. 1999, 40, 7857-7860.

72. Baguley, P.A.; Walton, J.C. Flight from the Tyranny of Tin: The Quest

for Practical Radical Sources Free from Metal Encumbrances. Angew. Chem. Int. Ed. Engl. 1998, 37, 3072-3082.

73. Fletcher, R.; Kizil, M.; Lampard, C.; Murphy, J.A.; Roome, S.J. Tetrathiafulvalene-mediated stereoselective synthesis of the tetracyclic core of Aspidosperma alkaloids. J. Chem. Soc., Perkin Trans. 1 1998, 2341-2352.

74. Callaghan, O.; Lampard, C.; Kennedy, A.R.; Murphy, J.A. A novel total synthesis of (+/-)-aspidospermidine. J. Chem. Soc., Perkin Trans. 1 1999, 995-1002.

75. de Luca, L.; Giacomelli, G.; Porcu, G.; Taddei, M. A New Supported Reagent for the Photochemical Generation of Radicals in Solution. Org. Lett. 2001, 3, 855-857.

76. Barton, D.H.R.; Crich, D.; Motherwell, W.B. The invention of new radical chain reactions. Part VIII. Radical chemistry of thiohydroxamic esters; A new method for the generation of carbon radicals from carboxylic acids. Tetrahedron 1985, 41, 3901-3924.

77. Clark, A.J.; Filik, R.P.; Haddleton, D.M.; Radigue, A.; Sanders, C.J.; Thomas, G.H.; Smith, M.E. Solid-Supported Catalysts for Atom-Transfer Radical Cyclization of 2-Haloacetamides. J. Org. Chem. 1999, 64, 8954-8957.

78. Clark, A.J.; Geden, J.V.; Thom, S. Solid-Supported Copper Catalysts for Atom-Transfer Radical Cyclizations: Assessment of Support Type and Ligand Structure on Catalyst Performance in the Synthesis of Nitrogen Heterocycles. J. Org. Chem. 2006, 71, 1471-1479.

79. Baldwin, J.E. Rules for ring closure. J. Chem. Soc. Chem. Commun. 1976, 734-736.

80. Baldwin, J.E.; Cutting, J.; Dupont, W.; Kruse, L.; Silberman, L.; Thomas, R.C. 5-Endo-trigonal reactions: a disfavoured ring closure. J. Chem. Soc. Chem. Commun. 1976, 736-738.

81. Clark, A.J.; Dell, C.P.; Ellard, J.M.; Hunt, N.A.; McDonagh, J.P. Efficient room temperature copper(I) mediated 5-endo radical cyclisations. Tetrahedron Lett. 1999, 40, 8619-8623.

82. Motoyama, Y.; Kamo, K.; Yuasa, A.; Nagashima, H. Catalytic atom-transfer radical cyclization by copper/bipyridine species encapsulated in polysiloxane. Chem. Commun. (Camb.) 2010, 46, 2256-2258.

83. Teduka, T.; Togo, H. I2-Mediated Photochemical Preparation of 2-Substituted 1,3-Dioxolanes and Tetrahydrofurans from Alcohols with Polymer-Supported Hypervalent Iodine Reagent, PSDIB. Synlett 2005, 2005, 923-926.

84. Kashiwagi, Y.; Ikezoe, H.; Ono, T. Oxidation of Alcohols with Nitroxyl

Radical Resins under Two-Phase Conditions. Synlett 2006, 2006, 69-72.

85. Sibi, M.P.; Chandramouli, S.V. Intermolecular free radical reactions on solid support. Allylation of esters. Tetrahedron Lett. 1997, 38, 8929-8932.

86. Enholm, E.J.; Gallagher, M.E.; Jiang, S.; Batson, W.A. Free Radical Allyl Transfers Utilizing Soluble Non-Cross-Linked Polystyrene and Carbohydrate Scaffold Supports. Org. Lett. 2000, 2, 3355-3357.

87. Enholm, E.J.; Gallagher, M.E. Free Radical Reactions on Soluble Supports from Ring-Opening Metathesis. Org. Lett. 2001, 3, 3397-3399.

88. Baytas, S.N.; Wang, Q.; Karst, N.A.; Dordick, J.S.; Linhardt, R.J. Solid-Phase Chemoenzymatic Synthesis of C-Sialosides. J. Org. Chem. 2004, 69, 6900-6903.

89. Zhu, X.; Ganesan, A. Intermolecular Conjugate Addition of Alkyl Radicals on Solid Phase. J. Comb. Chem. 1999, 1, 157-162.

90. Kumar, H.M.S.; Chakravarthy, P.P.; Rao, M.S.; Reddy, P.S.R.; Yadav, J.S. Free radical addition of haloalkanes to polymer bound olefins and its application to the solid-phase synthesis of pyrethroids. Tetrahedron Lett. 2002, 43, 7817-7819.

91. Yim, A.M.; Vidal, Y.; Viallefont, P.; Martinez, J. Solid-phase synthesis of alpha-amino acids by radical addition to adehydroalanine derivative. Tetrahedron Lett. 1999, 40, 4535-4538.Caddick, S.; Hamza, D.; Wadman, S.N.; Wilden, J.D. Solid-Phase Intermolecular Radical

92. Reactions 2: Synthesis of C-Glycopeptide Mimetics via a Novel Acrylate Acceptor. Org. Lett. 2002, 4, 1775-1777.

93. Miyabe, H.; Fujishima, Y.; Naito, T. Novel Synthesis of alpha-Amino Acid Derivatives through Triethylborane-Induced Solid-Phase Radical Reactions. J. Org. Chem. 1999, 64, 2174-2175.

94. Miyabe, H.; Nishimura, A.; Fujishima, Y.; Naito, T. Carbon-carbon bond construction on solid support: triethylborane-induced radical reactions of oxime ethers. Tetrahedron 2003, 59, 1901-1907.

95. Miyabe, H.; Ueda, M.; Naito, T. Carbon-Carbon Bond Construction Based on Radical Addition to C=N Bond. Synlett 2004, 2004, 1140-1157.

96. Miyabe, H. Development of Solid-Phase Radical Reactions Using Oxime Ethers as a Radical Acceptor. Yakugaku Zasshi 2000, 120, 667-676.

97. Miyabe, H. Development of Carbon Radical Addition to Imine Derivatives. Yakugaku Zasshi 2003, 123, 285-294.

98. Miyabe, H.; Konishi, C.; Naito, T. Stereocontrol in Solid-Phase Radical

Reactions: Radical Addition to Oxime Ether Anchored to Polymer Support. Org. Lett. 2000, 2, 1443-1445.

99. Miyabe, H.; Konishi, C.; Naito, T. Diastereoselective Solid-Phase Radical Addition to Oxime Ether Anchored to Polymer Support. Chem. Pharm. Bull. 2003, 51, 540-544.

100. Jeon, G.H.; Yoon, J.Y.; Kim, S.; Kim, S.S. Radical Reaction of Phenylsulfonyl Oxime Ethers on Solid Support: Application to the Synthesis of alpha-Amino Esters. Synlett 2000, 2000, 128-130.

101. Miyabe, H.; Tanaka, H.; Naito, T. Triethylborane-induced solid-phase radical cyclization of oxime ethers. Tetrahedron Lett. 1999, 40, 8387-8390

102. Miyabe, H.; Fujii, K.; Tanaka, H.; Naito, T. Solid-phase tandem radical addition-cyclisation reaction of oxime ethers. Chem. Commun. (Camb.) 2001, 831-832.

103. Miyabe, H.; Tanaka, H.; Naito, T. Solid-Phase Tandem Radical Addition-Cyclization Reaction: Triethylborane-Induced Reaction of Oxime Ethers Anchored to Polymer Support. Chem. Pharm. Bull. 2004, 52, 842-847.

104. Akamatsu, H.; Fukase, K.; Kusumoto, S. Solid-Phase Synthesis of Indol-2-ones by Microwave-Assisted Radical Cyclization. Synlett 2004, 2004, 1049-1053.

105. Tokimoto, H.; Fukase, K. New deprotection method of the 2,2,2-trichloroethoxycarbonyl (Troc) group with (Bu3Sn)2. Tetrahedron Lett. 2005, 46, 6831-6832.

106. Allin, S.M.; Bowman, W.R.; Elsegood, M.R.J.; McKee, V.; Karim, R.; Rahman, S.S. Synthetic applications of aryl radical building blocks for cyclisation onto azoles. Tetrahedron 2005, 61, 2689-2696.

107. Berteina, S.; de Mesmaeker, A.; Wendeborn, S. Functionalization via Radical Cyclization of Cyclohexenediol Derivatives Bound to Polystyrene. Synlett 1999, 1999, 1121-1123.

108. Du, X.; Armstrong, R.W. Synthesis of Benzofuran Derivatives on Solid Support via SmI2-Mediated Radical Cyclization. J. Org. Chem. 1997, 62, 5678-5679.

109. Du, X.; Armstrong, R.W. SmI2-mediated sequential radical cyclization/anionic capture of aryl iodides on solid support. Tetrahedron Lett. 1998, 39, 2281-2284.

110. Berteina, S.; Mesmaeker, A.D. Application of radical chemistry to solid support synthesis. Tetrahedron Lett. 1998, 39, 5759-5762.

111. Berteina, S.; Wendeborn, S.; de Mesmaeker, A. Radical and Palladium-Mediated Cyclization of Ortho-Iodo Benzyl Enamines: Application to

Solid Phase Synthesis. Synlett 1998, 1998, 1231-1233.

112. Chevet, C.; Jackson, T.; Santry, B.; Routledge, A. A Cleaner Approach to Solid-Supported Radical Chemistry: Application of Hypophosphite Salts to Intramolecular C-C Bond Formation on the Solid-Phase. Synlett 2005, 2005, 477-480.

113. Helliwell, P.A.; Bailey, V.A.; Chevet, C.; Clarke, D.B.; Lloyd, A.; Macarthur, R.; Routledge, A. A mass spectrometric investigation into microenvironmental effects in solid-supported radical chemistry. React. Funct. Polym. 2010, 70, 110-115.

114. Routledge, A.; Abell, C.; Balasubramanian, S. An Investigation into Solid-Phase Radical Chemistry-Synthesis of Furan Rings. Synlett 1997, 1997, 61-62.

115. Jia, G.; Iida, H.; Lown, J.W. Solid-Phase Synthesis of 1-Chloromethyl-1,2-dihydro-3H-benz[e]indole (seco-CBI) and a Polyamide Conjugate. Synlett 2000, 2000, 603-606.

116. Watanabe, Y.; Ishikawa, S.; Takao, G.; Toru, T. Radical cyclization on solid support: Synthesis of gamma-butyrolactones. Tetrahedron Lett. 1999, 40, 3411-3414.

117. Enholm, E.J.; Cottone, J.S. Highly Diastereoselective Radical Cyclizations on Soluble Ring Opening Metathesis Supports. Org. Lett. 2001, 3, 3959-3962.

118. Renaud, P.; Gerster, M. Use of Lewis Acids in Free Radical Reactions. Angew. Chem. Int. Ed. Engl. 1998, 37, 2562-2579.

119. Murakata, M.; Jono, T.; Mizuno, Y.; Hoshino, O. Construction of Chiral Quaternary Carbon Centers by Catalytic Enantioselective Radical-Mediated Allylation of alpha-Iodolactones Using Allyltributyltin in the Presence of a Chiral Lewis Acid. J. Am. Chem. Soc. 1997, 119, 11713-11714.

120. Porter, N.A.; Giese, B.; Curran, D.P. Acyclic stereochemical control in free-radical reactions. Acc. Chem. Res. 1991, 24, 296-304.

121. Nishida, M.; Hayashi, H.; Yamaura, Y.; Yanaginuma, E.; Yonemitsu, O.; Nishida, A.; Kawahara, N. A simple preparation of (R)-(2-cyclopentenyl)acetic acid and (R)-(2-cyclohexenyl)acetic acid using beta-diastereoselective radical cyclization in the presence of Lewis acid. Tetrahedron Lett. 1995, 36, 269-272.

122. Nishida, M.; Ueyama, E.; Hayashi, H.; Ohtake, Y.; Yamaura, Y.; Yanaginuma, E.; Yonemitsu, O.; Nishida, A.; Kawahara, N. Lewis Acid-Promoted Diastereoselective Radical Cyclization Using Chiral alpha,beta-Unsaturated Esters. J. Am. Chem. Soc. 1994, 116, 6455-6456.

123. Badone, D.; Bernassau, J.M.; Cardamone, R.; Guzzi, U.

Diastereoselective Chelation-Controlled Radical Cyclization of Chiral Oxazolidinone-Derived 2-Alkenamides and Modeling of the Transition State. Angew. Chem. Int. Ed. Engl. 1996, 35, 535-538.

124. Lamberto, M.; Corbett, D.F.; Kilburn, J.D. Thiol-mediated free radical cyclisations of isocyanides on solid support. Tetrahedron Lett. 2004, 45, 8541-8543.

125. Harrowven, D.C.; May, P.J.; Bradley, M. Sulfur-mediated radical cyclisation reactions on solid support. Tetrahedron Lett. 2003, 44, 503-506.

126. Beckwith, A.L.J.; Schiesser, C.H. Regio- and stereo-selectivity of alkenyl radical ring closure: A theoretical study. Tetrahedron 1985, 41, 3925-3941.

127. Spellmeyer, D.C.; Houk, K.N. Force-field model for intramolecular radical additions. J. Org. Chem. 1987, 52, 959-974.

128. Nicolaou, K.C.; Baran, P.S.; Zhong, Y.L.; Choi, H.S.; Yoon, W.H.; He, Y.; Fong, K.C. Total Synthesis of the CP Molecules CP-263,114 and CP-225,917 – Part 1: Synthesis of Key Intermediates and Intelligence Gathering. Angew. Chem. Int. Ed. Engl. 1999, 38, 1669-1675.

129. Nicolaou, K.C.; Baran, P.S.; Zhong, Y.L.; Fong, K.C.; He, Y.; Yoon, W.H.; Choi, H.S. Total Synthesis of the CP Molecules CP-225,917 and CP-263,114 – Part 2: Evolution of the Final Strategy. Angew. Chem. Int. Ed. Engl. 1999, 38, 1676-1678.

130. Nicolaou, K.C.; Zhong, Y.L.; Baran, P.S. New Synthetic Technology for the Rapid Construction of Novel Heterocycles – Part 1: The Reaction of Dess-Martin Periodinane with Anilides and Related Compounds. Angew. Chem. Int. Ed. Engl. 2000, 39, 622-625.

131. Nicolaou, K.C.; Baran, P.S.; Zhong, Y.L.; Barluenga, S.; Hunt, K.W.; Kranich, R.; Vega, J.A. Iodine(V) Reagents in Organic Synthesis. Part 3: New Routes to Heterocyclic Compounds via o-Iodoxybenzoic Acid-Mediated Cyclizations: Generality, Scope, and Mechanism. J. Am. Chem. Soc. 2002, 124, 2233-2244.

132. Caddick, S.; Hamza, D.; Wadman, S.N. Solid-phase intermolecular radical reactions 1. Sulfonyl radical addition to isolated alkenes and alkynes. Tetrahedron Lett. 1999, 40, 7285-7288.

133. Plourde Jr, R.; Johnson Jr, L.L.; Longo, R.K. Solid-Phase Free Radical Addition of Thiols to Allyl Ethers. Synlett 2001, 2001, 439-441.

134. Timár, Z.; Gallagher, T. Constructing the 2-(thiobenzyl)ethyl carbamate linker via thiyl radical addition. Tetrahedron Lett. 2000, 41, 3173-3176.

135. Dublanchet, A.C.; Lusinchi, M.; Zard, S.Z. Xanthates and solid-

phase chemistry. A new soluble polymer analogue of Wang resin. Tetrahedron 2002, 58, 5715-5721.

136. Back, T.G.; Zhai, H. Cyclizations and cycloadditions of acetylenic sulfones on solid supports. Chem. Commun. (Camb.) 2006, 21, 326-328.

137. Attardi, M.E.; Taddei, M. The Barton radical decarboxylation on solid phase. A versatile synthesis of peptides containing modified amino acids. Tetrahedron Lett. 2001, 42, 3519-3522.

138. Koley, D.; Colon, O.C.; Savinov, S.N. Chemoselective Nitration of Phenols with tert-Butyl Nitrite in Solution and on Solid Support. Org. Lett. 2009, 11, 4172-4175.

Chapter 2

ASYMMETRIC SYNTHESIS OF (7AS)-7A-METHYL-4,5,7, 7A-TETRAHYDRO-1H-INDENE-2,6-DIONE AND USEFUL DERIVATIVES THEREOF

Samuël Demin, Dirk Van Haver and Pierre J. De Clercq*

Ghent University, Department of Organic Chemistry, Laboratory for Organic Synthesis, Krijgslaan281 (S4), B-9000 Gent, Belgium

* Author to whom correspondence should be addressed: E-mail: pierre.declercq@UGent.be

Received: 7 September 2006; in revised form: 14 September 2006 / Accepted: 14 September 2006 Published: 18 September 2006

ABSTRACT

The enantioselective synthesis of the title compound, using Meyers' bicyclic lactam methodology, is described. This compound and a few of its derivatives are useful intermediates in natural product synthesis.

INTRODUCTION

In the context of the development of structural analogs of calcitriol, the hormonally active metabolite of vitamin D3 [1], we required the *cis*-fused perhydrindanone **7** (Scheme 1) [2]. We herein describe in detail the enantioselective synthesis of the potentially useful dione **5** and its further conversion to

RESULTS AND DISCUSSION

The synthesis of **7**, a *cis*-fused angularly substituted perhydrindane dione, in which one of the carbonyl groups is protected as cyclic ketal, involves two phases. First Meyers' methodology for the enantioselective synthesis of hydrindenones is applied in the preparation of **5** [3]. In a second phase, the cyclohexanone carbonyl group of **5** is protected as an ethylene ketal and the *cis*-fusion in **7** is obtained by stereoselective catalytic hydrogenation of **6**.

Meyers' approach for the asymmetric synthesis of angularly substituted hydrinden-2-ones proceeds in 3 stages: (i) the asymmetric introduction of the quaternary center using a chiral bicyclic lactam such as **1**; (ii) the reductive intramolecular alkylation in which the 1,4-diketone is generated; (iii) the intramolecular aldolisation of the latter to the hydrinden-2-one. We chose to introduce the second carbonyl group in **5** via oxidative cleavage of the exocyclic double bond in **4**. Following Meyers' methodology the latter is obtained from diastereomer **2a**.

To stereoselectively obtain the correct configuration at the 6-position in **2a**, the methyl group and the unsaturated side chain need to be introduced sequentially and in that precise order. Indeed, the prior introduction of the unsaturated side chain could eventually lead to the formation of spirocyclic derivatives [4]. On the other hand, the preferred *endo*-alkylation of Meyers' bicyclic lactam template has been well documented [5]. Hence, enantiomerically pure **(+)-1** was required as starting material [6].

Scheme 1. Synthetic pathway to tetrahydro-1*H*-inden-2,6-dione **5** and derivatives.

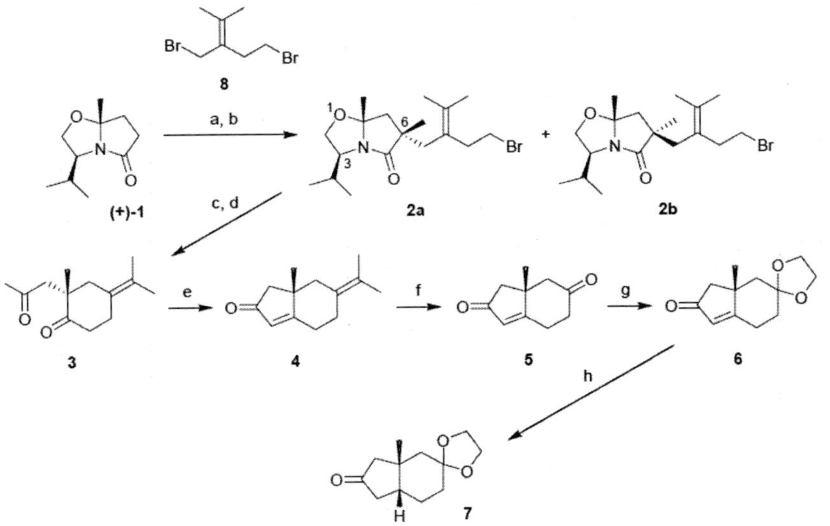

(a) (i) LDA, THF, –78 °C; (ii) MeI, –78 °C, 2 h (96%); (b) (i) LDA, DMPU, THF, –78 °C; (ii) **8**, –78 °C; 3 h (84%); (c) *t*-BuLi, KH, THF, –78 °C, 45 min; (d) *n*-Bu4NH2PO4, EtOH, 75 °C, 12 h (68% from **2a**); (e) KOH, EtOH, rt, 16 h (89%); (f) O3, CH2Cl2, –78 °C (73%); (g) HO(CH2)2OH, TsOH, toluene, reflux, 2 h (92%); (h) H2, Pd/C, EtOAc, rt, 2.5 h (100%).

The methylation of **1** led to a 9:1 diastereomeric mixture, which was not separated. After further deprotonation with LDA (DMPU, THF) and alkylation with the known dibromide **8** [7], a 7:3 mixture of **2a** and **2b** was obtained, which was separated by chromatography. The preferred formation of the *endo*-isomer **2a** was proven by 1H-NMR nOe signal-enhancement studies establishing the relative positions of the Me groups and H atoms indicated in Figure 1. The major isomer **2a** was subjected to

Meyers' protocol (KH, *t*-BuLi) yielding diketone **3**. Basic treatment (KOH, EtOH) led to hydrinden-2- one **4**. Selective cleavage of the exocyclic double bond in **4** (ozone, Me2S) gave diketone **5** in which the saturated carbonyl group was further protected to afford the ketal **6**. Finally, catalytic hydrogenation of the latter occurred exclusively from the convex side of the bicyclic molecule leading to the *cis*-fused perhydrindanone **7**, the structure of which was confirmed by 1H-NMR structural analysis: upon irradiation of the angular Me group a clear nOe enhancement of the signal of the bridgehead hydrogen was observed, confirming the *cis*-fusion of the hydrindane (Figure 1).

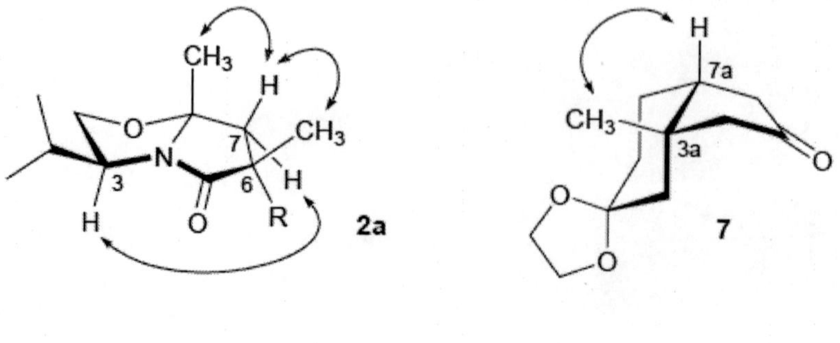

Figure 1. 1H-NMR nOe signal-enhancement studies of **2a** and **7**. **2a**

Conclusions

Angularly substituted perhydrindanones are useful intermediates in synthesis. The asymmetric synthesis of **7**, in which an additional carbonyl group in protected ketal form is present (overall yield 33%), following a sequence of 7 steps starting from commercially available **(+)-1** provides a useful example of the general applicability of Meyers' methodology.

General

Tetrahydrofuran (THF) was distilled from benzophenone ketyl. Dichloromethane (DCM) was distilled from CaH2. Toluene was distilled from sodium. TLC were run on glass plates precoated with silica gel (Merck, 60F-254). Column chromatography was performed on silica gel (Merck, 230-400 mesh). IR spectra were recorded on a Perkin–Elmer series 1600 FT-IR spectrometer. 1H-NMR and 13C-NMR spectra were recorded on a Bruker AM-500 spectrometer. Hydrogen chemical shifts δ are reported in ppm relative to CDCl3 (7.26 ppm) as an internal reference. *J* values are given in Hz. Carbon chemical shifts δ are reported in ppm relative to CDCl3 (77.16 ppm) as an internal reference. Mass spectra (EI) were recorded on a Hewlett–Packard 5898A spectrometer at 70 eV.

(3S,6S,7aR)-6-[2-(2-Bromoethyl)-3-methylbut-2-en-1-yl]-3-isopropyl-6,7a-dimethyltetrahydropyrrolo[2,1-b][1,3]oxazol-5(6H)-one **((+)-2a) [8]**

To *i*-Pr2NLi (LDA, 2 M solution in THF; 110 mL, 0.22 mol) was added dry THF (1100 mL) and the solution was cooled to –78 °C. (3*S*,7a*R*)-3-Isopropyl-7a-methyltetrahydropyrrolo[2,1-*b*][1,3]-oxazol-5(6*H*)-one ((**+)-1**; 20 g, 0.11 mol) was added dropwise and the mixture was stirred for 30 min. MeI (46.74 g, 20.5 mL, 0.33 mol) was then added dropwise, the reaction mixture was stirred for 2 h and then allowed to warm to rt. An aqueous saturated NH4Cl solution (50 mL) was added and the mixture was stirred for 1 h. H2O was added, the aqueous layer was extracted with Et2O (3×), the combined organic layers were dried over anhydrous MgSO4 and concentrated under reduced pressure. The residue was purified by column chromatography on silica gel (isooctane/EtOAc, 7:3) to afford a 9:1 mixture of the (6*R*/6*S*)-diastereomers of the methylated bicyclic lactam (20.8 g, 96%).

To LDA (2 M solution in THF; 7.5 mL, 20 mmol) was added dry THF (100 mL) and the solution was cooled to –78 °C. 1,3-Dimethyl-3,4,5,6-tetrahydro-2(1*H*)-pyrimidinone (DMPU; 11 mL) was added and the mixture was stirred for 10 min. The above mixture of methylated bicyclic lactams (2.0 g, 10.1 mmol) in dry THF (10 mL) was added dropwise and the reaction mixture was stirred for 2 h. 5-Bromo-3-(bromomethyl)-2-methylpent-2-ene (**8**; 5.12 g, 20 mmol) in dry THF (10 mL) was then added dropwise and stirring was continued for 1 h. An aqueous saturated NH4Cl solution (20 mL) was added and the mixture was allowed to warm to rt. H2O was added, the aqueous layer was extracted with Et2O, the combined organic layers were dried over anhydrous MgSO4 and concentrated under reduced pressure. The obtained 7:3 mixture of isomers of the dialkylated bicyclic lactam (3.1 g, 84%) was separated by column chromatography on silica gel (isooctane/ EtOAc, 7:3) to give the major (6*S*)- isomer **(+)-2a** as a solid: mp 45 °C; *Rf* (isooctane/EtOAc, 95:5) 0.22; [α]D rt +106.5 (*c* 1.15, CHCl3); IR (KBr pellet) ν 2963, 2871, 1708, 1463, 1376, 1381, 1298, 1252, 1218, 1176, 1130, 1037, 1006), 967, 903, 842, 808, 771, 684, 650 cm−1; 1H-NMR (500 MHz, CDCl3) δ 4.18 (1 H, dd, *J* = 8.8, 8.8 Hz), 3.77 (1 H, dd, *J* = 7.9, 7.9 Hz), 3.59–3.54 (1

H, m), 3.39–3.34 (1 H, m), 3.28–3.22 (1 H, m), 2.39 (2 H, s), 2.68–2.62 (1 H, m), 2.29–2.23 (1 H, m), 2.24 (1 H, d, *J* = 13.5 Hz), 1.88 (1 H, d, *J* = 13.3 Hz), 1.71 (3 H, s), 1.66–1.63 (1 H, m), 1.63 (3 H, s), 1.50 (3 H, s), 1.29 (3 H, s), 1.04 (3 H, d, *J* = 6.6 Hz), 0.88 (3 H, d, *J* = 6.6 Hz) ppm; 13C-NMR/DEPT (50 MHz, CDCl3) δ 183.7 (C), 132.9 (C), 126.8 (C), 96.8 (C), 70.6 (CH2), 61.6 (CH), 48.3 (C), 45.5 (CH2), 38.2 (CH2), 35.6 (CH2), 34.2 (CH), 30.9 (CH2), 27.5 (CH3), 25.8 (CH3), 21.4 (CH3), 20.7 (CH3), 20.6 (CH3), 18.9 (CH3) ppm; MS *m/z* (%) 372 (M+, 17), 359 (17), 358 (88), 357 (20), 356 (88), 341 (14), 331 (11), 330 (60), 328 (100), 326 (46), 310 (7), 300 (21). Anal. Calcd for C18H30BrNO2: C, 58.10; H, 8.06; N, 3.76. Found: C, 57.95; H, 8.25, N, 3.70.

(2S)-2-Methyl-4-(1-methylethylidene)-2-(2-oxopropyl) cyclohexanone **((+)-3)**

To a solution of dialkylated bicyclic lactam **(+)-2a** (1.0 g, 2.7 mmol) in dry THF (54 mL) were added *t*-BuLi (1.7 M solution in pentane; 3.32 mL, 5.6 mmol) and KH (0.324 g, 8.1 mmol) at –78 °C and the reaction mixture was stirred for 45 min. H2O (20 mL) was added and the mixture was allowed to warm to rt under stirring for 20 min. The solution was concentrated under reduced pressure to 1/3 of its volume and EtOH (20 mL) was added to obtain a homogeneous solution. A 1 M *n*-Bu4NH2PO4 solution (26.5 mL, 0.081 mmol) was added and the mixture was refluxed for 12 h. After cooling, H2O was added, the aqueous layer was extracted with Et2O (2×), the combined organic layers were dried over anhydrous MgSO4 and concentrated under reduced pressure. The residue was purified by column chromatography on silica gel (isooctane/EtOAc, 8:2) to give diketone **(+)-3** (0.388 g, 68%): *Rf* (isooctane/EtOAc, 8:2) 0.33; [α]D rt +20.5 (*c* 1.10, CHCl3); IR (KBr film) ν 2963, 2921, 1710, 1456, 1360, 1167, 1104 cm–1; 1H-NMR (500 MHz, CDCl3) δ 2.98 (1 H, AB, *J* = 17.7 Hz), 2.69–2.63 (2 H, m), 2.56–2.47 (2 H, m), 2.49 (1 H, AB, *J* = 17.7 Hz), 2.44–2.38 (1 H, m), 2.34 (1 H, d, *J* = 14.1 Hz), 2.10 (3 H, s), 1.71 (3 H, s), 1.70 (3 H, s), 1.03 (3 H, s) ppm; 13C-NMR/APT (125 MHz, CDCl3) δ 206.9 (C), 206.9 (C), 126.1 (C), 125.3 (C), 52.2 (CH2), 47.0 (C), 38.6 (CH2), 38.6 (CH2), 30.7 (CH3), 27.2 (CH2), 24.1 (CH3), 20.4 (CH3), 20.3 (CH3); MS *m/z* (%) 208 (M+, 1), 150 (49), 135 (34), 107 (43),93 (21), 79 (26), 67 (21), 43 (100).

(7aS)-7a-Methyl-6-(1-methylethylidene)-1,4,5,6,7,7a-hexahydro-2H-inden-2-one **((+)-4)**

To a solution of diketone **(+)-3** (2.5 g, 12 mmol) in EtOH (12 mL) was added KOH (0.64 g, 12 mmol) and the reaction mixture was stirred at rt for 16 h. H2O was added and the aqueous layer was extracted with Et2O (3×). The combined organic layers were washed with a saturated NaCl solution, dried over anhydrous MgSO4 and concentrated under reduced pressure. The residue was purified by column chromatography on silica gel (isooctane/EtOAc, 9:1) to give enone **(+)-4** (2.0 g, 89%) as yellow crystals: mp 84 °C; *Rf* (isooctane/ EtOAc, 8:2) 0.37; [α]D rt +53.15 (*c* 0.75, CHCl3); IR (KBr film) ν 2956, 2916, 2860, 2358, 1709, 1620, 1437, 1409, 1372, 1274, 1220, 1171, 840, 750, 670 cm−1; 1HNMR (500 MHz, CDCl3) δ 5.79 (1 H, d, *J* = 1.6 Hz), 2.96 (1 H, m), 2.89 (1 H, dd, *J* = 13.0, 2.3 Hz), 2.72 (1 H, ddd, *J* = 13.6, 4.6, 2.3 Hz), 2.36 (1 H, ddd, *J* = 13.5, 13.5, 6.3 Hz), 2.33 (1 H, AB, *J* = 18.4 Hz), 2.28 (1 H, AB, *J* = 18.4 Hz), 1.82–1.75 (2 H, m), 1.77 (3 H, s), 1.72 (3 H, s), 1.13 (3H, s) ppm; 13C-NMR/APT (125 MHz, CDCl3) δ 208.2 (C), 188.0 (C), 126.8 (C), 126.2 (C), 126.0 (CH), 51.1 (CH2), 45.2 (C), 43.9 (CH2), 30.6 (CH2), 28.9 (CH2), 24.2 (CH3), 20.5 (CH3), 20.4 (CH3) ppm; MS *m/z* (%) 190 (M+, 100), 175 (32), 162 (10), 147 (76), 133 (15), 119 (57), 105 (44), 91 (47), 79 (38), 55 (28), 41 (45).

(7aS)-7a-Methyl-4,5,7,7a-tetrahydro-1H-indene-2,6-dione **((–)-5)**

O3 was bubbled through a solution of alkene **(+)-4** (0.48 g, 2.5 mmol) in DCM (126 mL) at −78 °C for 8 min. Me2S (0.3 mL, 3.4 mmol) was added, the reaction mixture was allowed to warm to 0 °C, stirred for 1 h and then allowed to warm to rt. The solvent was evaporated and the residue was purified by column chromatography on silica gel (Et2O/isooctane, 9:1) to give ketone **(–)-5** (0.298 g, 73%): *Rf* (Et2O/ isooctane, 9:1) 0.23; [α]D rt −62.0 (*c* 1.10, CHCl3); IR (KBr film) ν 3070, 2953, 1710, 1678, 1622, 1432, 1337, 1296, 1198, 1090, 1070, 902 cm−1; 1H-NMR (500 MHz, CDCl3) δ 6.02 (1 H, d, *J* = 1.7 Hz), 3.06 (1 H, ddd, *J* = 14.6, 7.5, 1.4 Hz), 2.83 (1 H, dddd, *J* = 14.6, 13.1, 6.7, 1.7 Hz), 2.67–2.61 (2 H, m), 2.52–2.44 (2 H, m), 2.43 (1 H, AB, *J* = 18.6 Hz), 2.39 (1 H, AB, *J* = 18.6 Hz), 1.25 (3 H, s) ppm; 13C-NMR/APT (125 MHz, CDCl3) δ 207.2 (C), 206.4 (C), 181.2 (C), 128.3 (C), 54.2 CH2), 51.2 (CH2), 45.4 CH), 40.0 (CH2), 26.09 (CH2), 25.5 (CH3) ppm; MS *m/z*

(%) 164 (M+, 86), 149 (18), 136 (14), 121 (39), 107 (44), 93 (56), 79 (100), 55 (25), 53 (30).

(3a'S)-3a'-Methyl-3a',4',6',7'-tetrahydrospiro[1,3-dioxolane-2,5'-inden]-2'(3'H)-one **((+)-6)**

To a solution of ketone **(–)-5** (0.77 g, 4.7 mmol) in dry toluene (47 mL) were added HOCH2CH2OH (0.3 g, 0.27 mL, 4.7 mmol) and TsOH (90 mg, 0.47 mmol). The reaction mixture was refluxed for 2 h with azeotropic removal of the generated H2O (Dean–Stark). The solution was allowed to cool, concentrated under reduced pressure and an aqueous saturated NaHCO3 solution (10 mL) was added. The aqueous layer was extracted with Et2O (3 × 20 mL), the combined organic layers were dried over anhydrous MgSO4 and concentrated under reduced pressure. The residue was purified by column chromatography on silica gel (Et2O/isooctane, 9:1) to give ketal **(+)-6** (0.9 g, 92%) as whiteyellow crystals: mp 72 °C; *Rf* (Et2O/ isooctane, 9:1) 0.46; [α]D rt +6.28 (*c* 1.03, CHCl3); IR (KBr pellet)

2989, 2965, 2925, 1707, 1621, 1448, 1363, 1290, 1262, 1181, 1105, 1069, 1010, 952, 934, 907), 866, 844, 716 cm–1; 1H-NMR (500 MHz, CDCl3) δ 5.83 (1 H, s), 4.05–4.00 (2 H, m), 3.92 (2 H, t, *J* = 6.09 Hz), 2.69 (2 H, dd, *J* = 8.8, 3.2 Hz), 2.31 (1 H, AB, *J* = 18.3 Hz), 2.24 (1 H, AB, *J* = 18.3 Hz), 2.07– 2.00 (2 H, m), 1.70–1.62 (2 H, m), 1.37 (3 H, s) ppm; 13C-NMR/APT (125 MHz, CDCl3) δ 185.97 (C), 126.8 (CH), 107.9 (C), 107.9 (C), 64.9 (CH2), 64.0 (CH2), 53.3 (CH2), 47.3 (CH2), 43.5 (C), 35.7 (CH2), 26.3 (CH3), 25.2 (CH2) ppm; MS *m/z* (%) 208 (M+, 26), 193 (34), 152 (13), 121 (13), 107 (26), 86 (100), 79 (35), 55 (18), 53 (21). Anal. Calcd for C12H16O3: C, 69.21; H, 7.74. Found: C, 68.86; H, 7.95.

(3a'S,7a'R)-3a'-Methylhexahydrospiro[1,3-dioxolane-2,5'-inden]-2'(3'H)-one **((+)-7)**

To a solution of enone **(+)-6** (1.0 g, 4.8 mmol) in EtOAc (96 mL) was added Pd/C (10%; 0.5 g, 0.48 mmol) and the reaction mixture was shaken under 4 bar H2 pressure (Parr apparatus) at rt for 2.5 h. The mixture was filtered through Celite® and the solvent was evaporated under reduced pressure to afford saturated ketone **(+)-7** (1.0 g, 100%): *Rf* (isooctane/EtOAc, 9:1) 0.09; [α]D rt +43.4 (*c* 1.5, CHCl3); IR (KBr film) v 2931, 2884, 2250, 1736, 1453, 1407, 1364, 1256, 1211, 1145, 1098, 1068, 1018, 994, 911, 732, 648, 608 cm–1; 1H NMR (500 MHz, CDCl3) δ 3.93–3.90 (4 H, m), 2.73 (1 H, AB, *J* = 18.4 Hz), 2.47 (1 H, dd, *J* = 18.8, 7.7 Hz), 2.04 (1 H, dd, *J* = 18.8, 4.2 Hz), 1.95–1.93 (1 H, m), 1.87 (1 H, AB, *J* = 18.4 Hz), 1.85–1.80 (1 H, m), 1.71–1.66 (2 H, m), 1.59–1.48 (3 H, m), 1.12 (3 H, s) ppm; 13C NMR/APT (125 MHz, CDCl3) δ 219.4 (C), 108.4 (C), 64.2 (CH2), 63.9 (CH2), 49.9 (CH2), 43.0 (CH2), 42.0 (CH2), 40.8 (CH), 39.7 (CH2), 32.0 (C), 28.8 (CH3), 26.0 (CH2) ppm; MS *m/z* (%) 210 (M+, 3), 153 (9), 126 (28), 99 (100), 86 (30), 55 (18), 42 (15).

ACKNOWLEDGEMENTS

Financial assistance of the FWO-Vlaanderen (projects G.0150.02 and G.0036.00) is gratefully acknowledged. S. D. thanks the IWT for a scholarship.

REFERENCES AND NOTES

1. Proceedings of the 12th Workshop on Vitamin D; Bouillon, R.; Norman, A. W.; Pasqualini, J. R.; Eds.; *J. Steroid Biochem. Mol. Biol.* **2004**, *89–90*, 1–633, and the previous 11 volumes in this series; (b) *Vitamin D*; Feldman, D.; Glorieux, F. H.; Pike, J. W., Eds.; Academic Press: San Diego, **1997**.
2. Demin, S.; Van Haver, D.; Vandewalle, M.; De Clercq, P. J.; Bouillon, R.; Verstuyf, A. *Bioorg. Med. Chem. Lett.* **2004**, *14*, 3885–3888.
3. Snyder, L.; Meyers, A. I. *J. Org. Chem.* **1993**, *58*, 7507–7515.
4. Stork, G.; Danheiser, R. L.; Ganem, B. *J. Am. Chem. Soc.* **1973**, *95*, 3414–3415.
5. Meyers, A. I.; Seefeld, M. A.; Lefker, B. A.; Blake, J. F.; Williard, P. G. *J. Am. Chem. Soc.* **1998**, *120*, 7429–7438. (b) Meyers, A. I.; Seefeld, M. A.; Lefker, B. A. *J. Org. Chem.* **1996**, *61*, 5712–5713.
6. The enantiomeric purity of commercial **(+)-1**, purchased from Aldrich, was established by chiral GC (0.25-μm SUPELCO β-DEX 120 column, 30 m × 0.25 mm, 120 °C). Comparison with *Molecules* **2006**, *11* **713** racemic **(±)-1**, prepared from DL-valinol and levulinic acid according to Meyers, showed virtually 100% ee. See: Meyers, A. I.; Lefker, B. A. *Tetrahedron* **1987**, *43*, 5663–5676.
7. Posner, G. H.; Hamill, T. G. *J. Org. Chem.* **1988**, *53*, 6031–6035; (b)

Paquette, L. A.; Charumilind, P.; Kravetz, T. M.; Böhm, M. C.; Gleiter, R. *J. Am. Chem. Soc.* **1983,** *105,* 3126– 3135.

8. The ACD/I-Lab Web service (ACD/IUPAC Name Free 8.05) was used to generate the chemical names of compounds **1–8**

Chapter 3

ASYMMETRIC SYNTHESIS OF TERTIARY THIOLS AND THIOETHERS

Jonathan Clayden and Paul MacLellan

School of Chemistry, University of Manchester, Oxford Rd, Manchester M13 9PL, UK

ABSTRACT

Enantiomerically pure tertiary thiols provide a major synthetic challenge, and despite the importance of chiral sulfur-containing compounds in biological and medicinal chemistry, surprisingly few effective methods are suitable for the asymmetric synthesis of tertiary thiols. This review details the most practical of the methods available.

INTRODUCTION

Organosulfur compounds play key roles in many biological structures and functions: two of the 21 proteinogenic amino acids contain sulfur, and seven of the 10 best-selling drugs in the US in 2009 were organosulfur compounds (Figure 1) [1]. Glutathione plays a crucial role in primary metabolism, and (*R*)-thioterpineol (limonenethiol or "grapefruit mercaptan") is an important and extremely powerful flavour compound, providing the distinctive taste of grapefruit (Figure 2). The consequent need to prepare and manipulate enantiomerically pure organosulfur species has powered the development of asymmetric synthetic methods leading to various classes of organic sulfur compounds, with chirality residing at sulfur, at carbon, or at both [2-8].

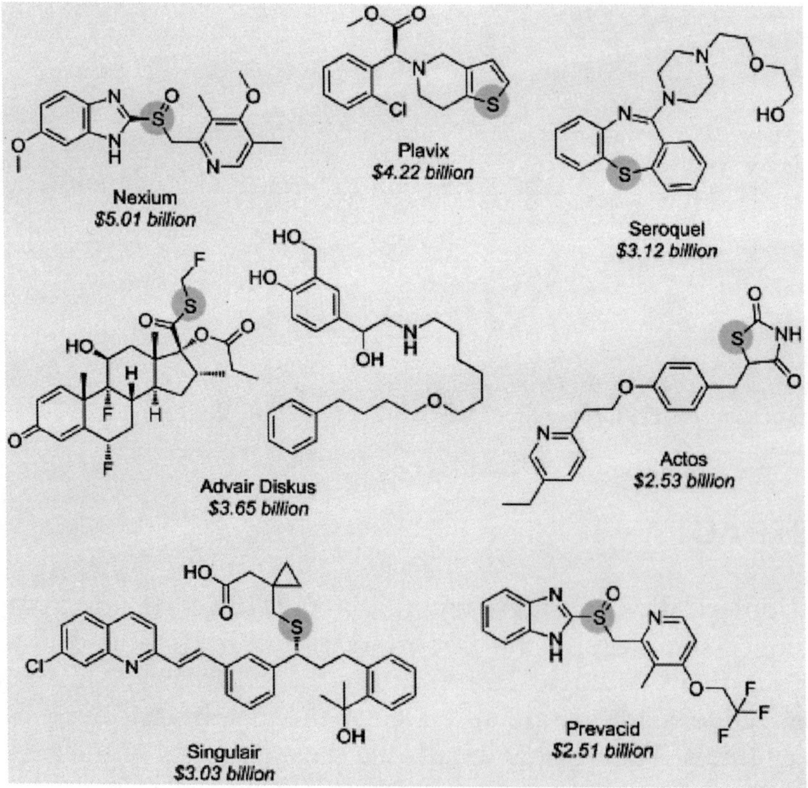

Nexium *$5.01 billion*
Plavix *$4.22 billion*
Seroquel *$3.12 billion*
Advair Diskus *$3.65 billion*
Actos *$2.53 billion*
Singulair *$3.03 billion*
Prevacid *$2.51 billion*

Figure 1: Seven out of the ten top selling drugs in the USA in 2009 contain sulfur. Figures in italics are total retail sales in dollars [1].

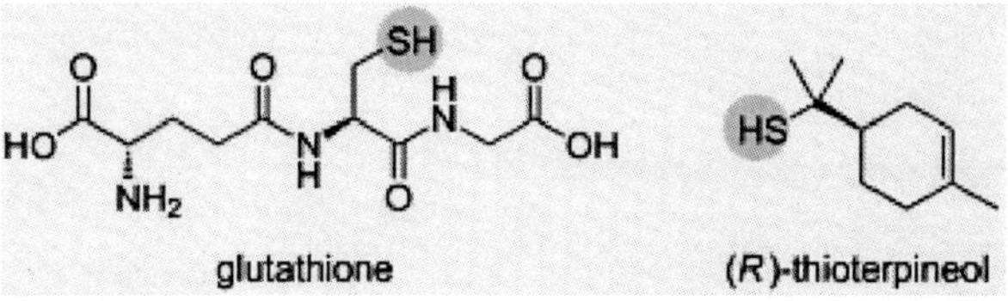

Figure 2: Naturally occurring organosulfur compounds glutathione and (*R*)-thioterpineol.

This field of asymmetric organosulfur chemistry is particularly well developed in connection with sulfur(IV) and sulfur(VI) species with chirality at sulfur - namely sulfoxides, sulfinates, sulfimines and sulfilimines [9]. Chiral sulfoxides and sulfonium ylids have themselves been extensively used as tools for asymmetric synthesis [3]. With regard to chiral sulfur(II) compounds - namely thiols and thioethers (sulfides) with chirality at carbon - methods available for their asymmetric preparation are abundant, and usually rely on stereospecific substitution reactions[9]. These reactions are well suited to the construction of *secondary* thiol derivatives. By contrast, few methods are suitable for the asymmetric preparation of simple *tertiary* thiols **1**. While the asymmetric synthesis of simple tertiary *alcohols* is generally achieved by controlling enantiofacial selectivity in nucleophilic attack on a prochiral ketone [10], comparable approaches to tertiary thiols **1** are not practical due to the instability of thioketones, their tendency to undergo nucleophilic attack at sulfur rather than carbon, and their intolerable stench [11]. As a result, enantiomerically pure tertiary thiols are a remarkably difficult class of compounds to make, despite the simplicity of their structure. This review will cover the methods available for the asymmetric synthesis of chiral tertiary thiols and their sulfur(II) derivatives.

Two general approaches to the preparation of a chiral tertiary thiol **1** may be envisaged, entailing disconnection of either a C–S or a C–C bond (Figure 3). The thiol could be installed by stereoselective attack of a sulfur-centred nucleophile on a substituted carbon centre (C–S bond formation). Alternatively, stereoselective alkylation, arylation or acylation of a secondary sulfur-based substrate could generate the quaternary centre (C–C bond formation).

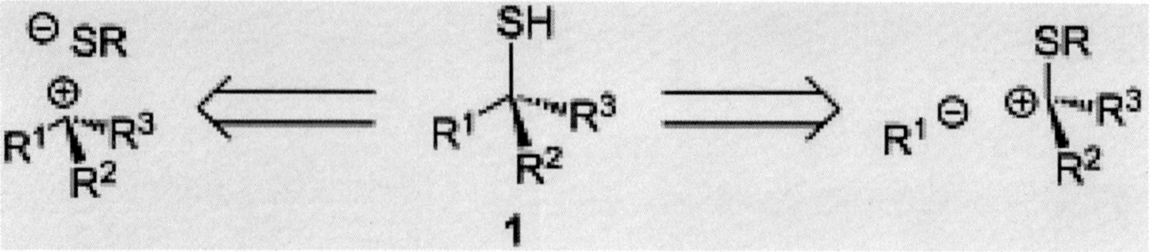

Figure 3: Methods for the synthesis of chiral tertiary thiol **1**.

CARBON–SULFUR BOND FORMATION

S_N2 displacement of a leaving group

Stereospecific nucleophilic attack on substituted carbon atoms is a simple and versatile way to construct stereocentres next to heteroatoms with overall inversion of stereochemistry. Sulfur nucleophiles are commonly used very effectively to accomplish reactions of this type [9]. However, S_N2 displacements are very sensitive to steric crowding at the reaction centre: S_N1 substitution and elimination reactions are almost always favoured over the S_N2 pathway in quaternary electrophiles, limiting the use of S_N2 reactions for the synthesis of tertiary thiols.

Sulfonate leaving groups

The efficiency of sulfonates as leaving groups has allowed the preparation of certain families of tertiary thiols and thioethers. For example, S_N2 displacement of a mesylate leaving group by thiophenol can be accomplished using α-hydroxy esters **2** (Scheme 1) [12].

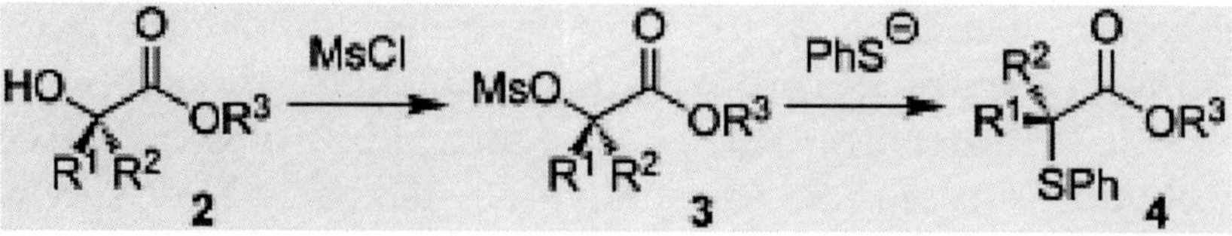

Scheme 1: Preparation of thioethers **4** from α-hydroxy esters.

The presence of the α-ester group of **3** promotes S_N2 reaction in two ways: The electron-withdrawing nature of the substituent inhibits S_N1 dissociation and carbocation formation, and the planar ester group poses minimal steric hindrance towards approach of the nucleophile.

Yields of substitution in α-aryl-α-hydroxy esters **5** are poor with a significantly decreased enantiomeric ratio in the product **6** (Scheme 2). The low yields can be attributed to the formation of β-phenylthioesters **7** by elimination followed by conjugate addition. Even in the presence of the adjacent carbonyl group, competing S_N1 dissociation of the benzylic leaving group leads to a loss of enantiomeric purity in some cases. The use of α,α-dialkyl hydroxy esters **8** is more successful: Thioethers **10** are formed in high yield and with almost complete stereospecificity (Scheme 3).

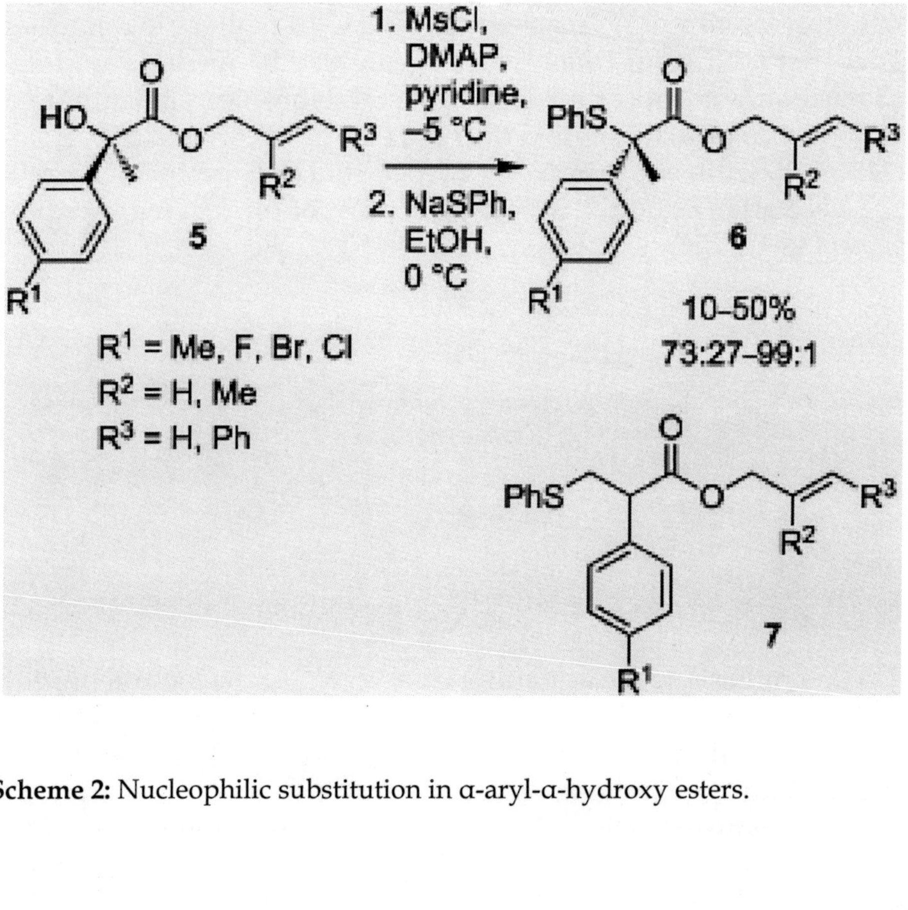

Scheme 2: Nucleophilic substitution in α-aryl-α-hydroxy esters.

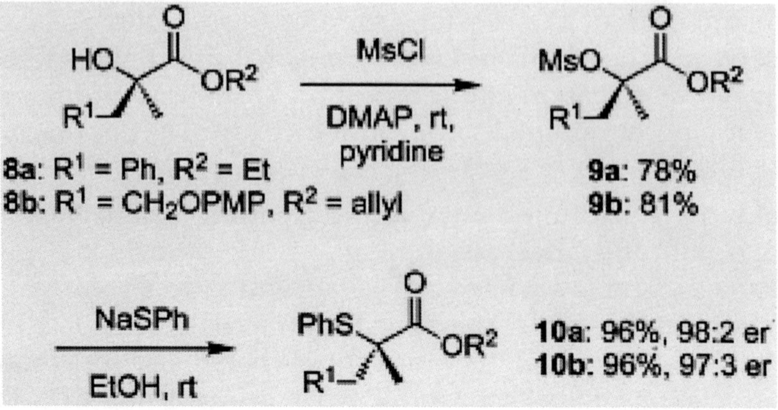

Scheme 3: Preparation of α,α-dialkylthioethers.

Using the same principles of low steric bulk and electronic inhibition of the S_N1 reaction pathway, α-(sulfonyloxy)nitriles, easily prepared from cyanohydrins, can also be made to undergo S_N2 reaction with sulfur nucleophiles [13]. However, substitution of the quaternary α-(sulfonyloxy)nitrile **11** is very slow, and conversion of**11** to **12** with potassium thioacetate in DMF proceeds in only 35% yield after 72 hours. S_N2 displacement of the leaving group by thioacetate in toluene proceeds faster at 55 °C (Scheme 4).

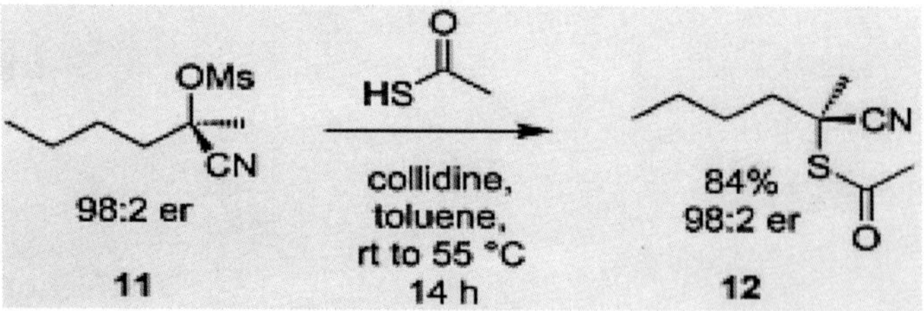

Scheme 4: Preparation of α-cyanothioacetate **12**.

The synthesis of the unnatural enantiomer of the natural product spirobrassinin has been achieved by substitution at a quaternary centre by a sulfur nucleophile [14]. Intramolecular cyclisation of a dithiocarbamate **13** allows isolation of **14** with complete inversion of stereochemistry, although no yield for this reaction was reported (Scheme 5).

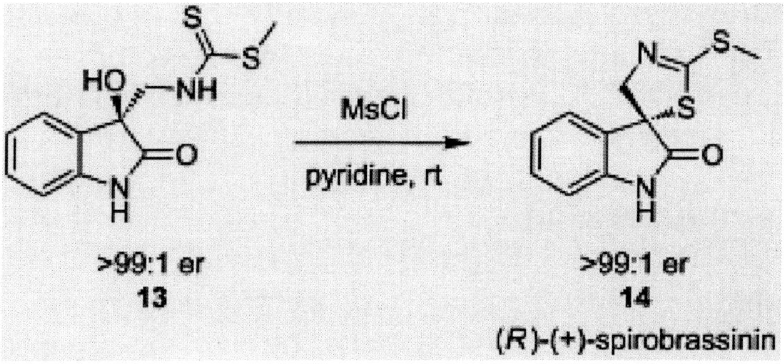

Scheme 5: Synthesis of (*R*)-(+)-spirobrassinin.

Peregrina and co-workers have developed a general method for the synthesis of α-functionalised β-amino acids[15-17] employing nucleophilic ring opening of cyclic sulfamidate **15** via an invertive S_N2 mechanism (Scheme 6)[16].

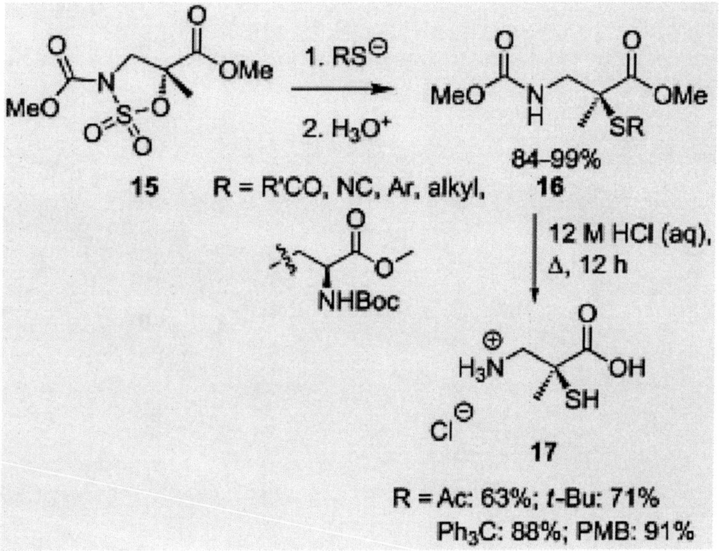

Scheme 6: Opening of cyclic sulfamidates with thiol nucleophiles.

A variety of nucleophiles are capable of ring opening the sulfamidate **15** in good yields, and simple functional group transformations allow isolation of the functionalised β-amino acids. Sulfur-based nucleophiles generally undergo reaction with sulfamidate **15** cleanly to give products **16** with complete inversion of configuration [16,17]. Acid catalysed cleavage of a thioether or thioacetate yields *tert*-thiols **17** in good yield.

Epoxide ring opening

S_N2 displacements from quaternary electrophiles require specific structural features to avoid competing racemisation: Excellent leaving groups and electron withdrawing, non-aromatic and small substituents at the reactive centre all favour invertive substitution. Under basic conditions, epoxides also undergo invertive reactions, even at quaternary centres. Sulfur-based nucleophiles have been employed in nucleophilic ring opening of epoxides in hindered systems [18-23], with most examples of the latter occurring at quaternary carbon atoms which are part of ring systems [18-22], and in particular in steroids [18-20]. For example, the thiol-containing androgen **19** is prepared in 71% yield by epoxide ring opening of **18** with potassium hydrogen sulfide (Scheme 7)[18].

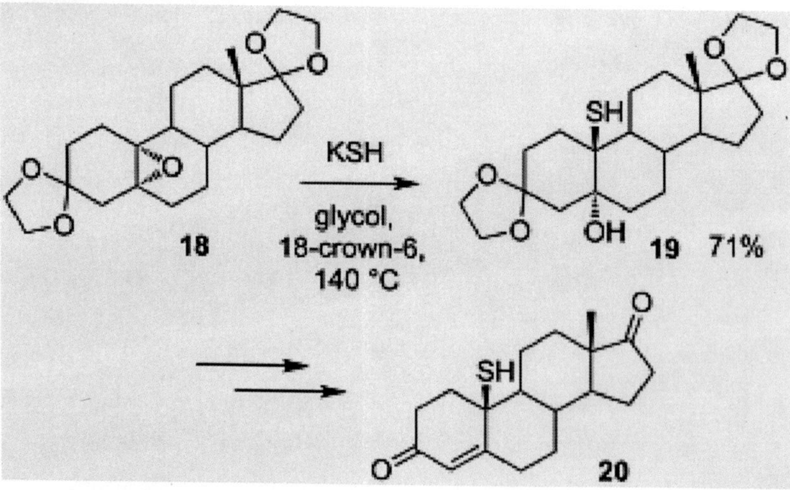

KSH

glycol, 18-crown-6, 140 °C

18

19 71%

20

Scheme 7: Synthesis of androgen **20**.

Epoxide ring opening by a sulfur nucleophile was also employed as the key step in the synthesis of (+)-BE-52440A (**22**) [22] (Scheme

8). Dimerisation of nanaomycin derivative **21** through a bridging sulfide involves a double regioselective invertive epoxide opening.

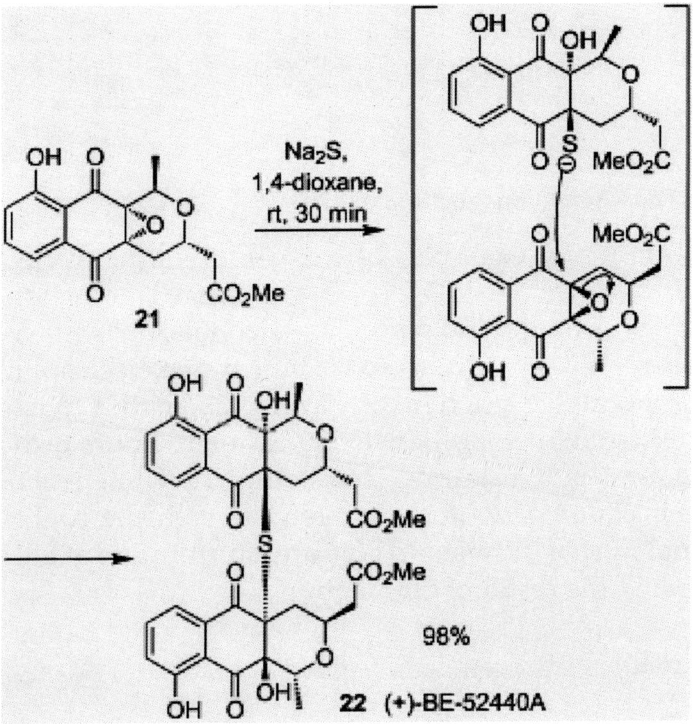

Scheme 8: Synthesis of (+)-BE-52440A.

Mitsunobu reactions

The Mitsunobu reaction offers an operationally straightforward method for activating simple alcohols to invertive substitution, and it is widely used for constructing new stereodefined carbon–heteroatom bonds [24,25]. The Mitsunobu reaction proceeds by reaction of a phosphine and DIAD or DEAD with an alcohol **23** to form an *O*-phosphinite leaving group (Scheme 9). S_N2 substitution to yield **24** is accompanied by formation of a phosphine oxide.

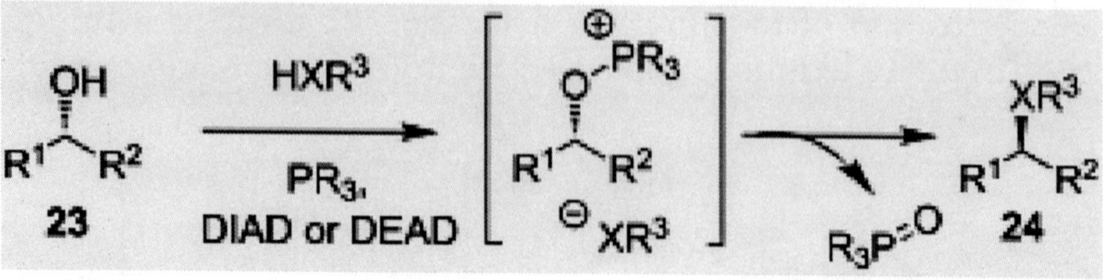

Scheme 9: The Mitsunobu reaction.

There are many examples of the use of sulfur nucleophiles in the Mitsunobu reaction [25-27]: The reaction is successful with a wide range of primary and secondary alcohols, but Mitsunobu-type reactions are very sensitive to steric bulk at the electrophilic carbon atom. For example, reaction of alcohol **25** with a phenol under standard Mitsunobu conditions at 50 °C for 16 hours provides only a trace amount of the desired chiral ether **26**, while reaction at 100 °C allows isolation of **26** in moderate yield (Scheme 10) [28]. A *gem*-diethyl analogue of **25** failed to give any substitution at all, the major product being the result of elimination.

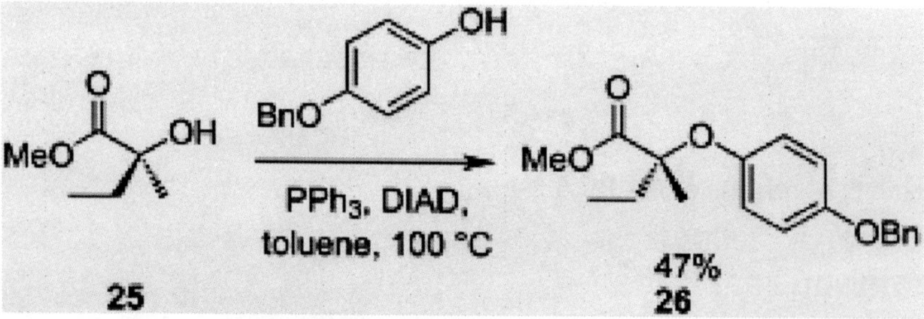

Scheme 10: Mitsunobu substitution at a quaternary centre.

Despite the proven reluctance of tertiary alcohols to undergo Mitsunobu reactions, there is one example of the displacement of a tertiary alcohol by a thiol under Mitsunobu conditions in the literature. La Clair reported the synthesis of the initially assigned structure of natural product hexacyclinol (**27**, Figure 4) [29].

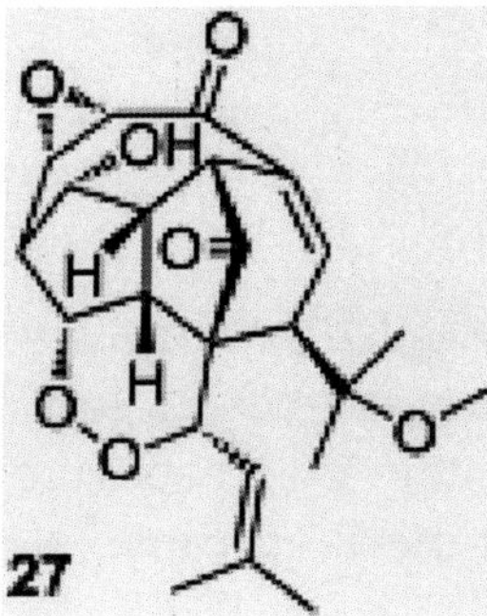

Figure 4: Initially assigned structure of hexacyclinol.

The reported synthesis entailed Mitsunobu reaction at a very hindered quaternary centre, accomplishing invertive substitution of **28** with thiophenol under mild reaction conditions to yield **29** in 94% yield (Scheme 11).

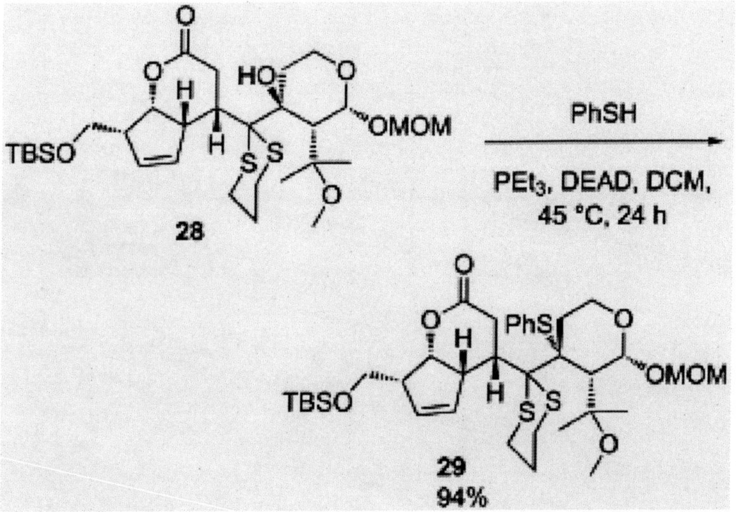

PhSH

PEt_3, DEAD, DCM, 45 °C, 24 h

28

29
94%

Scheme 11: Preparation of thioether **29**.

The paucity of literature precedents for Mitsunobu reactions of tertiary alcohols brings into sharp focus the remarkable success of this reaction under such mild conditions. However, this reaction and other unusual steps in the synthesis attracted scepticism among synthetic chemists [30], and many aspects of this synthesis were called into question when a revised structure of hexacyclinol was proposed by Rychnovsky [31]. A successful synthesis of the revised structure matched the published data for the natural compound [32].

A modification of the Mitsunobu reaction, that does allow the formation of sulfur-substituted tertiary carbon atoms with inversion of stereochemistry, was reported by Mukaiyama and co-workers [33,34]. The method employs benzoquinone derivatives instead of azodicarboxylates as the oxidising agent. A series of phosphinites **31** was prepared by treatment of alcohols **30** with chlorodiphenylphosphine in the presence of triethylamine and DMAP. S_N2 substitution of the phosphinites proceeds by oxidative activation of the leaving group with a 1,4-benzoquinone derivative, DBBQ (**35**) being most effective (Scheme 12). Addition of a thiol nucleophile to adduct **32** results in S_N2 inversion and isolation of the enantiomerically pure (94:6–99:1 er) tertiary thioether **33** and by-product **34**. Highest yields were obtained with BtzSH (**36**) and BoxSH (**37**).

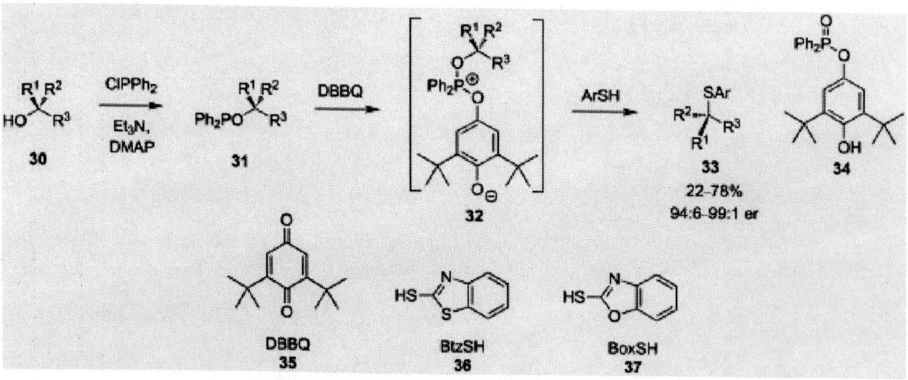

Scheme 12: Thioethers **33** prepared from phosphinites **31**.

The reaction proceeds well with many hindered substrates incorporating aromatic, alkyl and ester substituents with excellent stereospecificity. Enantiomerically pure thiols can also be made from the product: Aromatic thioether**38** is reduced with lithium aluminium hydride to give **39** in 95% yield (Scheme 13).

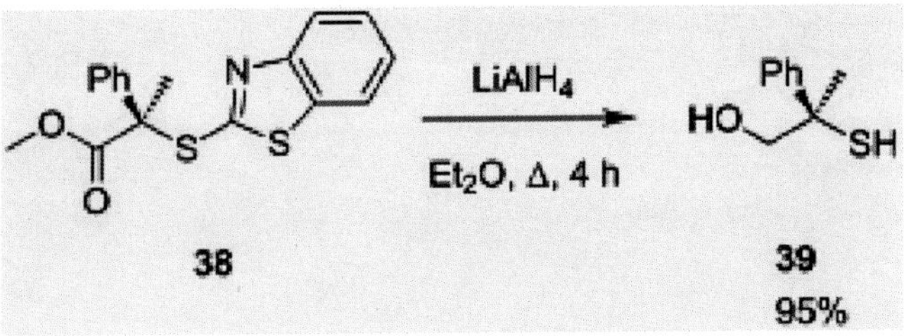

Scheme 13: Preparation of enantiomerically pure thiol **39**.

The methodology was later refined to allow substitution of alcohols in 1 step. Phenoxydiphenylphosphine was used to prepare the phosphinite intermediate and an azide was used as the oxidising agent instead of a benzoquinone derivative (Scheme 14) [35-37].

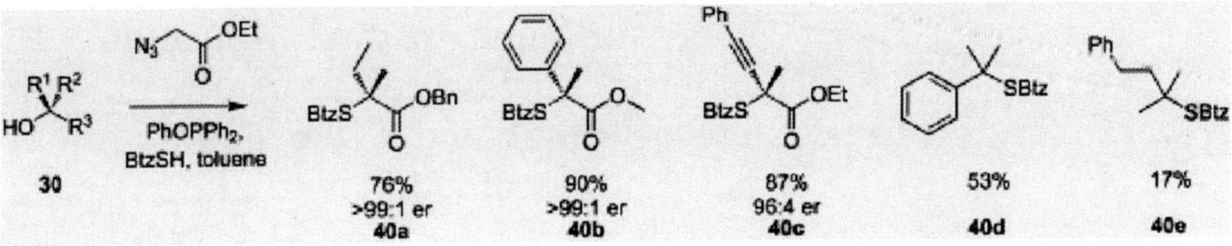

Scheme 14: Thioethers prepared by a modified Mitsunobu reaction.

These conditions allow the efficient preparation of chiral tertiary thioether products **40** from hindered tertiary alcohols. However, the yield of the simple alkyl-substituted thioether **40e** was a disappointing 17%. As with other methods, the preparation of simple optically pure chiral tertiary thiols by Mitsunobu-type procedures is not straightforward. Mitsunobu reactions also suffer from atom inefficiency due to the stoichiometric quantities of phosphorus-containing by-products produced by the reaction.

CONJUGATE ADDITION

The carbon–sulfur bond of enantiomerically pure tertiary organosulfur compounds **42** may be constructed by facially selective addition of a sulfur nucleophile to the disubstituted terminus of a conjugated alkene **41**(Scheme 15).

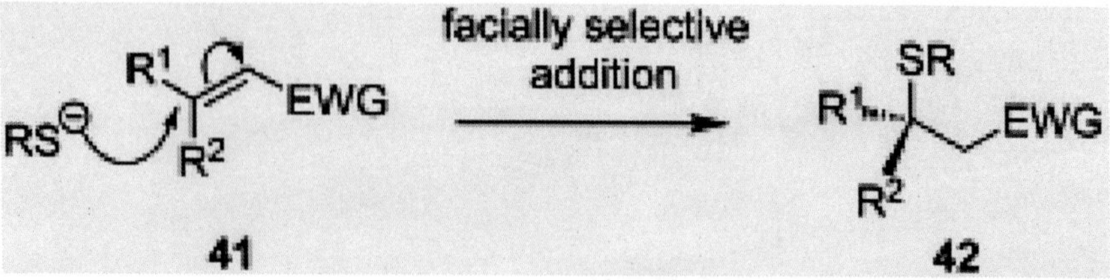

Scheme 15: Nucleophilic conjugate addition.

There are several examples of such facially selective addition of thiols to substituted sp^2 carbon atoms, including intramolecular conjugate addition in steroidal systems [38,39] and intermolecular addition directed by substrate stereochemistry [40,41] or a chiral auxiliary [42,43]. Despite these examples, there are very few general procedures for the preparation of chiral tertiary thiols and thioethers by this method. This may be attributed to three difficulties: The low reactivity of β,β-disubstituted Michael receptors, the difficulty in controlling π-facial stereoselectivity, and the equilibrium of stereoisomers through an addition/elimination mechanism.

The reduced reactivity of β,β-disubstituted Michael receptors towards nucleophilic addition is demonstrated by the catalytic asymmetric additions of thiols to α,β-unsaturated ketones [44]. (*R*)-$LaNa_3$tris(binaphthoxide) (LSB) catalyses asymmetric addition to cyclic enones (Scheme 16), and products of addition **43** are isolated in good yields and enantiomeric ratios.

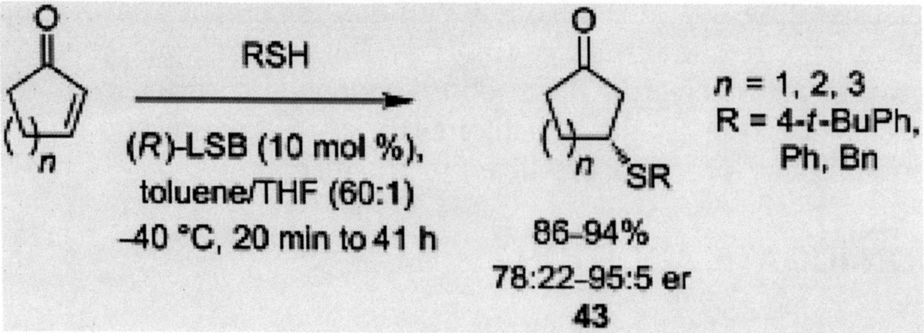

Scheme 16: Asymmetric addition to cyclic enones.

However, addition to the substituted substrate **44** is more difficult. A higher loading of the catalyst and higher temperature are required to produce the enantioenriched thioether **45** even in moderate yield and acceptable er (Scheme 17).

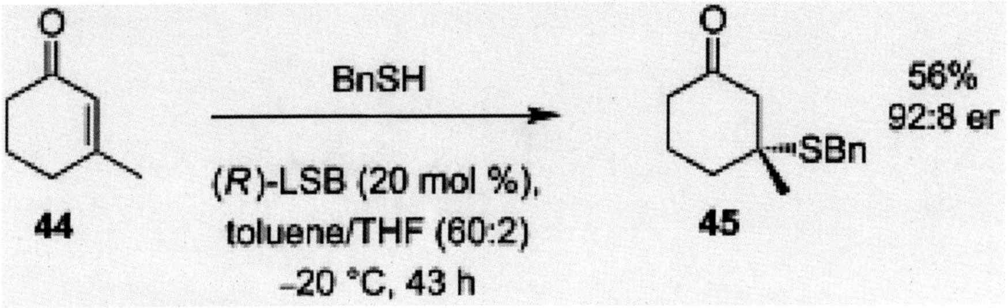

Scheme 17: Preparation of thioether **45**.

Despite the poor yield obtained with this substrate, similar conditions lead to catalytic kinetic resolution of the enantiomers of hindered enone **46** by addition, oxidation and elimination of a sulfenic acid under basic conditions (Scheme 18) [45].

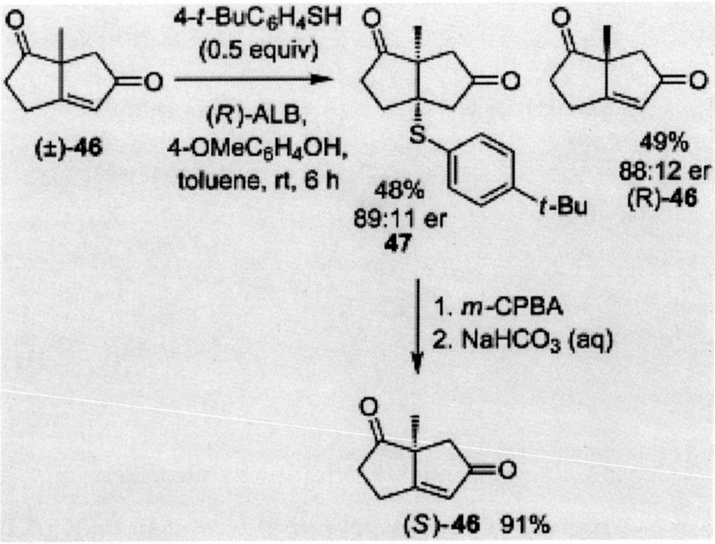

Scheme 18: Catalytic kinetic resolution of the enantiomers of enone **46**.

Xiao and co-workers developed an organocatalytic process for the addition of thiols to nitroalkenes [46]. Using thiourea organocatalyst **48**, conjugate addition of a variety of thiols to a range of nitroalkenes **49** proceeds to give**50** and hence **51** in good yield and good enantioselectivity (Scheme 19).

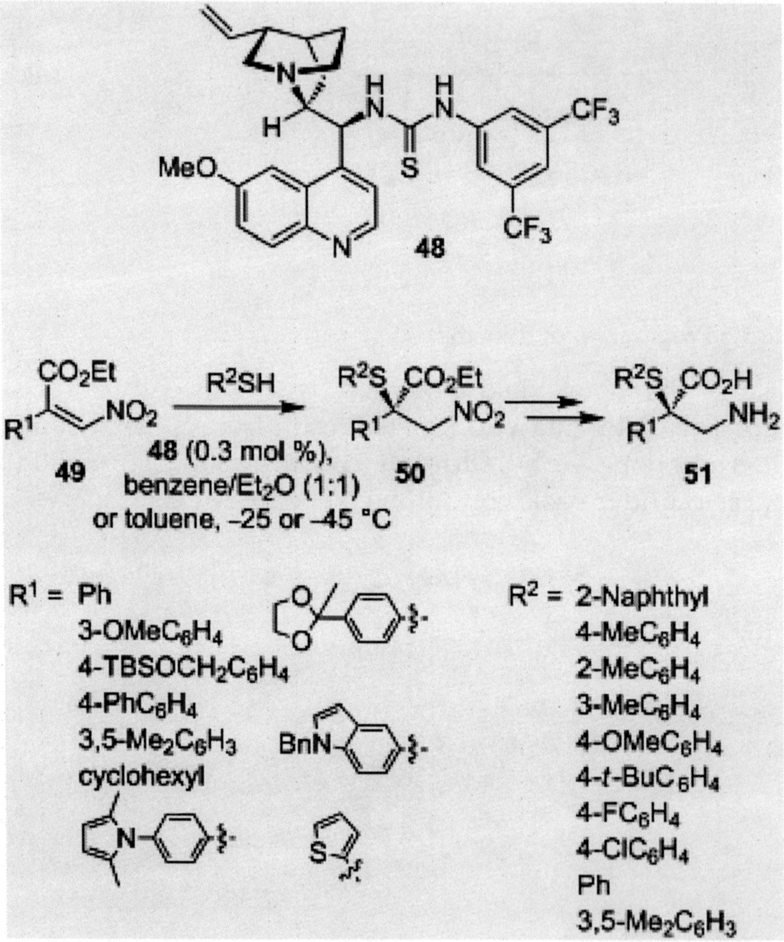

Scheme 19: Organocatalytic conjugate addition to nitroalkenes **49**.

α-Arylthio-β-amino acid **54** was prepared from Michael adduct **52** in three steps (Scheme 20) in good yield and with full conservation of enantiomeric purity.

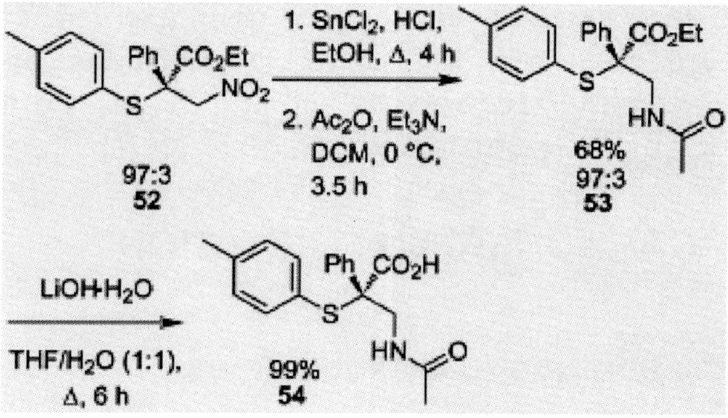

Scheme 20: Preparation of β-amino acid **54**.

Palomo and co-workers postulated that poor reactivity and facial selectivity in conjugate additions with sulfur could be minimised by an intramolecular approach [47,48]. A Lewis acid-promoted sulfur migration within *N*-enoyl oxazolidine-2-thione substrates **56** followed by hydrolysis gave optically pure tertiary thiols **57** directly (Scheme 21), with only the electron rich *para*-methoxyphenyl substituent performing poorly.

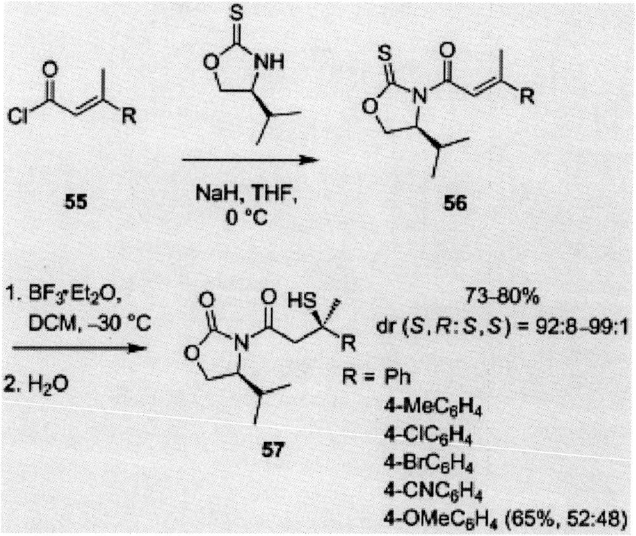

Scheme 21: Sulfur migration within oxazolidine-2-thiones **56**.

As with stereospecific nucleophilic displacement, only specific substrates are tolerated in conjugate additions, and further functionality is always required to activate the electrophilic alkene towards attack by the sulfur nucleophile.

CARBON–CARBON BOND FORMATION

Electrophilic addition to a-thioenolates

Diastereoselective alkylation of α-thioenolates has been used to prepare several types of enantiomerically pure tertiary thioethers [49-54]. The method of "self-regeneration of stereocentres" developed by Seebach [55] employs α-thiocarboxylic acids **58** condensed with pivaldehyde to generate 1,3-oxathiolan-4-ones **59** (Scheme 22) [49,50].

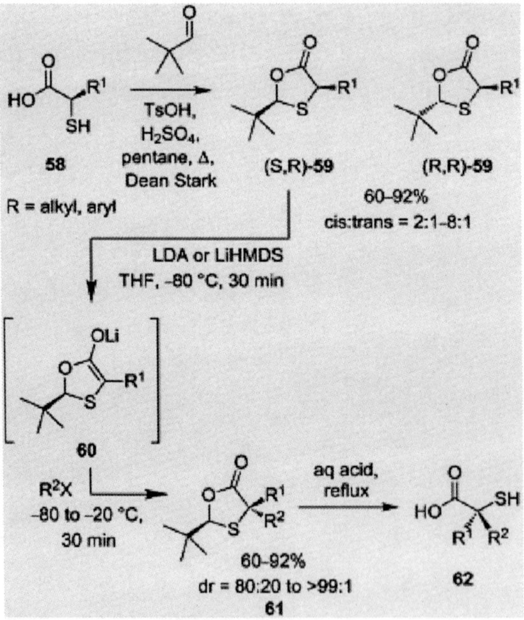

Scheme 22: Preparation of thiols **62** by self-regeneration of stereocentres.

The *cis* diastereomer is formed preferentially (2:1–8:1 selectivity), and can usually be purified by crystallisation. Enolate **60** is formed by deprotonation with a lithium base and treated with an electrophile. Electrophilic addition takes place diastereoselectively anti to the bulky *tert*-butyl group, producing oxathiolanone products **61** in excellent diastereoselectivities.

Enantiopure tertiary thiols **62** can be liberated from the oxathiolanone products by hydrolysis, and the method was employed by Townsend and co-workers in the synthesis of (5*R*)-thiolactomycin (**66**, Scheme 23) [51]. Diastereoselective addition of an aldehyde to oxathiolanone **64** affords **65** in 81% yield. Further synthetic manipulations allowed preparation of (5*R*)-thiolactomycin (**66**) in >99:1 er.

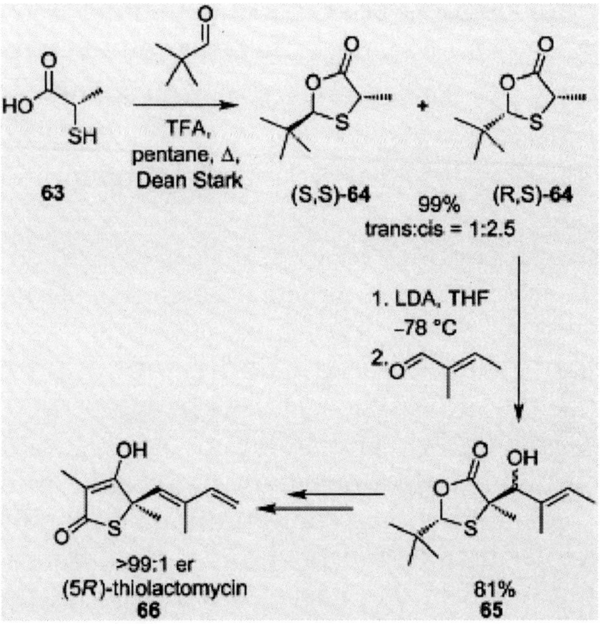

Scheme 23: Synthesis of (5*R*)-thiolactomycin.

Electrophilic attack on a-thioorganolithiums

Formation of carbon–carbon bonds adjacent to heteroatoms by deprotonation with an organolithium base and subsequent reaction with an electrophile has become an important and versatile method, especially when chiral ligands may be used to govern the enantioselectivity of the reaction pathway [56]. Tertiary organosulfur compounds can be made in this way by lithiation of a chiral secondary thiol derivative (Scheme 24).

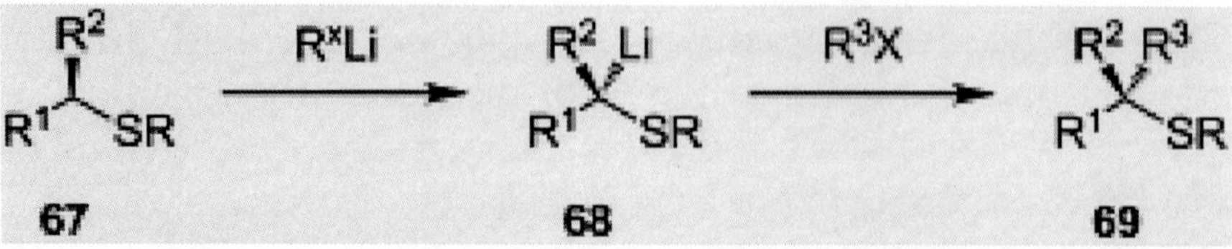

$R^1(R^2)CH$–SR (**67**) $\xrightarrow{R^xLi}$ $R^1(R^2)C(Li)$–SR (**68**) $\xrightarrow{R^3X}$ $R^1(R^2)C(R^3)$–SR (**69**)

Scheme 24: Preparation of tertiary thiols and thioethers via α-thioorganolithiums.

Three conditions must be met if this method is to be used for preparation of enantiomerically pure tertiary thiol derivatives:

1) The substrate **67** must be available in an enantiomerically pure form,
2) The lithiated intermediate **68** must be configurationally stable, and
3) Electrophilic addition to give **69** must proceed with either complete retention or complete inversion of stereochemistry.

Configurational stability in a-thioorganolithiums

In contrast to α-oxy and α-aminoorganolithiums, simple α-thioorganolithium compounds show high configurational lability even at –80 °C [57]. Beak noted configurational instability during the diastereoselective methylation of α-thioorganolithium **71** (Scheme 25) [58]. Equilibration of the diastereoisomers of **71** favoured axial methylation of the lithiated intermediate to give **72**.

Studies by both Hoffmann and co-workers [59,60] and by Reich and co-workers [61] showed that the rate determining step for racemisation of α-thio, α-seleno and α-telluroorganolithiums is rotation about the C–S, Se or Te bond [62]. Simple inversion in α-thio-substituted carbanions has a barrier as low as 0.5 kcal mol^{-1} [63] making inversion itself unlikely to comprise the rate determining step of the racemisation. Consistent with this explanation, the barrier to racemisation increases with greater steric bulk in the chalcogen substituent.

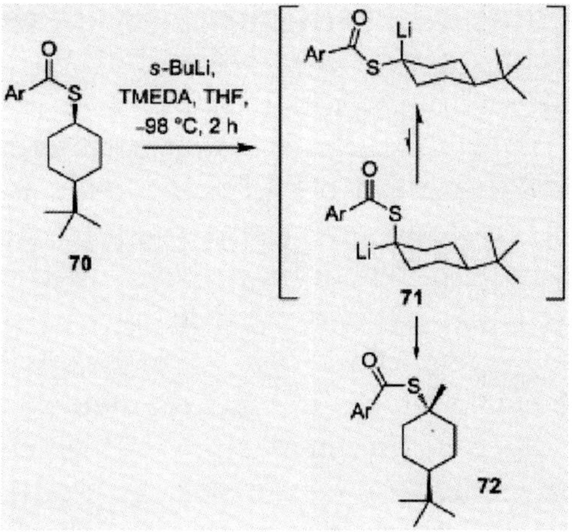

Scheme 25: Diastereoselective methylation of organolithium **71**.

Lithiation and alkylation of thiocarbamates

Hoppe and co-workers applied the observations of Hoffmann and Reich in comprehensive studies on configurationally stable α-lithiothiocarbamates. Stereoselective deprotonations of thiocarbamate **73** in the presence of (−)-sparteine **74** (Scheme 26) were attempted [64].

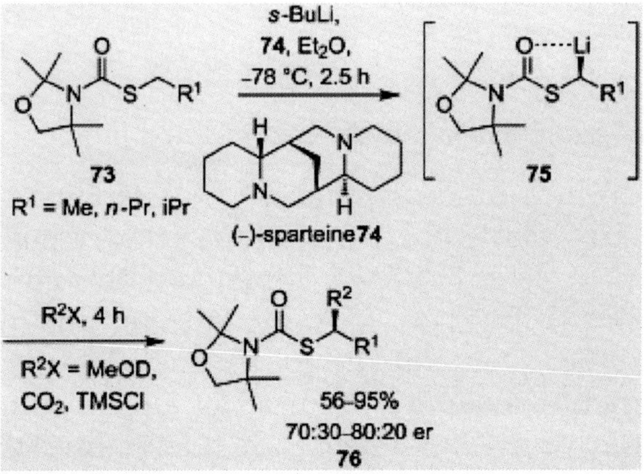

Scheme 26: Addition to lithiated thiocarbamate **75**.

Trapping the lithiated intermediate **75** with electrophiles gave products with poor enantioselectivities, but the results gave no indication of the configurational stability of the intermediate organolithium **75**. To establish the factors governing enantioselectivity in the electrophilic substitution of **73**, Hoppe made use of an extremely high kinetic H/D isotope effect observed for deprotonation in similar compounds [65]. Reaction of a test substrate demonstrated an isotope effect of $k_H/k_D \geq 100$. A racemic mixture of **77** was lithiated in the presence of (−)-sparteine and quenched with trimethylsilyl chloride (Scheme 27). Deuterated thiocarbamate **78** containing >99% deuterium was isolated from the reaction with an enantiomeric ratio of 67:33. The high deuterium content of the product indicated that both enantiomers of **77** must have been deprotonated, and thus that the lithio derivative of thiocarbamate **77**must be configurationally labile since the product is non-racemic.

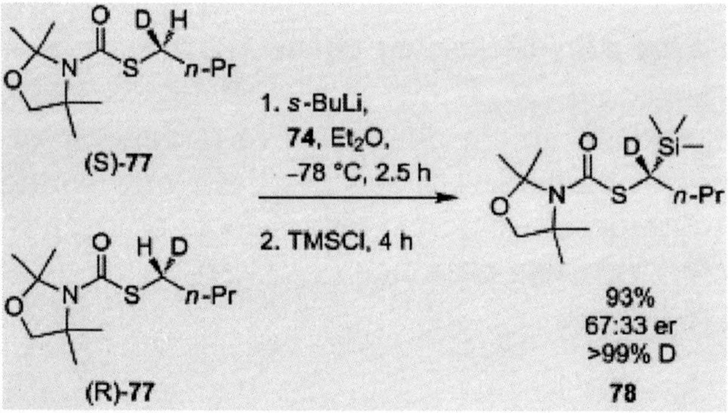

Scheme 27: Configurational lability in unhindered α-lithiothiocarbamates.

In order to induce greater configurational stability, thiocarbamate **79**, with increased steric bulk, was prepared. A sample of **79** with an enantiomeric ratio of 73:27 was lithiated with *sec*-butyllithium in diethyl ether and TMEDA at −78 °C for 2.5 hours and quenched with MeOD (Scheme 28). Deuterated thiocarbamate **78** was isolated with an enantiomeric ratio of 72:28, indicating almost full conservation of enantiomeric purity and hence demonstrating that the more congested organolithium **80** is now configurationally *stable* under the reaction conditions. Enantiomerically pure tertiary thiols can be liberated from the substitution products by cleavage of the carbamoyl functionality, providing a rare asymmetric route to simple unfunctionalised tertiary thiols.

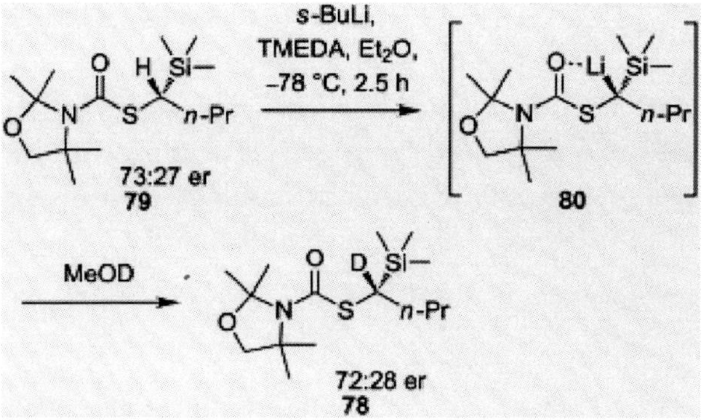

Scheme 28: Configurational stability in bulky α-lithiothiocarbamates.

The lithio derivative of benzylic thiocarbamate 81 was also configurationally stable in diethyl ether and TMEDA at −78 °C [66,67]. Reaction with a series of electrophiles allowed isolation of functionalised thiocarbamates 83 in excellent yield and enantiomeric ratio (Scheme 29).

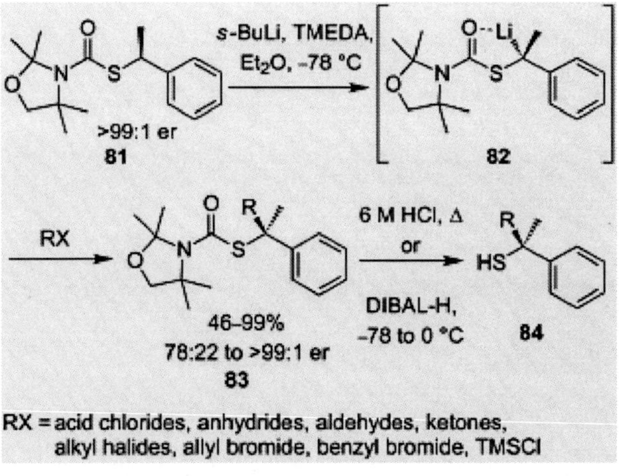

Scheme 29: Asymmetric functionalisation of secondary benzylic thiocarbamates.

Benzylic organolithiums are notorious for the lack of consistency with which they undergo electrophilic substitution [56], and 82 is remarkable in that all electrophiles, except protonating agents, react with complete *inversion* of configuration.

Enantiopure cyclohexenyl thiocarbamates also form configurationally stable α-thioallyllithium species [68-70]. Lithiation of thiocarbamate **85** followed by methylation results in the isolation of regioisomeric products arising from electrophilic substitution at either end of the lithioallyl system (Scheme 30).

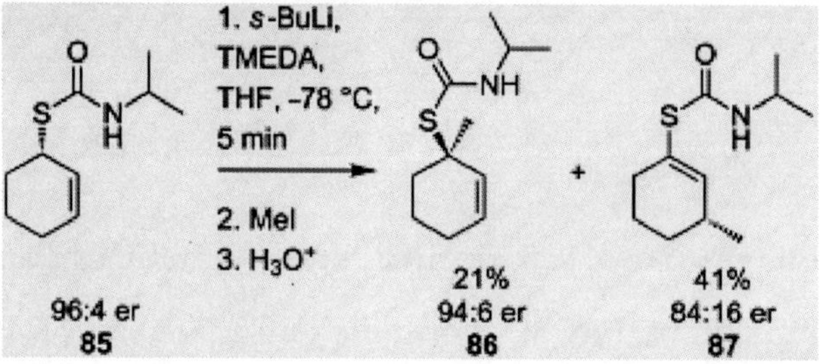

Scheme 30: Methylation of lithioallyl thiocarbamates.

In the case of the secondary thiocarbamate, the γ-substituted product **87** is preferentially formed, whilst with an*N,N*-diisopropylthiocarbamate selective formation of the α-substituted product **89** (Scheme 31) is observed.

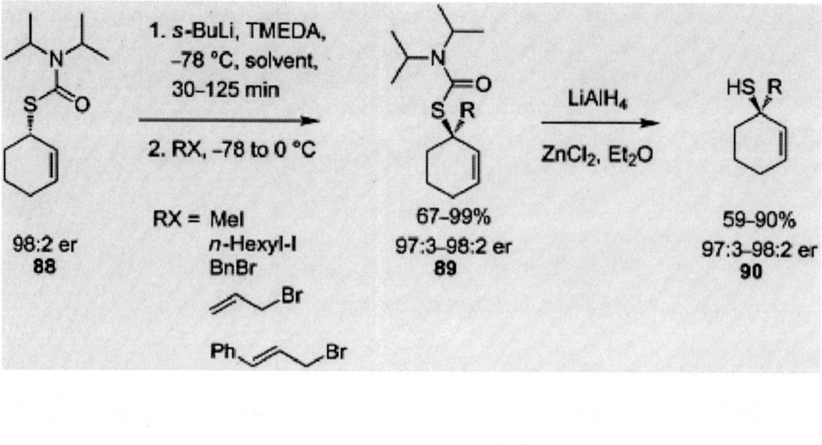

Scheme 31: Asymmetric preparation of tertiary allylic thiols.

Reduction of the functionalised thiocarbamate products **89** with lithium aluminium hydride affords enantiomerically pure thiols **90** in excellent yields and enantiomeric ratios.

Lithiation and rearrangement of thiocarbamates

Carbon–carbon bond formation in configurationally stable α-thioorganolithiums allows access to non-racemic tertiary thiols of varying structure and complexity. Hoppe's extensive studies of lithiated thiocarbamates have continued to display the synthetic utility of these species in the formation of secondary thiol derivatives. These were prepared by diastereoselective electrophilic additions in proline-derived systems [71,72] and asymmetric alkylation of thiocarbamates in the presence of a chiral ligand [73,74]. Stereospecific functionalisation of configurationally stable lithiated thiocarbamates with electrophiles is general for a range of structures, but until very recently it has not been applicable to *arylation* reactions.

However, we have discovered [75] a variant of a rearrangement reaction of lithiated ureas that we first reported in 2007 [76] in which a lithiated thiocarbamate undergoes an intramolecular aryl transfer. Treatment of an *N*-aryl *S*-benzyl thiocarbamate **92** leads to formation of a lithio derivative **93** comparable with those reported by Hoppe (Scheme 32). As in the case of lithiated *N*-aryl ureas [76-80] and *N*-aryl carbamates [81,82], the *N*-aryl ring is susceptible to attack by the anionic centre, and migration of the ring to the position α to sulfur occurs, presumably via an intermediate related to **94**. After work up the product is a tertiary thiocarbamate **95**, and simple mildly basic hydrolysis leads to the tertiary thiol **96**. The aryl migration is applicable to a range of products **96** (Table 1) and amounts to an intramolecular arylation, allowing the formation of otherwise inaccessible doubly benzylic tertiary thiols in enantiomerically enriched form from benzylic alcohols **91**. Similar to the rearrangement of lithiated ureas [76] (but interestingly in contrast to the rearrangement of lithiated *carbamates* [81,82]), aryl migration proceeds with retention of stereochemistry. Current work is continuing with the aim of expanding the utility of this new mode of reactivity displayed by thiocarbamates.

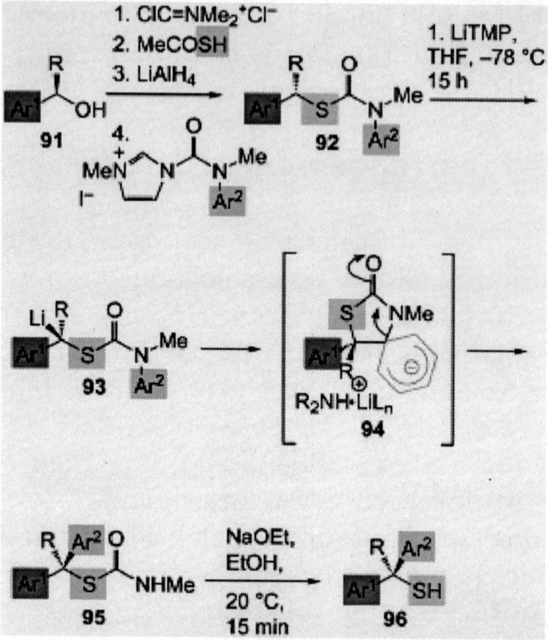

Scheme 32: Asymmetric preparation of thiols **96** by aryl migration in lithiated thiocarbamates.

Table 1: Scope of the thiocarbamate rearrangement [75].

SM, er	Ar¹	Ar²	R	95yield (%)	95er
(*S*)-**92a**, 98:2	Ph	4-MeC_6H_4	Me	83	96:4
(*S*)-**92b**, 97:3	Ph	2-OMeC_6H_4	Me	89	91:9
(*S*)-**92c**, 98:2	Ph	4-ClC_6H_4	Me	94	96:4
(*S*)-**92d**, 98:2	Ph	3-ClC_6H_4	Me	69	96:4
(*S*)-**92e**, 98:2	Ph	4-CNC_6H_4	Me	78	97:3
(*S*)-**92f**, 99:1	Ph	1-naphthyl	Me	98	96:4
(*S*)-**92g**, 98:2	Ph	3-FC_6H_4	Me	0	-
(*R*)-**92h**, 99:1	Ph	4-MeC_6H_4	*n*-Pr	74	98:2

(*R*)-**92i**, 96:4	3-CF_3	Ph	Me	85	67:33
(±)-**92j**, 50:50	4-OMe	4-MeC_6H_4	Me	86	50:50
(*R*)-**92k**, 79:21	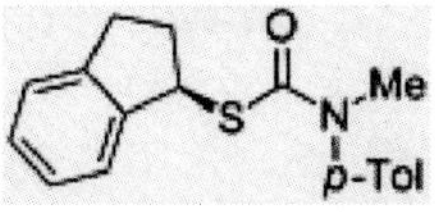			87	74:26

CONCLUSION

Of the methods available for the synthesis of tertiary thiols, the majority (for example, those involving epoxide opening or electrophilic attack on thio-substituted enolates) lead to thiols carrying further functionality in some form. Tertiary thiols without further functionality are still challenging, with organolithium-based alkylations and arylation (by rearrangement) of thiocarbamates offering the best prospects. Further developments in this area - particularly with regard to carbolithiation reactions [79] - are to be expected.

ACKNOWLEDGEMENTS

We are grateful to the Royal Society of Chemistry for a Briggs Scholarship (to PM) and for the Hickinbottom Fellowship (to JC).

REFERENCES

1. 2009 Top 200 Branded Drugs by Retail Dollars. Drug Topics, 2010;http://drugtopics.modernmedicine.com/drugtopics/data/articlestandard//drugtopics/252010/674961/article.pdf. Return to citation in text: [1] [2]
2. Cremlyn, R. J. *An Introduction to Organosulfur Chemistry;* John Wiley & Sons Ltd.: Chichester, U.K., 1996. Return to citation in text: [1]
3. Toru, T.; Bolm, C., Eds. *Organosulfur Chemistry in Asymmetric Synthesis;* Wiley-VCH: Weinheim, Germany, 2008. Return to citation in text: [1] [2]
4. Fernández, I.; Khiar, N. *Chem. Rev.* **2003,** *103,* 3651–3706. doi:10.1021/cr990372u Return to citation in text: [1]

5. Pellissier, H. *Tetrahedron* **2006,** *62,* 5559–5601. doi:10.1016/j.tet.2006.03.093 Return to citation in text: [1]
6. Mellah, M.; Voituriez, A.; Schulz, E. *Chem. Rev.* **2007,** *107,* 5133–5205. doi:10.1021/cr068440h Return to citation in text: [1]
7. Carreño, M. C.; Hernández-Torres, G.; Ribagorda, M.; Urbano, A. *Chem. Commun.* **2009,** 6129–6144.doi:10.1039/B908043K Return to citation in text: [1]
8. Wojaczyńska, E.; Wojaczyński, J. *Chem. Rev.* **2010,** *110,* 4303–4356. doi:10.1021/cr900147h Return to citation in text: [1]
9. Procter, D. J. *J. Chem. Soc., Perkin Trans. 1* **2001,** 335–354. doi:10.1039/b002081hReturn to citation in text: [1] [2] [3]
10. Shibasaki, M.; Kanai, M. *Chem. Rev.* **2008,** *108,* 2853–2873. doi:10.1021/cr078340rReturn to citation in text: [1]
11. Voss, J. *J. Sulfur Chem.* **2009,** *30,* 167–207. doi:10.1080/17415990802673017 Return to citation in text: [1]
12. Weaver, J. D.; Morris, D. K.; Tunge, J. A. *Synlett* **2010,** 470–474. doi:10.1055/s-0029-1219186 Return to citation in text: [1]
13. Effenberger, F.; Gaupp, S. *Tetrahedron: Asymmetry* **1999,** *10,* 1765–1775. doi:10.1016/S0957-4166(99)00154-8 Return to citation in text: [1]
14. Monde, K.; Taniguchi, T.; Miura, N.; Nishimura, S.-I.; Harada, N.; Dukor, R. K.; Nafie, L. A. *Tetrahedron Lett.* **2003,** *44,* 6017–6020. doi:10.1016/S0040-4039(03)01513-2 Return to citation in text: [1]
15. Avenoza, A.; Busto, J. H.; Corzana, F.; Jiménez-Osés, G.; Peregrina, J. M. *Chem. Commun.* **2004,** 980–981.doi:10.1039/b400282b Return to citation in text: [1]
16. Avenoza, A.; Busto, J. H.; Jiménez-Osés, G.; Peregrina, J. M. *J. Org. Chem.* **2006,** *71,* 1692–1695.doi:10.1021/jo051632cReturn to citation in text: [1] [2] [3]
17. Avenoza, A.; Busto, J. H.; Jiménez-Osés, G.; Peregrina, J. M. *Org. Lett.* **2006,** *8,* 2855–2858.doi:10.1021/ol060993r Return to citation in text: [1] [2]
18. Bednarski, P. J.; Nelson, S. D. *J. Med. Chem.* **1989,** *32,* 203–213. doi:10.1021/jm00121a037 Return to citation in text: [1] [2] [3] [4]
19. Kitagawa, I.; Ueda, Y.; Kawasaki, T.; Mosettig, E. *J. Org. Chem.* **1963,** *28,* 2228–2232.doi:10.1021/jo01044a018 Return to citation in text: [1] [2] [3]
20. Komeno, T.; Kishi, M. *Tetrahedron* **1971,** *27,* 1517–1526. doi:10.1016/S0040-4020(01)98017-2 Return to citation in text: [1] [2] [3]
21. Caine, D.; Crews, E.; Salvino, J. M. *Tetrahedron Lett.* **1983,** *24,* 2083–

2086. doi:10.1016/S0040-4039(00)81850-X Return to citation in text: [1] [2]

22. Tatsuta, K.; Suzuki, Y.; Toriumi, T.; Furuya, Y.; Hosokawa, S. *Tetrahedron Lett.* **2007,** *48,* 8018–8021.doi:10.1016/j.tetlet.2007.09.039 Return to citation in text: [1] [2] [3]
23. López, I.; Rodríguez, S.; Izquierdo, J.; González, F. V. *J. Org. Chem.* **2007,** *72,* 6614–6617.doi:10.1021/jo0709955 Return to citation in text: [1]
24. Mitsunobu, O. *Synthesis* **1981,** 1–28. doi:10.1055/s-1981-29317 Return to citation in text: [1]
25. Kumara Swamy, K. C.; Bhuvan Kumar, N. N.; Balaraman, E.; Pavan Kumar, K. V. P. *Chem. Rev.* **2009,** *109,*2551–2651. doi:10.1021/cr800278z Return to citation in text: [1] [2]
26. Walker, K. A. M. *Tetrahedron Lett.* **1977,** *18,* 4475–4478. doi:10.1016/S0040-4039(01)83541-3 Return to citation in text: [1]
27. Kotsuki, H.; Matsumoto, K.; Nishizawa, H. *Tetrahedron Lett.* **1991,** *32,* 4155–4158. doi:10.1016/S0040-4039(00)79890-X Return to citation in text: [1]
28. Shi, Y.-J.; Hughes, D. L.; McNamara, J. M. *Tetrahedron Lett.* **2003,** *44,* 3609–3611. doi:10.1016/S0040-4039(03)00728-7 Return to citation in text: [1]
29. La Clair, J. J. *Angew. Chem., Int. Ed.* **2006,** *45,* 27692773. doi:10.1002/anie.200504033 Return to citation in text: [1]
30. Marris, E. *Nature* **2006,** *442,* 492–493. doi:10.1038/442492c Return to citation in text: [1]
31. Rychnovsky, S. D. *Org. Lett.* **2006,** *8,* 2895–2898. doi:10.1021/ol0611346 Return to citation in text: [1]
32. Porco, J. A., Jr.; Su, S.; Lei, X.; Bardhan, S.; Rychnovsky, S. D. *Angew. Chem., Int. Ed.* **2006,** *45,* 57905792.doi:10.1002/anie.200602854 Return to citation in text: [1]
33. Ikegai, K.; Pluempanupat, W.; Mukaiyama, T. *Chem. Lett.* **2005,** *34,* 638–639. doi:10.1246/cl.2005.638 Return to citation in text: [1]
34. Ikegai, K.; Pluempanupat, W.; Mukaiyama, T. *Bull. Chem. Soc. Jpn.* **2006,** *79,* 780–790. doi:10.1246/bcsj.79.780 Return to citation in text: [1]
35. Kuroda, K.; Maruyama, Y.; Hayashi, Y.; Mukaiyama, T. *Chem. Lett.* **2008,** *37,* 836–837. doi:10.1246/cl.2008.836 Return to citation in text: [1]
36. Kuroda, K.; Maruyama, Y.; Hayashi, Y.; Mukaiyama, T. *Bull. Chem. Soc. Jpn.* **2009,** *82,* 381–392.doi:10.1246/bcsj.82.381 Return to citation in text: [1]

37. Mukaiyama, T.; Kuroda, K.; Maruyama, Y. *Heterocycles* **2010,** *80,* 63–82. doi:10.3987/REV-09-SR(S)1 Return to citation in text: [1]
38. Childers, W. E.; Furth, P. S.; Shih, M.-J.; Robinson, C. H. *J. Org. Chem.* **1988,** *53,* 5947–5951.doi:10.1021/jo00260a026 Return to citation in text: [1]
39. Lesuisse, D.; Gourvest, J.-F.; Benslimane, O.; Canu, F.; Delaisi, C.; Doucet, B.; Hartmann, C.; Lefrançois, J.-M.; Tric, B.; Mansuy, D.; Philibert, D.; Teutsch, G. *J. Med. Chem.* **1996,** *39,* 757–772. doi:10.1021/jm950539l Return to citation in text: [1]
40. Hanessian, S.; Reddy, B. *Tetrahedron* **1999,** *55,* 3427–3443. doi:10.1016/S0040-4020(98)01152-1 Return to citation in text: [1]
41. McDevitt, J. P.; Lansbury, P. T., Jr. *J. Am. Chem. Soc.* **1996,** *118,* 3818–3828. doi:10.1021/ja9525622 Return to citation in text: [1]
42. Ohata, K.; Terashima, S. *Tetrahedron Lett.* **2006,** *47,* 2787–2791. doi:10.1016/j.tetlet.2006.02.058 Return to citation in text: [1]
43. Ohata, K.; Terashima, S. *Bioorg. Med. Chem. Lett.* **2007,** *17,* 4070–4074. doi:10.1016/j.bmcl.2007.04.067 Return to citation in text: [1]
44. Emori, E.; Arai, T.; Sasai, H.; Shibasaki, M. *J. Am. Chem. Soc.* **1998,** *120,* 4043–4044. doi:10.1021/ja980397v Return to citation in text: [1]
45. Emori, E.; Iida, T.; Shibasaki, M. *J. Org. Chem.* **1999,** *64,* 5318–5320. doi:10.1021/jo9904922 Return to citation in text: [1]
46. Lu, H.-H.; Zhang, F.-G.; Meng, X.-G.; Duan, S.-W.; Xiao, W.-J. *Org. Lett.* **2009,** *11,* 3946–3949.doi:10.1021/ol901572x Return to citation in text: [1]
47. Palomo, C.; Oiarbide, M.; Dias, F.; López, R.; Linden, A. *Angew. Chem., Int. Ed.* **2004,** *43,* 3307–3310.doi:10.1002/anie.200453889 Return to citation in text: [1]
48. Palomo, C.; Oiarbide, M.; López, R.; González, P. B.; Gómez-Bengoa, E.; Saá, J. M.; Linden, A.*J. Am. Chem. Soc.* **2006,** *128,* 15236–15247. doi:10.1021/ja0654027 Return to citation in text: [1]
49. Strijtveen, B.; Kellogg, R. M. *J. Org. Chem.* **1986,** *51,* 3664–3671. doi:10.1021/jo00369a020 Return to citation in text: [1] [2]
50. Strijtveen, B.; Kellogg, R. M. *Tetrahedron* **1987,** *43,* 5039–5054. doi:10.1016/S0040-4020(01)87682-1 Return to citation in text: [1] [2]
51. McFadden, J. M.; Frehywot, G. L.; Townsend, C. A. *Org. Lett.* **2002,** *4,* 3859–3862. doi:10.1021/ol026685k Return to citation in text: [1] [2]
52. Alibés, R.; Bayón, P.; de March, P.; Figueredo, M.; Font, J.; Marjanet, G. *Org. Lett.* **2006,** *8,* 1617–1620.doi:10.1021/ol060173e Return to citation in text: [1]

53. Arpin, A.; Manthorpe, J. M.; Gleason, J. L. *Org. Lett.* **2006,** *8,* 1359–1362. doi:10.1021/ol060106k Return to citation in text: [1]

54. Tiong, E. A.; Gleason, J. L. *Org. Lett.* **2009,** *11,* 1725–1728. doi:10.1021/ol802643kReturn to citation in text: [1]

55. Seebach, D.; Naef, R.; Calderari, G. *Tetrahedron* **1984,** *40,* 1313–1324. doi:10.1016/S0040-4020(01)82417-0 Return to citation in text: [1]

56. Clayden, J. *Organolithiums: Selectivity for Synthesis;* Tetrahedron Organic Chemistry Series, Vol. 23; Pergamon: Oxford, U.K., 2002. Return to citation in text: [1] [2]

57. McDougal, P. G.; Condon, B. D.; Laffosse, M. D., Jr.; Lauro, A. M.; VanDerveer, D. *Tetrahedron Lett.* **1988,** *29,*25472550. doi:10.1016/S00404039(00)86108-0 Return to citation in text: [1]

58. Beak, P.; Becker, P. D. *J. Org. Chem.* **1982,** *47,* 3855–3861. doi:10.1021/jo00141a010 Return to citation in text: [1]

59. Ruhland, T.; Dress, R. K.; Hoffmann, R. W. *Angew. Chem., Int. Ed. Engl.* **1993,** *32,* 14671468.doi:10.1002/anie.199314671 Return to citation in text: [1]

60. Hoffmann, R. W.; Dress, R. K.; Ruhland, T.; Wenzel, A. *Chem. Ber.* **1995,** *128,* 861–870.doi:10.1002/cber.19951280903 Return to citation in text: [1]

61. Reich, H. J.; Dykstra, R. R. *Angew. Chem., Int. Ed. Engl.* **1993,** *32,* 1469–1470. doi:10.1002/anie.199314691 Return to citation in text: [1]

62. Aggarwal, V. K. *Angew. Chem., Int. Ed. Engl.* **1994,** *33,* 175177. doi:10.1002/anie.199401751 Return to citation in text: [1]

63. Wiberg, K. B.; Castejon, H. *J. Am. Chem. Soc.* **1994,** *116,* 10489–10497. doi:10.1021/ja00102a016 Return to citation in text: [1]

64. Kaiser, B.; Hoppe, D. *Angew. Chem., Int. Ed. Engl.* **1995,** *34,* 323–325. doi:10.1002/anie.199503231 Return to citation in text: [1]

65. Hoppe, D.; Paetow, M.; Hintze, F. *Angew. Chem., Int. Ed. Engl.* **1993,** *32,* 394–396. doi:10.1002/anie.199303941 Return to citation in text: [1]

66. Hoppe, D.; Kaiser, B.; Stratmann, O.; Fröhlich, R. *Angew. Chem., Int. Ed. Engl.* **1997,** *36,* 2784–2786.doi:10.1002/anie.199727841 Return to citation in text: [1]

67. Stratmann, O.; Kaiser, B.; Fröhlich, R.; Meyer, O.; Hoppe, D. *Chem.–Eur. J.* **2001,** *7,* 423–435.doi:10.1002/1521-3765(20010119)7:2<423::AID-CHEM423>3.0.CO;2-Y Return to citation in text: [1]

68. Marr, F.; Fröhlich, R.; Hoppe, D. *Org. Lett.* **1999,** *1,* 2081–2083. doi:10.1021/ol991134o Return to citation in text: [1]

69. Marr, F.; Hoppe, D. *Org. Lett.* **2002,** *4,* 4217–4220. doi:10.1021/ol0266828 Return to citation in text: [1]

70. Marr, F.; Fröhlich, R.; Wibbeling, B.; Diedrich, C.; Hoppe, D. *Eur. J. Org. Chem.* **2002,** 29702988.doi:10.1002/10990690(200209)2002:17<2970::AID-EJOC2970>3.0.CO;2-J Return to citation in text: [1]

71. Sonawane, R. P.; Fröhlich, R.; Hoppe, D. *Chem. Commun.* **2006,** 3101–3103. doi:10.1039/b604029b Return to citation in text: [1]

72. Sonawane, R. P.; Mück-Lichtenfeld, C.; Fröhlich, R.; Bergander, K.; Hoppe, D. *Chem.-Eur. J.* **2007,** *13,* 6419–6429. doi:10.1002/chem.200601853 Return to citation in text: [1]

73. Sonawane, R. P.; Fröhlich, R.; Hoppe, D. *Adv. Synth. Catal.* **2006,** *348,* 1847–1854.doi:10.1002/adsc.200606177 Return to citation in text: [1]

74. Lange, H.; Bergander, K.; Frölich, R.; Kehr, S.; Nakamura, S.; Shibata, N.; Toru, T.; Hoppe, D. *Chem.–Asian J.***2008,** *3,* 88–101. doi:10.1002/asia.200700262 Return to citation in text: [1]

75. MacLellan, P.; Clayden, J. *Chem. Commun.* **2011,** *47,* 3395–3397. doi:10.1039/C0CC04912C Return to citation in text: [1] [2]

76. Clayden, J.; Dufour, J.; Grainger, D. M.; Helliwell, M. *J. Am. Chem. Soc.* **2007,** *129,* 7488–7489.doi:10.1021/ja071523a Return to citation in text: [1] [2] [3]

77. Clayden, J.; Hennecke, U. *Org. Lett.* **2008,** *10,* 3567–3570. doi:10.1021/ol801332n Return to citation in text: [1]

78. Bach, R.; Clayden, J.; Hennecke, U. *Synlett* **2009,** 421–424. doi:10.1055/s-0028-1087543 Return to citation in text: [1]

79. Clayden, J.; Donnard, M.; Lefranc, J.; Minassi, A.; Tetlow, D. J. *J. Am. Chem. Soc.* **2010,** *132,* 6624–6625.doi:10.1021/ja1007992 Return to citation in text: [1] [2]

80. Tetlow, D. J.; Hennecke, U.; Raftery, J.; Waring, M. J.; Clarke, D. S.; Clayden, J. *Org. Lett.* **2010,** *12,* 5442–5445. doi:10.1021/ol102155h Return to citation in text: [1]

81. Clayden, J.; Farnaby, W.; Grainger, D. M.; Hennecke, U.; Mancinelli, M.; Tetlow, D. J.; Hillier, I. H.; Vincent, M. A.*J. Am. Chem. Soc.* **2009,** *131,* 3410–3411. doi:10.1021/ja808959e Return to citation in text: [1] [2]

82. Fournier, A. M.; Brown, R. A.; Farnaby, W.; Miyatake-Ondozabal, H.; Clayden, J. *Org. Lett.* **2010,** *12,* 2222–2225. doi:10.1021/ol100627c Return to citation in text: [1] [2]

Chapter 4

ENVIRON-ECONOMIC SYNTHESIS AND CHARACTERIZATION OF SOME NEW 1,2,4-TRIAZOLE DERIVATIVES AS ORGANIC FLUORESCENT MATERIALS AND POTENT FUNGICIDAL AGENTS

Harshita Sachdeva, Rekha Saroj, Sarita Khaturia, and Diksha Dwivedi

Department of Chemistry, Faculty of Engineering and Technology, Mody Institute of Technology and Science, Lakshmangarh, Rajasthan 332311, India

ABSTRACT

A multicomponent one-pot clean cyclocondensation reaction of 4-chloro-2-nitro aniline/amino acids and aromatic aldehydes/indole-2,3-diones with thiosemicarbazide in water yielding triazole/spiro indole-triazole derivatives in high yields and shorter reaction time and displaying excellent florescent property is reported. The developed MCR may provide a valuable practical tool for the synthesis of new drugs containing the title core fragment. All the newly synthesized compounds have been characterized by IR, ^{1}HNMR, ^{13}CNMR, and fluorescence study and also been screened for antimicrobial activity.

INTRODUCTION

Multicomponent and domino reactions are efficient and effective methods in the sustainable and diversity-oriented synthesis of heterocycles and such reactions have attracted enormous interest in recent years [1]. Thiosemicarbazide and its derivatives are an important class of synthetic compounds, having large variety of applications due to their wide spectrum of biological activities [2] including antiviral [3] and antitumoral [4] as well as parasiticidal activity against Plasmodium falciparum, Plasmodium berghei [5], Trypanosoma cruzi [6–8] Trypanosoma brucei rhodesiense [9], and Toxoplasma gondii [10].

The 1,2,4-triazoles and their derivatives are found to be associated with various biological activities such as anticonvulsant [11], antifungal [12], anticancer [13], antiinflammatory [14], and antibacterial properties [15]. Also several compounds containing 1,2,4-triazole rings are well known as drugs; for example, fluconazole is used as an antimicrobial drug, while vorozole, letrozole, and anastrozole are nonsteroidal drugs used for the treatment of cancer.

The increasing diversity of small molecule libraries is an important source for the discovery of new drug candidates. In terms of this trend, the literature survey showed that indole derivatives possess anticancer [16, 17], antioxidant [18], antibacterial [19], antifungal [20, 21], antiviral [22, 23], and antihypertensive activities [24]. Indole-3-carbon atom in the form of spiro carbon atom exhibits enhanced biological activities [25, 26].

The important biological activities of triazole derivatives impelled us to take up the synthesis of these new combinational heterocycles which are likely to have augmented diverse biological activity. The developed MCR may provide a valuable practical tool for the synthesis of novel physiologically active agents containing the title core fragment. Several methods [27–30] for the synthesis of 1,2,4-triazole derivatives are reported in the literature but all these methods require the presence of organic solvents and also require long reaction time. Further, there are only few reports [31–33] available in the literature regarding the synthesis of triazole derivatives under environmentally benign conditions. Earlier we also reported [33] the synthesis of spiro indole-triazoles under microwaves using montmorillonite as an inorganic solid support.

In recent years, organic research is mainly focused on the development of greener and environ-economic methods which involve use of alternative reaction media to replace volatile and hazardous solvents like benzene, toluene, and methanol, commonly used in organic synthesis. Nowadays, many organic transformations have been carried out in water [34–38]. It is a unique solvent because it is readily available, inexpensive, nontoxic, safer, and environmentally benign.

Recently, triazole-based fluorescent sensors have been developed for the selective detection of platinum ions in aqueous solutions [38] and synthesis of a pyrenyl-appended triazole-based calyxarene [39] as a fluorescent sensor for Cd^{+2} and Zn^{+2} is carried out. Prompted by rapidly expanding applications of organic fluorescent materials for electroluminescence (EL), dyelasers, sensors, probes, and phototherapeutic agents, and in view of the florescent properties associated with triazole derivatives, development of new fluorescent organic compounds with high functionality has been the subject of intense study for more than a decade [40–43].

In this context, we have synthesized new 1,2,4-triazole derivatives (4/5) using water as a green solvent under the umbrella of green chemistry and synthesized compounds have been evaluated for antibacterial and antifungal activity and also fluorescent properties.

RESULT AND DISCUSSION

In continuation of our work to develop greener and expeditious protocols for the synthesis of heterocyclic compounds [44–48], herein we wish to report a highly efficient procedure for the synthesis of new substituted-2H-1,2,4-triazole phenol derivatives (4) and spiro (indole-triazole) propanoic acid derivatives (5) using water (containing 1-2 mL of alcohol) by the reaction of 4-chloro-2-nitro aniline (1)/amino acids (1′) and aromatic aldehydes (2)/1H indole-

2,3-diones (2′), respectively, with thiosemicarbazide (3) yielding triazole (4)/spiro indole-triazole (5) derivatives in high yields and shorter reaction displaying excellent florescent property (Scheme 1).

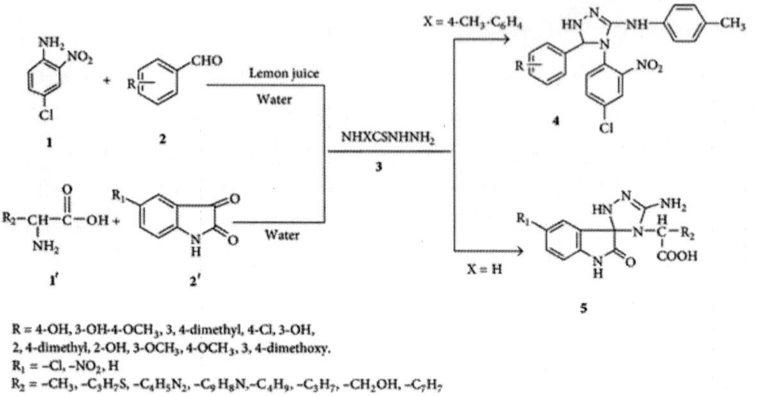
Lemon juice
Water
NHXCSNHNH2
X = 4-CH3-C6H4
X = H
R = 4-OH, 3-OH-4-OCH3, 3, 4-dimethyl, 4-Cl, 3-OH, 2, 4-dimethyl, 2-OH, 3-OCH3, 4-OCH3, 3, 4-dimethoxy.
R1 = -Cl, -NO2, H
R2 = -CH3, -C3H7S, -C4H5N2, -C9H8N, -C4H9, -C3H7, -CH2OH, -C7H7

Scheme 1 Synthesis of triazole derivatives 4 and 5

One of the tools used to combine economic aspect of new reactions with environmental aspects is the multicomponent reaction strategy. Compound 4 was synthesized by the one-pot multicomponent reaction of 1, 2, and 3 in the presence of lemon juice as a natural acidic catalyst in aqueous medium. There are few reports [49, 50] of synthesis of heterocyclic compounds using lemon juice. For the present work, we have used extract of Citrus limonum species of lemon as natural catalyst for the synthesis of triazole derivatives. The main ingredients of lemon juice are moisture (85%), carbohydrates (11.2%), citric acid (5–7%), protein (1%), vitamin-C (0.5%), fat (0.9%), minerals (0.3%), fibers (1.6%), and some other organic acids. As lemon juice is acidic in nature (pH ≈2-3) and percentage of citric acid (5–7%) is more than other acids, it works as acid catalyst [49] for synthesis of triazole derivatives. Using this methodology these reactions were completed in shorter reaction times (1-2 hrs) with yields of the product ranging from 78 to 83%. To our satisfaction, we found that the use of 2 mL of lemon juice resulted in quantitative yield of the corresponding triazole derivatives (4) within 1 to 2 hrs.

Mechanism of formation of compound 4 involves acid-catalyzed nucleophilic attack of 4-chloro-2-nitro aniline on carbonyl carbon forming arylimino derivatives preferentially . The nucleophilic attack of the amino electrons of thiosemicarbazide at the electrophilic benzylidene carbon accompanied by the migration of a hydrogen atom forms an intermediate thiol derivative which on cyclisation and desulfurization with the loss of H_2S gives the suggested triazole derivatives (4) (Scheme 2).

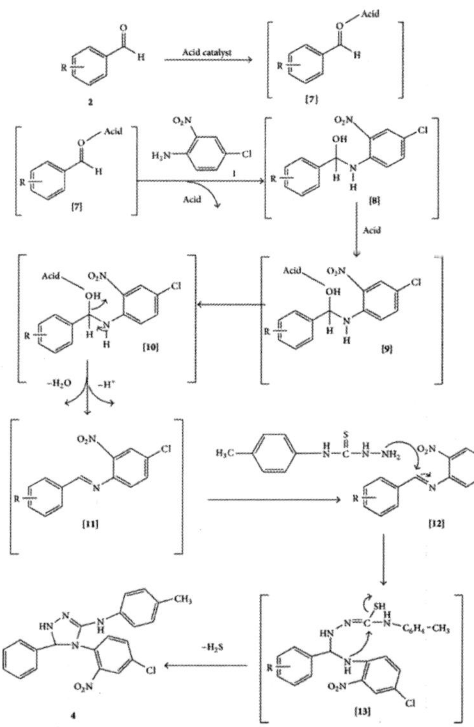

Scheme 2 Plausible mechanism of formation of compound 4

On the other hand, synthesis of spiro indole triazole (5) was carried out in the absence of lemon juice by the reaction of 1′, 2′, and 3 in the presence of water containing small amount of alcohol as a solvent. The progress of reaction was monitored by TLC. After completion of reaction, mixture was cooled to room temperature and poured on to crushed ice; the product formed was filtered and recrystallized from ethanol. Formation of spiro indole-triazoles (5) (reported by us) [31] took place in the absence of natural acid, formation of which is confirmed by spectral analyses.

The third carbonyl group of isatin being more electrophilic forms 3-aryl imino derivatives preferentially . The formation of final compound 5 involves the nucleophilic attack of thiosemicarbazide at the electrophilic indolyl carbon accompanied by the migration of hydrogen atom to form an intermediate thiol derivative . This is followed by cyclization and desulphurization with the loss of hydrogen sulphide to give the suggested spiro compound 5(Scheme 3).

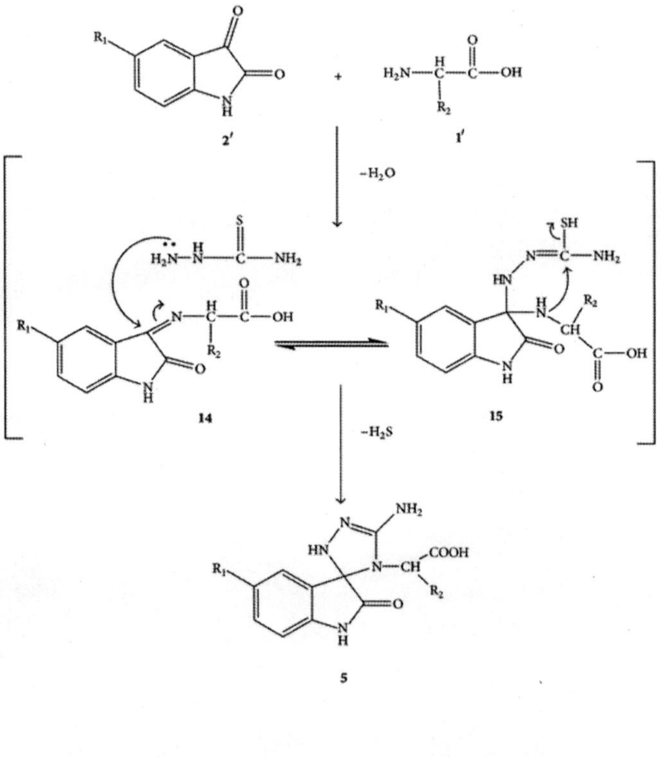

Scheme 3 Plausible mechanism of formation of compound 5

In summary, it can be stated that the present green synthetic protocol is highly efficient as it avoids the use of hazardous solvents at any stage of the reaction. The scope of the method was further studied by reacting differently substituted 1H-indole-2,3diones with amino acids and thiosemicarbazide. The identity of the products (Tables 3 and 4) obtained was confirmed by their IR,^{1}H NMR, and ^{13}C NMR spectral data.

FLUORESCENCE STUDY

A fluorescence study was carried out with newly synthesized triazole derivatives (4a–j) and (5a–l) in order to relate the fluorescence properties to the nature of donating and acceptor groups attached to these moieties. Fluorescence spectra of 5.7×10^{-5} M solutions of compounds were measured; excitation and emission maxima are also reported. Emission spectra of compounds (4a–j) and (5a–p) were run in DMSO. The resulting triazoles formed by the reaction of 4-Cl, 2-NO_2 aniline/amino acids and aromatic aldehydes/ indole-2,3-diones with thiosemicarbazide result in a bathochromic (red) shift of emission maxima (Tables 1 and 2). The combination of 4-chlorophenyl-substituted triazole to 4-chloro-2-nitro aniline in compound 4d results in the shift of λem. from 326 to 390.98 nm owing to the electron's withdrawing nature of Cl group. Further, combination of imidazole/indole-substituted triazole in 5j and 5l to 5-chloro-1 H indole-2,3-dione also results in red shift. For amino substituents, the bathochromic shift of the fluorescence as well of the absorption band is due to the large electron releasing ability of the nitrogen in amine group (5a–k) (Figures 1, 2,3, and 4).

Table 1 Fluorescence data of compounds (4a–j).

Entry	Molar concentration	4-Cl, 2-NO_2 aniline (reactants) λem·(nm)	Products λem·(nm)
4a	5×10^{-5}	326	376.969
4b	6×10^{-5}	326	362.008
4c	6×10^{-5}	326	355.83
4d	5×10^{-5}	326	390.98
4e	6×10^{-5}	326	354.83
4f	6×10^{-5}	326	360.01
4g	6×10^{-5}	326	353.63
4h	6×10^{-5}	326	356.07
4i	6×10^{-5}	326	354.5
4j	6×10^{-5}	326	354.26

Table 2 Fluorescence data of compounds (5a–l).

Entry	Molar concentration	1*H*-indole-2,3-diones (H, 5-Cl, 5-NO_2) reactants λem·(nm)	Products λem·(nm)
5a	7×10^{-5}	296	362.008
5b	6×10^{-5}	296	360.108
5c	6×10^{-5}	296	418.053
5e	6×10^{-5}	296	415.915
5f	7×10^{-5}	296	386.943
5g	7×10^{-5}	296	413.066
5h	6×10^{-5}	296	412.116
5j	5×10^{-5}	291	422.090
5k	5×10^{-5}	291	371.982
5l	5×10^{-5}	291	435.864

Table 3 Physical characterization data of compounds (4a–j).

Entry	Product	Time (hr.)	Yield (%)	M.P. (°C)	Color
4a	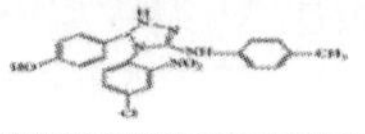	1	82	190	Yellow
4b	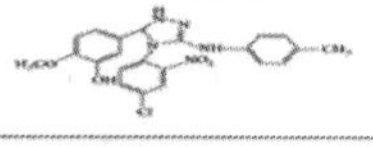	2	80	220–222	Pale yellow
4c	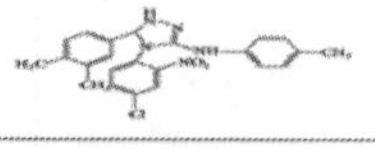	1.5	81	140	Orange
4d	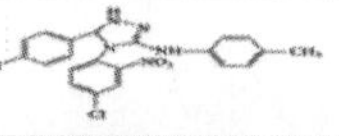	1	78	180	Orange
4e	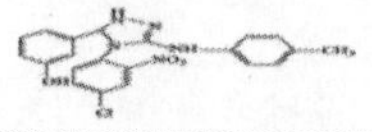	1.5	78	200	Orange
4f	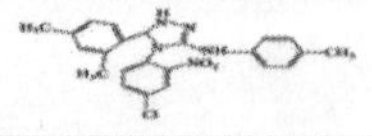	1	80	195	Orange
4g	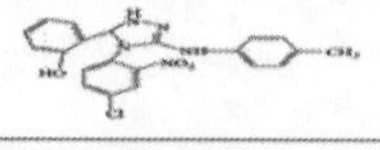	1	80	220	Orange
4h	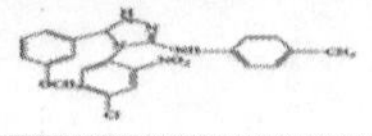	1	81	210	Yellow
4i	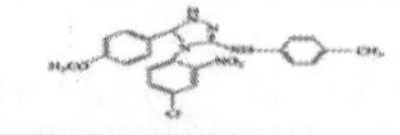	1.2	82	240	Yellow
4j	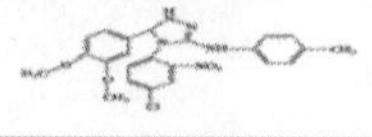	1.5	83	160	Orange

Table 4 Physical characterization data of compounds (5a–p).

Entry	Product	Time (hr)	Yield (%)	M.P. (°C)	Color
5a	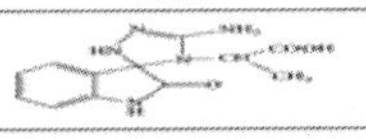	1.5	80	105	Red
5b		1.5	72	92	Brown
5c		2	76	140	Dark brown
5d		2	72	125	Dark brown
5e		1.5	77	170	Orange
5f		1	72	195	Orange
5g		1.5	75	90	Brown
5h		1.5	75	140–142	Red
5i		1.5	71	152	Yellow
5j		2	70	120–122	Orange
5k	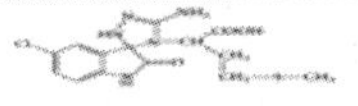	1.5	70	220	Red
5l		1.5	69	220	Red
5m		1.5	62	210	Pale yellow
5n	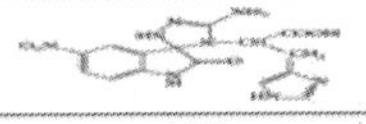	1.5	69	207	Pale yellow
5o		2	70	194	Pale yellow
5p	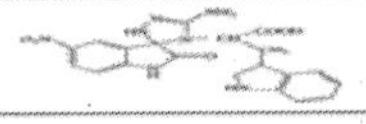	2	70	125	Pale yellow

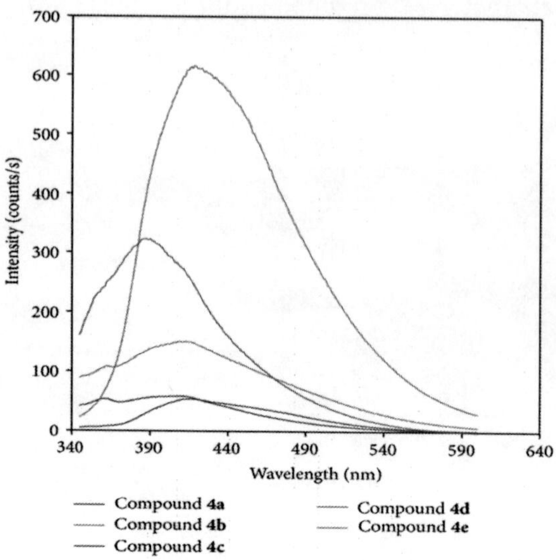

Figure 1 Fluorescence spectra of compounds (4a–e).

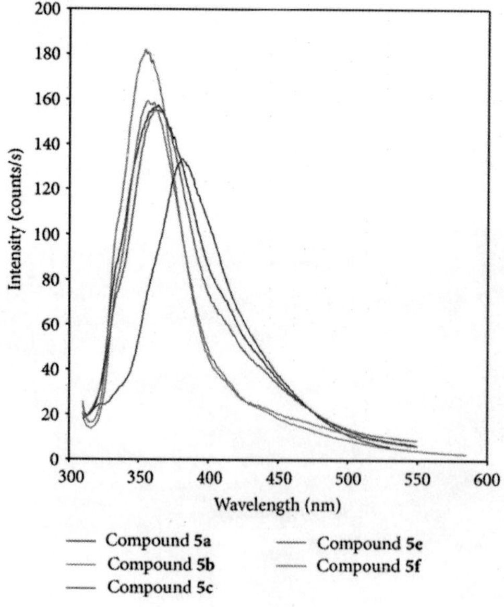

Figure 2 Fluorescence spectra of compounds (4f–j).

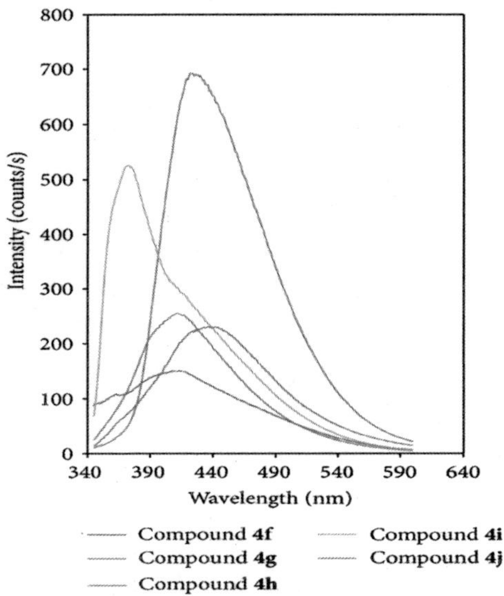

Figure 3: Fluorescence spectra of compounds (5a–f).

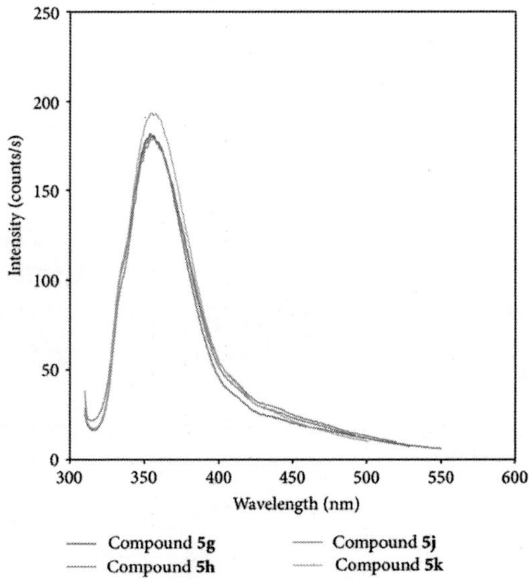

Figure 4 Fluorescence spectra of compounds (5g–k).

MATERIALS AND METHODS

General

Reagents and solvents were obtained from commercial sources and used without further purification. Melting points were determined on a Toshniwal apparatus. The ^{1}H NMR and ^{13}C NMR spectra of synthesized compounds have been carried out at SAIF, Punjab University, Chandigarh. ^{1}H NMR spectra were recorded on Bruker Avance II 400 NMR spectrometer using DMSO-d_6 and CDCl3 as solvent and tetramethylsilane (TMS) as internal reference standard. The purity of compounds was checked on thin layers of silica gel in various nonaqueous solvent systems, for example, benzene : ethyl acetate (8 : 2). IR spectra were recorded in KBr on a Perkin Elmer Infrared L1600300 spectrum Two Li-Ta spectrophotometer and fluorescence studies were carried out on RF-5301 PC spectrofluorophotometer, Shimadzu at FET, MITS, Laxmangarh, Sikar, Rajasthan, India.

General Procedure for the Synthesis of Compounds 4a–j

An equimolar mixture of substituted aromatic aldehyde (0.01 mol), 4-chloro-2-nitro aniline (0.01 mol), and methyl phenyl thiosemicarbazide in water (15 mL containing 1-2 mL of ethanol) in the presence of lemon juice (2 mL) was mixed in round bottom flask and the mixture was refluxed for the time needed to complete the reaction (as monitored by TLC). After completion of the reaction, mixture was cooled to room temperature and the solid mass was filtered and recrystallized from ethanol.

General Procedure for the Synthesis of Compounds 5a–p

An equimolar mixture of substituted 1H-indole-2,3-dione (0.012 mol), substituted amino acids (0.012 mol) and thiosemicarbazide in water (15 mL containing 1-2 mL of ethanol) was mixed in round bottom flask and the mixture was refluxed for the time needed to complete the reaction (as monitored by TLC). The initial syrupy reaction mixture solidifies within 2-3 hours. After completion of the reaction, mixture was cooled to room temperature and the solid mass was poured onto crushed ice, filtered, and recrystallized from ethanol.

ANTIMICROBIAL ACTIVITY

Synthesized compounds 4a–j and 5a–p were evaluated for antibacterial and antifungal activity.

Antibacterial Activity

Synthesized compounds (4a–j) were screened for their antibacterial activity against Gram-positive bacteria Bacillus licheniformis, Staphylococcus aureus, and Micrococcus luteus and Gram-negative bacteria Pseudomonas aeruginosa and Escherichia coli by the agar well diffusion method. 5 mL aliquot of nutrient broth was inoculated with the test organism and incubated at 37°C for 24 hours. Sterile nutrient agar plates were also prepared and holes of 5 mm diameter were cut using a sterile cork borer ensuring proper distribution. The test organisms after 24 hours of incubation were spread onto separate agar plates. The chemical compounds were dissolved in DMSO at a particular concentration or poured into appropriately labelled holes using a pipette in aseptic conditions. A hole containing DMSO served as a control. The plates were left at room temperature for two hours to allow the diffusion of the sample followed by incubation at 37°C for 24 hours in inverted position. The antimicrobial activity was determined by measuring the diameter of the zone (mm) showing complete inhibition with respect to control (DMSO) and reference compounds streptomycin and erythromycin. It has been observed that all the compounds tested showed good to excellent activity against tested bacteria.

Compound 4g shows good activity against bacteria Staphylococcus aureus at 250 ppm concentration. On the other hand, compound 4j shows excellent activity against bacteria Pseudomonas aeruginosa, Staphylococcus aureus, and Micrococcus luteus at 500 ppm concentration. Compound 4b shows good activity against bacteria Staphylococcus aureus, Micrococcus luteus, and Escherichia coli and compound 4d also shows good activity against bacteria Pseudomonas aeruginosa, Bacillus licheniformis, and Micrococcus luteus at 500 ppm concentration (Table 5, Figure 5).

Table 5 Antibacterial evaluation of the synthesized compounds (4a–j).

Compounds	Zone of inhibition (mm) (250 ppm)					(500 ppm)				
	A	B	C	D	E	A	B	C	D	E
4a	6	8	7	9	9	9	6	10	7	—
4b	7	6	9	8	—	9	9	11	10	11
4c	—	9	9	—	8	4	9	4	—	9
4d	6	—	—	8	—	10	10	—	10	—
4e	8	9	6	9	6	9	4	6	10	3
4f	—	—	9	7	8	9	—	9	9	10
4g	7	8	10	—	7	—	10	—	—	11
4h	7	—	9	9	—	3	—	10	6	—
4i	9	7	—	8	9	—	8	9	—	10
4j	9	9	—	9	—	10	9	11	11	—
DMSO	—	—	—	—	—	—	—	—	—	—
Streptomycin	13	10	14	10	11	10	10	14	14	12
Erythromycin	12	11	—	12	—	12	10	—	11	13

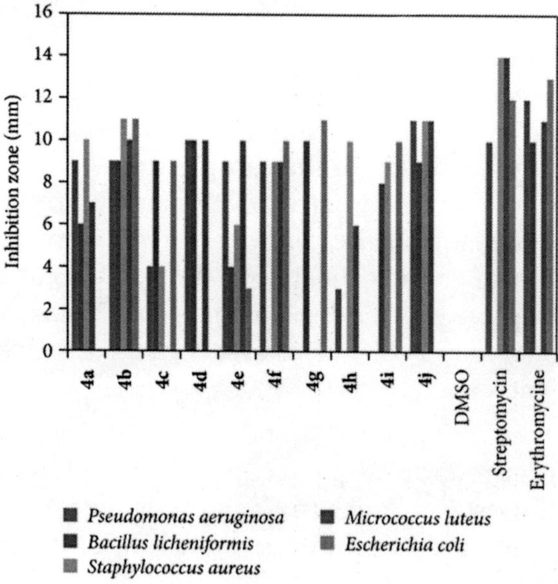

Figure 5 antibacterial evaluations of the synthesized compounds (4a–j) at 500 ppm

Antifungal Activity

Antifungal activity was performed by cup plate method at concentration 500 ppm and 250 ppm against Aspergillus niger, Penicillum sp. fusarium oxysporum, Alternaria brassicicola, Chaetomium orium, and Lycopodium sp. by measuring the zone of inhibition in mm. Sabour and dextrose agar were employed as culture medium, by pouring the sterile agar into Petri dishes in aseptic conditions. 0.1 mL of each standardized test organism culture was spread onto agar plates. The test compounds (250 and 500 ppm), the standard drug solutions, and the solvent control (DMSO) were placed in the cavity separately. Then the plates were maintained at −4°C for 1 hour to allow the diffusion of solution into the medium. All the fungal plates were incubated at 28°C for 42–72 hrs and the zone of inhibition was measured in mm. The antimicrobial activity was determined by measuring the diameter of the zone (mm) showing complete inhibition with respect to control (DMSO) and reference compounds streptomycin and erythromycin.

Table 6 Antifungal evaluation of the synthesized compounds (4a–j).

	Zone of inhibition (mm)											
Compounds	Concentration (250 ppm)						Concentration (500 ppm)					
	A	B	C	D	E	F	A	B	C	D	E	F
4a	7	9	8	8	8	8	8	8	10	10	9	10
4b	6	7	9	—	—	—	—	9	9	10	—	11
4c	—	9	—	5	9	5	—	8	9	—	12	9
4d	5	—	10	—	—	8	9	—	9	—	11	—
4e	—	8	8	—	8	—	7	—	11	—	—	—
4f	7	7	9	9	—	6	—	8	9	11	—	12
4g	6	6	.10	8	—	—	9	7	10	9	9	10
4h	—	9	—	—	9	—	8	9	11	—	11	11
4i	8	8	—	—	8	8	—	6	—	—	10	—
4j	4	—	8	8	—	—	9	—	11	8	9	—
DMSO	—	—	—	—	—	—	—	—	—	—	—	—
Streptomycin	11	11	13	12	13	10	11	—	12	13	15	14
Erythromycin	14	13	10	11	12	13	14	10	—	14	14	12

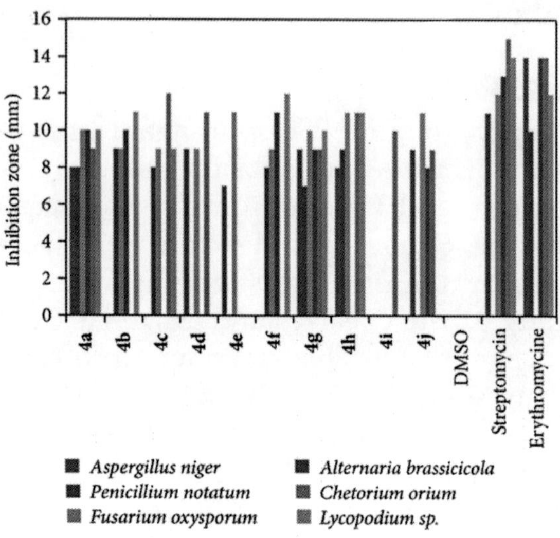

Figure 6 Antifungal evaluation of the synthesized compounds (4a–j) at 500 ppm.

Compounds 4d and 4g show good activity against fungus Fusarium oxysporum at 250 ppm due to presence of –Cl and –OH substituents. Compound 4c shows excellent activity against fungusChaetomium orium and 4f shows excellent activity against fungus Alternaria brassicicola andLycopodium sp. at 500 ppm. Compound 4h shows excellent activity against fungus Fusarium oxysporum, Chaetomium orium, and Lycopodium sp. at 500 ppm. Compound 4a shows good activity against fungus Fusarium oxyforum, Alternaria brassicicola, and Lycopodium sp. at 500 ppm concentration due to the presence of –OH substituent (Table 6, Figure 6).

Antibacterial Activity

Compound 5j shows good activity against bacteria Pseudomonas aeruginosa and 5k shows good activity against bacteria Staphylococcus aureus, respectively at 250 ppm concentration. Compound5m shows excellent activity against bacteria Staphylococcus aureus at 250 ppm concentration due to presence of $-NO_2$ substituent. On the other hand, compounds 5b, 5g, and 5n show excellent activity against bacteria Escherichia coli at 500 ppm concentration (Table 7, Figure 7).

Table 7 Antibacterial evaluation of the synthesized compounds (5a–p).

Compounds	Zone of inhibition (mm) (250 ppm)					Zone of inhibition (mm) (500 ppm)				
	A	B	C	D	E	A	B	C	D	E
5a	8	8	7	9	9	9	6	10	7	—
5b	6	6	9	8	—	—	9	11	10	11
5c	9	9	9	—	8	4	9	4	—	9
5d	—	—	—	8	—	10	10	—	10	4
5e	8	9	6	9	6	—	4	6	10	3
5f	7	—	9	7	8	9	—	9	9	10
5g	—	8	10	—	7	—	10	—	—	11
5h	7	—	9	9	—	3	—	10	6	5
5i	9	7	—	8	9	—	8	9	—	10
5j	11	9	—	9	—	10	9	11	11	—
5k	9	—	11	—	8	8	8	12	2	10
5l	9	7	—	9	9	—	9	—	10	10
5m	—	—	12	8	—	9	—	11	9	—
5n	9	8	8	7	9	—	—	—	9	11
5o	10	8	10	9	7	8	8	12	4	9
5p	9	7	9	6	—	9	9	11	8	—
DMSO	—	—	—	—	—	—	—	—	—	—
Streptomycin	13	10	14	10	11	10	—	14	14	12
Erythromycin	12	11	—	12	—	12	10	—	11	13

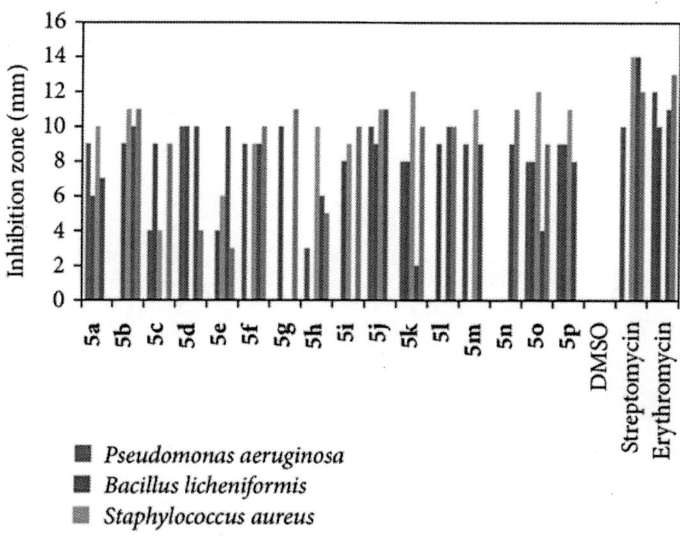

Figure 7 Antibacterial evaluation of the synthesized compounds 5a–p at 500 ppm.

ANTIFUNGAL ACTIVITY

Compound 5d shows good activity against Fusarium oxysporum at 250 ppm concentration. Compounds 5b, 5h, and 5m show good activity against Lycopodium sp. at 500 ppm concentration. Compounds 5d and 5h show good activity against Chaetomium orium at 500 ppm concentration. Compounds 5f and 5p also show good activity against Alternaria brassicicola at 500 ppm concentration; compounds 5e, 5h, and 5j also show good activity against Fusarium oxysporum at 500 ppm concentration (Table 8, Figure 8).

Table 8 Antifungal evaluation of the synthesized compounds (5a–p).

Compounds	Zone of inhibition (mm)											
	Concentration (250 ppm)						Concentration (500 ppm)					
	A	B	C	D	E	F	A	B	C	D	E	F
5a	7	9	8	9	8	8	8	8	10	10	9	10
5b	8	7	9	—	—	—	—	9	9	10	—	11
5c	6	9	—	8	9	5	—	8	9	—	12	9
5d	4	—	10	—	—	8	9	—	9	—	11	—
5e	—	8	8	—	8	—	7	—	11	—	—	—
5f	7	7	9	9	—	6	—	8	9	11	—	12
5g	6	6	10	8	—	—	9	7	10	9	9	10
5h	8	9	—	—	9	—	8	9	11	—	11	11
5i	8	8	—	—	8	8	—	6	—	—	10	—
5j	4	—	8	8	—	—	9	—	11	8	9	—
5k	7	—	9	10	—	—	8	9	—	7	—	—
5l	—	9	—	—	9	6	—	8	10	10	9	—
5m	9	10	9	9	—	6	—	—	9	9	9	11
5n	—	9	8	7	8	8	—	—	9	—	9	9
5o	9	—	—	—	—	7	8	—	—	10	7	10
5p	8	8	9	7	7	7	7	9	9	11	12	9
DMSO	—	—	—	—	—	—	—	—	—	—	—	—
Streptomycin	11	11	13	12	13	10	11	—	12	13	15	14
Erythromycin	14	13	10	11	12	13	14	10	—	14	14	12

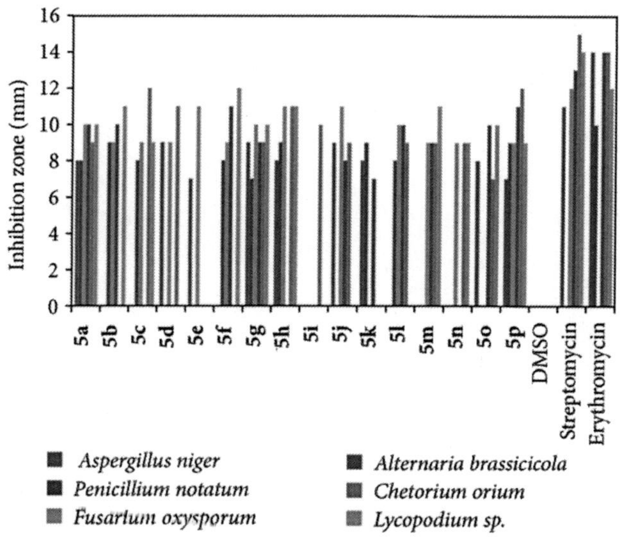

Figure 8 Antifungal evaluation of the synthesized compounds (5a–p) at 500 ppm.

SPECTROSCOPIC CHARACTERIZATION DATA

Spectroscopic Characterization Data of 4a–j

4-(5-(p-Tolylamino)-4-(4-chloro-2-nitrophenyl)-3,4-dihydro-2H-1,2,4-triazol-3-yl)phenol (4a).

m.p. 190°C; IR (KBr): 3476, 3355, 3169, 2980, 1604, 1580, 1369,722, 699 cm^{-1}; ^{1}H NMR (δ ppm DMSO-d_6): 2.35 (s, 3H, CH_3), 5.0 (s, 1H, OH), 5.04 (s, 1H, CH), 6.34–7.15 (m, 4H, aromatic), 6.46–6.95 (m, 4H, aromatic), 6.69–7.25 (m, 3H, aromatic), 7.01 (s, 1H, N–H), 8.97 (s, 1H, N–H). ^{13}C NMR (400 MHz, DMSO-d_6): 156.92 (C–OH), 154.98 (C=N), 141.48–129.72 (aromatic carbons), 128.08–116.20 (aromatic carbons), 135.82–112.14 (aromatic carbons), 66.92 (C–H), 24.34 (CH_3) ppm. Anal.calcd for $C_{21}H_{18}N_5O_3Cl$: C, 59.51; H, 4.28; N, 16.52. Found: C, 59.68; H, 4.29; N, 16.53.

5-(5-(p-Tolylamino)-4-(4-chloro-2-nitrophenyl)-3,4-dihydro-2H-1,2,4-triazol-3-yl)-2-methoxyphenol (4b).

m.p. 220–222°C; IR (KBr): 3472, 3354, 3140, 2960, 1619, 1506, 1368, 1272, 1204, 1045, 701, 648 cm^{-1}; ^{1}HNMR (δ ppm DMSO-d_6): 2.38 (s, 3H, CH_3), 3.73 (s, 3H, OCH_3), 5.01 (s, 1H, OH), 5.06 (s, 1H, CH), 6.34–6.81 (m, 4H, aromatic), 6.42–6.51 (m, 3H, aromatic), 6.69–7.98 (m, 3H, aromatic), 7.21 (s, 1H, N–H), 8.98 (s, 1H, N–Hs). ^{13}CNMR (400 MHz, DMSO-d_6): 156.55 (C–OH), 154.98 (C=N), 149.16 ($COCH_3$), 141.48–135.82 (aromatic carbons), 129.72–123.22 (aromatic carbons), 124.12–112.14 (aromatic carbons), 66.92 (C–H), 56.20 ($COCH_3$) 24.32 (CH_3) ppm. Anal.calcd for $C_{22}H_{20}ClN_5O_4$: C, 58.22; H, 4.44; N, 15.43. Found: C, 58.05; H, 4.27; N, 15.42.

4-(4-Chloro-2-nitrophenyl)-4,5-dihydro-5-(3,4-dimethylphenyl)-N-p-tolyl-1H-1,2,4-triazol-3-amine (4c).

m.p. 140°C; IR (KBr): 3442, 3317, 3110, 2980, 1645, 1600, 1510, 1350, 1260, 1091, 745, 660 cm^{-1}; ^{1}HNMR (δ ppm DMSO-d_6): 2.35 (s, 3H, CH_3), 2.35 (s, 3H, CH_3), 2.35 (s, 3H, CH_3), 5.10 (s, 1H, CH), 6.34–7.18 (m, 4H, aromatic), 6.46–6.95 (m, 3H, aromatic), 6.67–7.20 (m, 3H, aromatic) 7.10 (s, 1H, N–H), 9.01 (s, 1H, N–H). ^{13}CNMR (400 MHz, DMSO-d_6): 154.92 (C=N), 149.10–135.86 (aromatic carbons), 129.76–116.22 (aromatic carbons), 124.18–112.12 (aromatic carbons), 66.98 (C–H), 24.34 (CH_3), 18.12 (CH_3), 17.88 (CH_3) ppm. Anal.calcd for $C_{23}H_{22}N_5O_2Cl$: C, 63.37; H, 5.09; N, 16.07. Found: C, 63.55; H, 5.10; N, 16.09.

4-(4-Chloro-2-nitrophenyl)-5-(4-chlorophenyl)-4,5-dihydro-N-p-tolyl-1H-1,2,4-triazol-3-amine (4d).

m.p. 180°C; IR (KBr): 3440, 3327, 3114, 2981, 1645, 1600, 1511, 1462, 1388, 1291, 1090, 781, 642 cm^{-1};^{1}HNMR (δ ppm DMSO-d_6): 2.35 (s, 3H, CH_3), 5.06 (s, 1H, CH), 6.34–7.15 (m, 4H, aromatic), 6.46–6.95 (m, 4H, aromatic), 6.67–7.20 (m, 3H, aromatic) 7.01 (s, 1H, N–H), 8.83 (s, 1H, N–H). ^{13}CNMR (400 MHz, DMSO-d_6): 154.94 (C=N), 149.10–135.88 (aromatic carbons), 129.78–123.22 (aromatic carbons), 116.28–112.18 (aromatic carbons), 66.92 (C–H), 24.38 (CH_3), ppm. Anal.calcd for $C_{21}H_{17}N_5O_2Cl_2$: C, 57.03; H, 3.87; N, 15.83. Found: C, 57.21; H, 3.86; N, 15.84.

3-(5-(p-Tolylamino)-4-(4-chloro-2-nitrophenyl)-3,4-dihydro-2H-1,2,4-triazol-3-yl)phenol (4e).

m.p. 200°C; IR (KBr): 3476, 3355, 3169, 2980, 1604, 1580, 1369, 722, 699 cm^{-1}; ^{1}H NMR (δ ppm DMSO-d_6): 2.35 (s, 3H, CH_3), 5.0 (s, 1H, OH), 5.04 (s, 1H, CH), 6.34–7.15 (m, 4H, aromatic), 6.46–6.95 (m, 4H, aromatic), 6.67–7.20 (m, 3H, aromatic), 7.23 (s, 1H, N–H), 9.03 (s, 1H, N–H). ^{13}C NMR (400 MHz, DMSO-d_6): 156.24 (C–OH), 154.94 (C=N), 145.10–135.86 (aromatic carbons), 129.72–123.28 (aromatic carbons), 124.18–112.12 (aromatic carbons), 66.94 (C–H), 24.32 (CH_3), ppm. Anal.calcd for $C_{21}H_{18}ClN_5O_3$: C, 59.51; H, 4.28; N, 16.52. Found: C, 59.34; H, 4.27; N, 16.51.

4-(4-Chloro-2-nitrophenyl)-4,5-dihydro-5-(2,4-dimethylphenyl)-N-p-tolyl-1H-1,2,4-triazol-3-amine (4f).

m.p. 195°C; IR (KBr): 3477, 3330, 3169 2980, 1604, 1600, 1508, 1369, 1091, 722, 699 cm^{-1}; ^{1}H NMR (δppm DMSO-d_6): 2.35 (s, 3H, CH_3), 2.38 (s, 3H, CH_3), 2.32 (s, 3H, CH_3), 5.08 (s, 1H, CH), 6.34–7.15 (m, 4H, aromatic), 6.46–6.95 (m, 3H, aromatic), 6.67–7.20 (m, 3H, aromatic), 7.67 (s, 1H, N–H), 8.99 (s, 1H, N–H). ^{13}C NMR (400 MHz, DMSO-d_6): 154.96 (C=N), 140.42–129.74 (aromatic carbons), 135.82–123.22 (aromatic carbons), 124.16–112.16 (aromatic carbons), 66.92 (C–H), 24.36 (CH_3), 24.14 (CH_3), 16.14 (CH_3). ppm Anal.calcd for $C_{23}H_{22}N_5O_2Cl$: C, 63.37; H, 5.09; N, 16.07. Found: C, 63.54; H, 5.08; N, 16.06.

2-(5-(p-Tolylamino)-4-(4-chloro-2-nitrophenyl)-3,4-dihydro-2H-1,2,4-triazol-3-yl)phenol (4g).

m.p. 220°C; IR (KBr): 3476, 3354, 3158, 1630, 1504, 1341, 1203, 723, 699 cm^{-1}; ^{1}H NMR (δ ppm DMSO-d_6): 2.35 (s, 3H, CH_3), 5.01 (s, 1H, OH), 5.04 (s, 1H, CH), 6.34–7.15 (m, 4H, aromatic), 6.46–6.95 (m, 4H, aromatic), 6.67–7.20 (m, 3H, aromatic), 7.0 (s, 1H, N–H), 8.93 (s, 1H, N–H). ^{13}C NMR (400 MHz, DMSO-d_6): 156.24 (C–OH), 154.92 (C=N), 141.42–135.86 (aromatic carbons), 136.32–124.12 (aromatic carbons),

128.02–112.16 (aromatic carbons), 66.94 (C–H), 24.32 (CH_3), ppm Anal.calcd for $C_{21}H_{18}N_5O_3Cl$: C, 59.51; H, 4.28; N, 16.52. Found: C, 59.33; H, 4.26; N, 16.50.

4-(4-Chloro-2-nitrophenyl)-4,5-dihydro-5-(3-methoxyphenyl)-N-p-tolyl-1H-1,2,4-triazol-3-amine (4h).

m.p. 210°C; IR (KBr): 3476, 3356, 2924, 1629, 1562, 1505, 1369, 1341, 1251, 1203, 1158, 723, 700 cm^{-1}; ^{1}H NMR (δ ppm DMSO-d_6): 2.35 (s, 3H, CH_3), 3.73 (s, 3H, OCH_3), 5.04 (s, 1H, CH), 6.34–7.15 (m, 4H, aromatic), 6.46–6.95 (m, 4H, aromatic), 6.67–7.98 (m, 3H, aromatic), 7.09 (s, 1H, N–H), 9.01 (s, 1H, N–H). ^{13}C NMR (400 MHz, DMSO-d_6): 154.42 (C=N), 149.16 ($COCH_3$), 145.62–136.32 (aromatic carbons), 137.72–128.08 (aromatic carbons), 124.16–112.12 (aromatic carbons), 66.94 (C–H), 56.24 ($COCH_3$), 24.34 (CH_3), ppm, Anal.calcd for $C_{22}H_{20}N_5O_3Cl$: C, 60.34; H, 4.60; N, 15.99. Found: C, 60.16; H, 4.62; N, 15.97.

4-(4-Chloro-2-nitrophenyl)-4,5-dihydro-5-(4-methoxyphenyl)-N-p-tolyl-1H-1,2,4-triazol-3 amine (4i).

m.p. 240°C; IR (KBr): 3477, 3352, 2934, 1627, 1564, 1507, 1364, 1345, 1254, 1201, 1150, 727, 700 cm^{-1}; ^{1}H NMR (δ ppm DMSO-d_6): 2.35 (s, 3H, CH_3), 3.73 (s, 3H, OCH_3), 5.04 (s, 1H, CH), 6.34–7.15 (m, 4H, aromatic), 6.63–6.95 (m, 4H, aromatic), 6.67–7.98 (m, 3H, aromatic), 7.0 (s, 1H, N–H), 8.93 (s, 1H, N–H). ^{13}C NMR (400 MHz, DMSO-d_6): 156.24 (O–CH_3), 154.44 (CH_3), 150.14 ($COCH_3$), 145.64–135.82 (aromatic carbons), 136.34–124.12 (aromatic carbons), 128.02–112.18 (aromatic carbons), 66.98 (C–H), 55.90 ($COCH_3$), 55.22 (CH_3), 24.32 (CH_3) ppm, Anal.calcd for $C_{22}H_{20}N_5O_3Cl$: C, 60.34; H, 4.60; N, 15.98. Found: C, 60.17; H, 4.62; N, 15.96.

4-(4-Chloro-2-nitrophenyl)-4,5-dihydro-5-(3,4-dimethoxyphenyl)-N-p-tolyl-1H-1,2,4 triazol-3-amine (4j).

m.p. 160°C; IR (KBr): 3476, 3354, 2922, 1627, 1508, 1464, 1339, 1203, 1134, 1044 793, 699 cm^{-1}; ^{1}H NMR (δ ppm DMSO-d_6): 2.35 (s, 3H,

CH_3), 3.73 (s, 3H, OCH_3), 3.73 (s, 3H, OCH_3), 5.04 (s, 1H, CH), 6.34–7.81 (m, 4H, aromatic), 6.46–6.95 (m, 3H, aromatic), 6.67–7.98 (m, 3H, aromatic), 7.0 (s, 1H, N–H), 8.93 (s, 1H, N–H). ^{13}C NMR (400 MHz, DMSO-d_6): 154.96 (C=N), 150.14 (2×$COCH_3$), 149.54 (O–CH_3), 147.62 (O–CH_3), 141.48–129.74 (aromatic carbons), 124.18–120.08 (aromatic carbons), 123.28–112.14 (aromatic carbons), 66.92 (C–H), 56.24 (CH_3), 56.20 (2×$COCH_3$), 24.34 (CH_3) ppm Anal.calcd for $C_{23}H_{22}N_5O_4Cl$: C, 59.04; H, 4.74; N, 14.97. Found: C, 59.22; H, 4.76; N, 14.95.

Spectroscopic Characterization Data of 5a-p

2-(5-Amino-2-oxo-1, 2-dihydrospiro (indole-3, 3-(1, 2, 4) triazol)-4(2H)-yl) propanoic acid (5a).

m.p. 105°C; IR (KBr): 3419, 3159, 2980, 1700, 1674, 1593, 1465, 1060, 1091 cm^{-1}; ^{1}H NMR (δ ppm DMSO-d_6): 1.23 (d, 3H, CH_3), 2.00 (s, 2H, NH_2 of triazole ring), 3.67 (q, 1H, CH), 7.0 (s, 1H, N–H of triazole ring), 6.88–7.52 (m, 4H, aromatic), 8.0 (s, 1H, NH of indole ring), 11.0 (s, 1H, OH). ^{13}C NMR (400 MHz, DMSO-d_6): 174.12 (C=O, acid), 168.24 (C=O, indole), 162.14 (C=N), 141.16–130.74 (aromatic carbons), 76.56 (N–C–NH), 46.22 (CH), 14.64 (CH_3) ppm. Anal.calcd for $C_{12}H_{13}N_5O_3$: C, 52.36; H, 4.76; N, 25.44. Found: C, 52.53; H, 4.79; N, 25.48.

2-(5-Amino-2-oxo-1, 2-dihydrospiro (indole-3, 3-(1, 2, 4) triazol)-4(2H)-yl)-4-(methylsulfanyl) butanoic acid (5b).

m.p. 98°C; IR (KBr): 3420, 3150, 2982, 1702, 1675, 1590, 1460, 1062, 780, 672 cm^{-1}; ^{1}H NMR (δ ppm DMSO-d_6): 2.00 (s, 2H, NH_2 of triazole ring), 2.01 (q, 2H, CH_2), 2.09 (s, 3H, CH_3), 2.44 (t, 2H, CH_2), 3.49 (t, 1H, CH), 7.1 (s, 1H, N–H of triazole ring), 6.86–7.54 (m, 4H, aromatic), 8.1 (s, 1H, NH of indole ring), 11.1 (s, 1H, OH). ^{13}C NMR (400 MHz, DMSO-d_6): 174.18 (C=O, acid), 168.26 (C=O, indole), 162.12 (C=N), 142.14–122.14 (aromatic carbons), 75.54 (N–C–NH), 51.22 (CH), 30.08 (CH_2), 28.14 (CH_2), 17.64 (CH_3) ppm. Anal.calcd for $C_{14}H_{17}N_5O_3S$: C, 50.14; H, 5.11; N, 20.88. Found: C, 50.36; H, 5.14; N, 20.84.

2-(5-Amino-2-oxo-1, 2-dihydrospiro (indole-3, 3-(1, 2, 4) triazol)-4(2H)-yl)-3-(1H-imidazole-4-yl)propanoic acid (5c).

m.p. 140°C; IR (KBr): 3422, 3152, 2982, 1708, 1675, 1593, 1530, 1420, 1091 cm^{-1}; ^{1}H NMR (δ ppm DMSO-d_6): 2.01 (s, 2H, NH_2 of triazole ring), 2.78 (dd, 1H, CH), 3.88 (t, 1H, CH), 6.80 (s, 1H, CH), 7.0 (s, 1H, N–H of triazole ring), 7.44 (s, 1H, CH), 6.87–7.53 (m, 4H, aromatic), 8.2 (s, 1H, NH of indole ring), 11.0 (s, 1H, OH), 13.4 (s, 1H, NH). ^{13}C NMR (400 MHz, DMSO-d_6): 174.18 (C=O, acid), 168.26 (C=O, indole), 162.12 (C=N), 142.14–119.62 (aromatic carbons), 75.54 (N–C–NH), 45.22 (CH), 27.12 (CH_2) ppm. Anal.calcd for $C_{15}H_{15}N_7O_3$: C, 52.78; H, 4.43; N, 28.73. Found: C, 52.58; H, 4.45; N, 28.76.

2-(5-Amino-2-oxo-1, 2-dihydrospiro (indole-3, 3-(1, 2, 4) triazol)-4(2H)-yl)-3-(1H-indole-3-yl)propanoic acid (5d).

m.p. 185°C; IR (KBr): 3425, 3155, 2984, 1708, 1677, 1597, 1530, 1420, 1094 cm^{-1}; ^{1}H NMR (δppm DMSO-d_6): 2.02 (s, 2H, NH_2 of triazole ring), 2.65 (dd, 2H, CH_2), 3.88 (t, 1H, CH), 7.1 (s, 1H, N–H of triazole ring), 6.80 (s, 1H, CH), 6.88–7.52 (m, 4H, aromatic), 7.18–7.20 (m, 4H, aromatic), 8.3 (s, 1H, NH of indole ring), 10.1 (s, 1H, NH), 11.1 (s, 1H, OH). ^{13}C NMR (400 MHz, DMSO-d_6): 174.18 (C=O, acid), 168.26 (C=O, indole), 162.12 (C=N), 142.14–110.92 (aromatic carbons), 75.54 (N–C–NH), 46.22 (CH), 28.12 (CH_2) ppm. Anal.calcd for $C_{20}H_{18}N_6O_3$: C, 61.63; H, 4.65; N, 21.53. Found: C, 61.43; H, 4.68; N, 21.55.

2-(5-Amino-2-oxo-1, 2-dihydrospiro (indole-3, 3-(1, 2, 4) triazol)-4(2H)-yl)-3-methylpentanoic acid (5e).

m.p. 170°C; IR (KBr): 3427, 3150, 2980, 1701, 1672, 1592, 1533, 1424, 1094 cm^{-1}; ^{1}H NMR (δ ppm DMSO-d_6): 0.96 (t, 3H, CH_3), 1.06 (d, 3H, CH_3), 1.29 (m, 2H, CH_2), 2.03 (s, 2H, NH_2 of triazole ring), 2.21 (m, 2H, CH_2), 3.48 (d, 1H, CH), 7.0 (s, 1H, N–H of triazole ring), 6.88–7.52 (m, 4H, aromatic), 8.1 (s, 1H, NH of indole ring), 11.1 (s, 1H, OH). ^{13}C NMR (400 MHz, DMSO-d_6): 174.18 (C=O, acid), 168.26 (C=O, indole), 162.12 (C=N), 142.14–120.12 (aromatic carbons), 75.54 (N–C–NH), 55.22 (CH), 33.22 (CH), 25.4 (CH_2), 15.3 (CH_3), 11.14 (CH_3) ppm.

Anal.calcd for $C_{15}H_{19}N_5O_3$: C, 56.77; H, 6.03; N, 22.07. Found: C, 56.57; H, 6.06; N, 22.08.

2-(5-Amino-2-oxo-1, 2-dihydrospiro (indole-3, 3-(1, 2, 4) triazol)-4(2H)-yl)-3-methylbutanoic acid (5f).

m.p. 195°C; IR (KBr): 3433, 3155, 2988, 1709, 1678, 1595, 1533, 1427, 1094 cm^{-1}; ^1H NMR (δ ppm DMSO-d_6): 1.01 (d, 3H, CH_3), 1.03 (d, 3H, CH_3), 2.1 (s, 2H, NH_2 of triazole ring), 2.39 (m, 1H, CH), 3.49 (d, 1H, CH), 7.1 (s, 1H, N–H of triazole ring), 6.87–7.54 (m, 4H, aromatic), 8.2 (s, 1H, NH of indole ring), 11.2 (s, 1H, OH). ^{13}C NMR (400 MHz, DMSO-d_6): 174.12 (C=O, acid), 168.24 (C=O, indole), 162.16 (C=N), 142.12–120.16 (aromatic carbons), 75.58 (N–C–NH), 58.28 (CH), 27.14 (CH_2), 17.14 (CH_3), 17.12 (CH_3) ppm Anal.calcd for $C_{14}H_{17}N_5O_3$: C, 55.44; H, 5.65; N, 23.09. Found: C, 55.62; H, 5.68; N, 23.07.

2-(5-Amino-2-oxo-1, 2-dihydrospiro (indole-3, 3-(1, 2, 4) triazol)-4(2H)-yl)-3-hydroxypropanoic acid (5g).

m.p. 90°C; IR (KBr): 3480, 3433, 3151, 2986, 1702, 1673, 1595, 1531, 1422, 1091 cm^{-1}; ^1H NMR (δppm DMSO-d_6): 2.0 (s, 1H, OH), 2.2 (s, 2H, NH_2 of triazole ring), 3.58 (t, 1H, CH), 3.76 (dd, 2H, CH_2), 7.0 (s, 1H, N–H of triazole ring), 6.89–7.52 (m, 4H, aromatic), 8.1 (s, 1H, NH of indole ring), 11.1 (s, 1H, OH). ^{13}C NMR (400 MHz, DMSO-d_6): 174.12 (C=O, acid), 168.22 (C=O, indole), 162.16 (C=N), 142.16–120.18 (aromatic carbons), 75.54 (N–C–NH), 58.22 (CH_2), 51.22 (CH) ppm. Anal.calcd for $C_{12}H_{13}N_5O_4$: C, 49.48; H, 4.50; N, 24.04. Found: C, 49.66; H, 4.52; N, 24.07.

2-(5-Amino-2-oxo-1, 2-dihydrospiro (indole-3, 3-(1, 2, 4) triazol)-4(2H)-yl)-3-phenylpropanoic acid (5h).

m.p. 140–142°C; IR (KBr): 3431, 3155, 2988, 1709, 1677, 1592, 1534, 1422, 1091 cm^{-1}; ^1H NMR (δ ppm DMSO-d_6): 2.3 (s, 2H, NH_2 of triazole ring), 2.78 (dd, 2H, CH_2), 3.89 (t, 1H, CH), 7.2 (s, 1H, N–H of triazole ring), 6.88–7.52 (m, 4H, aromatic), 7.08–7.12 (m, 5H, aromatic), 8.0 (s, 1H, NH of indole ring), 11.0 (s, 1H, OH). ^{13}C NMR (400 MHz, DMSO-d_6): 174.18 (C=O, acid), 168.26 (C=O, indole), 162.12 (C=N), 142.14–122.14 (aromatic carbons), 75.54 (N–C–NH), 45.32 (CH), 34.22

(CH_2) ppm. Anal.calcd for $C_{18}H_{17}N_5O_3$: C, 61.53; H, 4.88; N, 19.93. Found: C, 61.33; H, 4.86; N, 19.95.

2-(5-Amino-5-chloro-2-oxo-1, 2-dihydrospiro (indole-3, 3-(1, 2, 4) triazol)-4(2H)-yl)-3-phenylpropanoic acid (5i).

m.p. 153°C; IR (KBr): 3433, 3151, 2986, 1702, 1673, 1595, 1531, 1422, 1093 cm^{-1}; ^1HNMR (δppm DMSO-d_6): 2.1 (s, 2H, NH_2 of triazole ring), 3.03 (dd, 2H, CH_2), 3.88 (t, 1H, CH), 7.3 (s, 1H, N–H of triazole ring), 7.08–7.21 (m, 5H, aromatic), 7.05–7.46 (m, 3H, aromatic), 8.2 (s, 1H, NH of indole ring), 11.1 (s, 1H, OH). ^{13}CNMR (400 MHz, DMSO-d_6): 174.18 (C=O, acid), 168.26 (C=O, indole), 162.12 (C=N), 142.14–120.12 (aromatic carbons), 75.54 (N–C–NH), 45.38 (CH), 34.22 (CH_2) ppm. Anal.calcd for $C_{18}H_{16}N_5O_3Cl$: C, 56.04; H, 4.18; N, 18.15. Found: C, 56.24; H, 4.21; N, 18.17.

2-(5-Amino-5-chloro-2-oxo-1, 2-dihydrospiro (indole-3, 3-(1, 2, 4) triazol)-4(2H)-yl)-3-(1H-imidazole-4-yl) propanoic acid (5j).

m.p. 180–182°C; IR (KBr): 3431, 3155, 2989, 1707, 1676, 1598, 1530, 1428, 1091, 721 cm^{-1}; ^1HNMR (δ ppm DMSO-d_6): 2.0 (s, 2H, NH_2 of triazole ring), 3.04 (dd, 2H, CH_2), 3.89 (t, 1H, CH), 6.80 (d, 1H, CH), 7.1 (s, 1H, N–H of triazole ring), 7.44 (d, 1H, CH), 7.05–7.47 (m, 3H, aromatic), 8.1 (s, 1H, NH of indole ring), 11.1 (s, 1H, OH), 13.4 (t, 1H, NH). ^{13}CNMR (400 MHz, DMSO-d_6): 174.18 (C=O, acid), 168.26 (C=O, indole), 162.12 (C=N), 142.14–120.12 (aromatic carbons), 75.54 (N–C–NH), 45.38 (CH), 27.12 (CH_2) ppm. Anal.calcd for $C_{15}H_{14}N_7O_3Cl$: C, 47.94; H, 3.76; N, 26.09. Found: C, 47.74; H, 3.78; N, 26.12.

2-(5-Amino-5-chloro-2-oxo-1, 2-dihydrospiro (indole-3, 3-(1, 2, 4) triazol)-4(2H)-yl)-5-(methylsulfanyl) butanoic acid acid (5k).

m.p. 220°C; IR (KBr): 3434, 3152, 2982, 1701, 1672, 1597, 1532, 1426, 1095, 726 cm^{-1}; ^1HNMR (δ ppm DMSO-d_6): 2.1 (s, 2H, NH_2 of triazole ring), 2.01 (q, 2H, CH_2), 2.08 (s, 3H, CH_3), 2.43 (t, 2H, CH_2), 3.49 (t, 1H, CH), 7.0 (s, 1H, N–H of triazole ring), 7.05–7.46 (m, 3H, aromatic),

8.0 (s, 1H, NH of indole ring), 10.1 (t, 1H, NH). 11.1 (s, 1H, OH). ^{13}C NMR (400 MHz, DMSO-d_6): 174.12 (C=O, acid), 168.22 (C=O, indole), 162.18 (C=N), 142.18–120.18 (aromatic carbons), 75.58 (N–C–NH), 51.22 (CH), 27.12 (CH_2), 25.14 (CH_2), 17.14 (CH_3) ppm. Anal. calcd for $C_{15}H_{18}N_5O_3SCl$: C, 46.93; H, 4.73; N, 18.24. Found: C, 46.73; H, 4.70; N, 18.27.

2-(5-Amino-5-chloro-2-oxo-1, 2-dihydrospiro (indole-3, 3-(1, 2, 4) triazol)-4(2H)-yl)-3-(2,3-dihydro-1H-indole-3-yl)propanoic acid (5l).

m.p. 230°C; IR (KBr): 3431, 3150, 2980, 1703, 1670, 1595, 1531, 1424, 1097, 729 cm^{-1}; ^{1}H NMR (δ ppm DMSO-d_6): 1.93 (t, 2H, CH_2), 2.2 (s, 2H, NH_2 of triazole ring), 3.00 (t, 1H, CH), 3.22 (dd, 2H, CH_2), 3.48 (t, 1H, CH), 7.1 (s, 1H, N–H of triazole ring), 6.35–6.91 (m, 4H, aromatic), 7.06–7.46 (m, 3H, aromatic), 8.3 (s, 1H, NH of indole ring), 10.1 (s, 1H, NH), 11.1 (s, 1H, OH). ^{13}C NMR (400 MHz, DMSO-d_6): 174.18 (C=O, acid), 168.26 (C=O, indole), 162.12 (C=N), 142.14–120.12 (aromatic carbons), 75.54 (N–C–NH), 49.32 (CH), 32.12 (CH_2) ppm. Anal.calcd for $C_{20}H_{19}N_6O_3Cl$: C, 56.28; H, 4.49; N, 19.69. Found: C, 56.48; H, 4.46; N, 19.66.

2-(5-Amino-5-nitro-2-oxo-1, 2-dihydrospiro (indole-3, 3-(1, 2, 4) triazol)-4(2H)-yl)-3-phenylpropanoic acid (5m).

m.p. 210°C; IR (KBr): 3433, 3152, 2982, 1701, 1675, 1591, 1532, 1422, 1365, 1092 cm^{-1}; ^{1}H NMR (δ ppm DMSO-d_6): 2.1 (s, 2H, NH_2 of triazole ring), 3.03 (dd, 2H, CH_2), 3.88 (t, 1H, CH), 7.3 (s, 1H, N–H of triazole ring), 7.08–7.21 (m, 5H, aromatic), 7.09–7.46 (m, 3H, aromatic), 8.2 (s, 1H, NH of indole ring), 11.1 (s, 1H, OH). ^{13}C NMR (400 MHz, DMSO-d_6): 174.12 (C=O, acid), 168.24 (C=O, indole), 162.14 (C=N), 142.12–120.10 (aromatic carbons), 75.52 (N–C–NH), 45.34 (CH), 34.28 (CH_2) ppm. Anal.calcd for $C_{18}H_{16}N_6O_5$: C, 54.54; H, 4.07; N, 21.20. Found: C, 54.37; H, 4.09; N, 21.21.

2-(5-Amino-5-nitro-2-oxo-1, 2-dihydrospiro (indole-3, 3-(1, 2, 4) triazol)-4(2H)-yl)-3-(1H-imidazole-4-yl) propanoic acid (5n).

m.p. 207°C; IR (KBr): 3431, 3155, 2986, 1704, 1672, 1594, 1536, 1422, 1366 1092 cm^{-1}; ^{1}HNMR (δ ppm DMSO-d_6): 2.0 (s, 2H, NH_2 of triazole ring), 3.04 (dd, 2H, CH_2), 3.89 (t, 1H, CH), 6.80 (d, 1H, CH), 7.1 (s, 1H, N–H of triazole ring), 7.44 (d, 1H, CH), 7.04–7.45 (m, 3H, aromatic), 8.1 (s, 1H, NH of indole ring), 11.3 (s, 1H, OH), 13.4 (t, 1H, NH). ^{13}CNMR (400 MHz, DMSO-d_6): 174.14 (C=O, acid), 168.26 (C=O, indole), 162.12 (C=N), 142.18–120.12 (aromatic carbons), 75.52 (N–C–NH), 45.35 (CH), 27.22 (CH_2) ppm Anal.calcd for $C_{15}H_{14}N_8O_5$: C, 46.63; H, 3.65; N, 29.01. Found: C, 46.43; H, 3.67; N, 29.03.

2-(5-Amino-5-nitro-2-oxo-1, 2-dihydrospiro (indole-3, 3-(1, 2, 4) triazol)-4(2H)-yl)-5-(methylsulfanyl)butanoic acid (5o).

m.p. 194°C; IR (KBr): 3434, 3153, 2984, 1702, 1672, 1599, 1533, 1421, 1362 1097 cm^{-1}; ^{1}HNMR (δ ppm DMSO-d_6): 1.80 (q, 2H, CH_2), 2.10 (s, 2H, NH_2 of triazole ring), 2.30 (s, 3H, CH_3), 2.50 (t, 2H, CH_2), 3.49 (t, 1H, CH), 7.1 (s, 1H, N–H of triazole ring), 7.78–7.98 (m, 3H, aromatic), 8.2 (s, 1H, NH of indole ring), 11.1 (s, 1H, OH). ^{13}CNMR (400 MHz, DMSO): 174.18 (C=O, acid), 168.22 (C=O, indole), 162.12 (C=N), 142.14–132.76 (aromatic carbons), 75.54 (N–C–NH), 51.22 (CH), 35.02 (CH_2), 27.14 (CH_2), 25.12 (CH_2), 17.64 (CH_3) ppm. Anal. calcd for $C_{15}H_{18}N_6O_5S$: C, 45.68; H, 4.60; N, 21.31. Found: C, 44.40; H, 4.26; N, 22.07.

2-(5-Amino-5-nitro-2-oxo-1, 2-dihydrospiro (indole-3, 3-(1, 2, 4) triazol)-4(2H)-yl)-3-(2,3-dihydro-1H-indole-3-yl)propanoic acid (5p).

m.p. 185°C; IR (KBr): 3431, 3152, 2982, 1706, 1678, 1594, 1531, 1429, 1365 1093 cm^{-1}; ^{1}HNMR (δ ppm DMSO-d_6): 1.93 (t, 2H, CH_2), 2.2 (s, 2H, NH_2 of triazole ring), 3.00 (t, 1H, CH), 3.47 (dd, 2H, CH_2), 3.60 (t, 1H, CH), 7.1 (s, 1H, N–H of triazole ring), 6.35–6.91 (m, 4H, aromatic), 7.78–7.98 (m, 3H, aromatic), 8.1 (s, 1H, NH of indole ring), 10.1 (s, 1H, NH), 11.1 (s, 1H, OH). ^{13}CNMR (400 MHz, DMSO): 174.18 (C=O, acid), 168.26 (C=O, indole), 162.12 (C=N), 142.14–117.14 (aromatic carbons), 75.54 (N–C–NH), 49.32 (CH), 32.12 (CH_2) ppm. Anal.calcd for $C_{20}H_{19}N_7O_5$: C, 54.92; H, 4.38; N, 22.42. Found: C, 44.40; H, 4.26; N, 22.07.

CONCLUSION

The use of water as a green solvent offers a convenient, nontoxic, and inexpensive reaction medium for the environ-economic synthesis of triazole derivatives. This procedure is simpler, economical, milder, and faster, including cleaner reactions, high yields of products, and a simple experimental and work-up procedure, which makes it a useful and attractive process and is also consistent with the green chemistry theme which affords good yields. Synthesized compounds are found to be excellent fluorescent materials and potent fungicidal agents.

ACKNOWLEDGMENTS

The authors are thankful to the Dean of FET, MITS, for providing necessary research facilities in the department. Financial assistance from FET, MITS is gratefully acknowledged. They are also thankful to SAIF Punjab University, Chandigarh, for the spectral and elemental analyses.

REFERENCE

1. T. J. J. Müller, "Multicomponent reactions," Beilstein Journal of Organic Chemistry, vol. 7, pp. 960–961, 2011.
2. R. P. Tenório, C. S. Carvalho, C. S. Pessanha, et al., "Synthesis of thiosemicarbazone and 4-thiazolidinone derivatives and their in vitro anti-Toxoplasma gondii activity," Bioorganic & Medicinal Chemistry Letters, vol. 15, pp. 2575–2578, 2005.
3. M. C. Pirrung, S. V. Pansare, K. Das Sarma, K. A. Keith, and E. R. Kern, "Combinatorial optimization of isatin-β-thiosemicarbazones as anti-poxvirus agents," Journal of Medicinal Chemistry, vol. 48, no. 8, pp. 3045–3050, 2005.
4. W.-X. Hu, W. Zhou, C.-N. Xia, and X. Wen, "Synthesis and anticancer activity of thiosemicarbazones," Bioorganic & Medicinal Chemistry

Letters, vol. 16, pp. 2213–2218, 2006.

5. R. B. de Oliveira, E. M. de Souza-Fagundes, R. P. P. Soares, A. A. Andrade, A. U. Kretti, and C. L. Zani, "Synthesis and antimalarial activity of semicarbazone and thiosemicarbazone derivatives," European Journal of Medicinal Chemistry, vol. 43, pp. 183–188, 2008.
6. A. Pérez-Rebolledo, L. R. Teixeira, A. A. Batista, et al., "4-Nitroacetophenone-derived thiosemicarbazones and their copper(II) complexes with significant in vitro anti-trypanosomal activity," European Journal of Medicinal Chemistry, vol. 43, pp. 939–948, 2008.
7. G. Aguirre, L. Boiani, H. Cerecetto et al., "In vitro activity and mechanism of action against the protozoan parasite Trypanosoma cruzi of 5-nitrofuryl containing thiosemicarbazones,"Bioorganic and Medicinal Chemistry, vol. 12, no. 18, pp. 4885–4893, 2004.
8. X. Du, C. Guo, E. Hansell et al., "Synthesis and structure-activity relationship study of potent trypanocidal thio semicarbazone inhibitors of the trypanosomal cysteine protease cruzain,"Journal of Medicinal Chemistry, vol. 45, no. 13, pp. 2695–2707, 2002.
9. N. Fujii, J. P. Mallari, E. J. Hansell et al., "Discovery of potent thiosemicarbazone inhibitors of rhodesain and cruzain," Bioorganic and Medicinal Chemistry Letters, vol. 15, no. 1, pp. 121–123, 2005.
10. R. P. Tenório, A. J. S. Góes, J. G. de Lima, A. R. de Faria, A. J. Alves, and T. M. Aquino, "Thiosemicarbazones: preparation methods, synthetic applications and biological importance," Química Nova, vol. 28, pp. 1030–1037, 2005.
11. J. M. Kane, B. M. Baron, M. W. Dudley, S. M. Sorensen, M. A. Staeger, and F. P. Miller, "2,4-Dihydro-3H-1,2,4-triazol-3-ones as anticonvulsant agents," Journal of Medicinal Chemistry, vol. 33, no. 10, pp. 2772–2777, 1990.
12. S. Rollas, N. Kalyoncuoglu, D. Sur-Altiner, and Y. Yegenoglu, "5-(4-Aminophenyl)-4-substituted-2,4-dihydro-3H-1,2,4-triazole-3-thione s: synthesis and antibacterial and antifungal activities," Pharmazie, vol. 48, no. 4, pp. 308–309, 1993.
13. B. E. Gilbert and V. Knight, "Biochemistry and clinical applications of ribavirin," Antimicrobial Agents and Chemotherapy, vol. 30, no. 2, p. 201, 1986.
14. P. C. Wade, B. Richard Vogt, T. P. Kissick, L. M. Simpkins, D. M. Palmer, and R. C. Millonig, "1-Acyltriazoles as antiinflammatory agents," Journal of Medicinal Chemistry, vol. 25, no. 3, pp. 331–333, 1982.

15. F. Malbec, R. Milcent, P. Vicart, and A. M. Bure, "Synthesis of new derivatives of 4-amino-2,4-dihydro-1,2,4-triazol-3-one as potential antibacterial agents," Journal of Heterocyclic Chemistry, vol. 21, no. 6, pp. 1769–1774, 1984.
16. L. I. Kruse, D. L. Ladd, P. B. Harrsch et al., "Synthesis, tubulin binding, antineoplastic evaluation, and structure-activity relationship of oncodazole analogues," Journal of Medicinal Chemistry, vol. 32, no. 2, pp. 409–417, 1989.
17. S. Ram, D. S. Wise, L. L. Wotring, J. W. McCall, and L. B. Townsend, "Synthesis and biological activity of certain alkyl 5-(alkoxycarbonyl)-1H-benzimidazole-2-carbamates and related derivatives: a new class of potential antineoplastic and antifilarial agents," Journal of Medicinal Chemistry, vol. 35, no. 3, pp. 539–547, 1992.
18. E. R. Cole, G. Crank, and A. Salam-Sheikh, "Antioxidant properties of benzimidazoles.,"Journal of Agricultural and Food Chemistry, vol. 22, no. 5, p. 918, 1974.
19. C. Hubschwerlen, P. Pflieger, J. L. Specklin et al., "Pyrimido[1,6-a] benzimidazoles: a new class of DNA gyrase inhibitors," Journal of Medicinal Chemistry, vol. 35, no. 8, pp. 1385–1392, 1992.
20. P. Venkatesan, "Albendazole," Journal of Antimicrobial Chemotherapy, vol. 41, pp. 145–147, 1998.
21. N. S. Pawar, D. S. Dalal, S. R. Shimpi, and P. P. Mahulikar, "Studies of antimicrobial activity ofN-alkyl and N-acyl 2-(4-thiazolyl)-1H-benzimidazoles," European Journal of Pharmaceutical Sciences, vol. 21, pp. 115–118, 2004.
22. T. Ishida, T. Suzuki, S. Hirashima et al., "Benzimidazole inhibitors of hepatitis C virus NS5B polymerase: identification of 2-[(4-diarylmethoxy)phenyl]-benzimidazole," Bioorganic and Medicinal Chemistry Letters, vol. 16, pp. 1859–1863, 2006.
23. L. Garuti, M. Roberti, and G. Gentilomi, "Synthesis and antiviral assays of some benzimidazole nucleosides and acyclonucleosides," Farmaco, vol. 56, no. 11, pp. 815–819, 2001.
24. K. Kubo, Y. Kohara, Y. Yoshimura et al., "Nonpeptide angiotensin II receptor antagonists. Synthesis and biological activity of potential prodrugs of benzimidazole-7-carboxylic acids,"Journal of Medicinal Chemistry, vol. 36, no. 16, pp. 2343–2349, 1993.
25. K. Stratmann, R. E. Moore, R. Bonjouklian et al., "Welwitindolinones, unusual alkaloids from the blue-green algae Hapalosiphon welwitschii and Westiella intricata. Relationship to fischerindoles and hapalindoles," Journal of the American Chemical Society, vol. 116, no. 22, pp. 9935–9942, 1994.

26. J. W. Skiles and D. Menil, "Spiro indolinone beta-lactams, inhibitors of poliovirus and rhinovlrus 3C-proteinases," Tetrahedron Letters, vol. 31, p. 7277, 1990.
27. A. Hasan, N. F. Thomas, and S. Gapil, "Synthesis, characterization and antifungal evaluation of 5-substituted-4-amino-1,2,4-triazole-3-thioesters," Molecules, vol. 16, no. 2, pp. 1297–1309, 2011.
28. S. Veda and H. Nagasawa, "Facile synthesis of 1,2,4-triazoles via a copper-catalyzed tandem addition-oxidative cyclization," Journal of the American Chemical Society, vol. 131, pp. 15080–15081, 2009.
29. G. M. Castanedo, P. S. Seng, N. Blaquiere, S. Trapp, and S. T. Staben, "Rapid synthesis of 1,3,5-substituted 1,2,4-triazoles from carboxylic acids, amidines, and hydrazines," Journal of Organic Chemistry, vol. 76, no. 4, pp. 1177–1179, 2011.
30. Y. Xu, M. Mc Laughlin, E. N. Bolton, and R. A. Reamer, "Practical synthesis of functionalized 1,5-disubstituted 1,2,4-triazole derivatives," Journal of Organic Chemistry, vol. 75, pp. 8666–8669, 2010.
31. A. Dandia, P. Sarawgi, K. Arya, and S. Khaturia, "Mild and ecofriendly tandem synthesis of 1,2,4-triazolo[4,3-a]pyrimidines in aqueous medium," Arkivoc, no. 16, pp. 83–92, 2006.
32. A. Jha, Y. L. N. Murthy, G. Durga, and T. T. Sundari, "Microwave-assisted synthesis of 3,5-Dibenzyl-4-amino-1,2,4-triazole and its diazo ligand, metal complexes along with anticancer activity," E-Journal of Chemistry, vol. 7, no. 4, pp. 1571–1577, 2010.
33. A. Dandia, H. Sachdeva, and R. Singh, "Montmorillonite catalysed synthesis of novel spiro[3H-indole-3,3'- [3H-1,2,4] triazol]-2(1H) ones in dry media under microwave irradiation," Journal of Chemical Research, vol. 2000, no. 6, pp. 272–275, 2000.
34. A. Dandia, R. Singh, S. Bhaskaran, and D. S. Shriniwas, "Versatile three-component procedure for combinatorial synthesis of biologically relevant scaffold spiro[indole-thiazolidinones] under aqueous conditions," Green Chemistry, vol. 13, pp. 1852–1859, 2011.
35. T. K. Huang, R. Wang, S. Lin, and X. Lu, "Montmorillonite K-10: an efficient and reusable catalyst for the synthesis of quinoxaline derivatives in water," Catalysis Communications, vol. 9, pp. 1143–1147, 2008.
36. G. P. Chiusoli, M. Costa, L. Cucchia, B. Gabriele, G. Salerno, and L. Veltri, "Carbon dioxide effect on palladium-catalyzed sequential reactions with carbon monoxide, acetylenic compounds and water," Journal of Molecular Catalysis A, vol. 204-205, pp. 133–142, 2003.

37. M. Lautens and M. Yoshida, "Rhodium-catalyzed addition of arylboronic acids to alkynyl aza-heteroaromatic compounds in water," Journal of Organic Chemistry, vol. 68, p. 762, 2003.

38. H. Kim, S. Lee, J. Lee, and J. Tae, "Rhodamine triazole-based fluorescent probe for the detection of Pt^{2+}," Organic Letters, vol. 12, no. 22, pp. 5342–5345, 2010.

39. S. Y. Park, J. H. Yoon, C. S. Hong et al., "A pyrenyl-appended triazole-based calix[4]arene as a fluorescent sensor for Cd^{2+} and Zn^{2+}," Journal of Organic Chemistry, vol. 73, no. 21, pp. 8212–8218, 2008.

40. J. D. Cheon, T. Mutai, and K. Araki, "Preparation of a series of novel fluorophores, N-substituted 6-amino and 6,6''-diamino-2,2':6',2''-terpyridine by palladium-catalyzed amination," Tetrahedron Letters, vol. 47, no. 29, pp. 5079–5082, 2006.

41. S. Coe, W. K. Woo, M. Bawendi, and V. Bulović, "Electroluminescence from single monolayers of nanocrystals in molecular organic devices," Nature, vol. 420, no. 6917, pp. 800–803, 2002.

42. G. Hughes and M. R. Bryce, "Electron-transporting materials for organic electroluminescent and electrophosphorescent devices," Journal of Materials Chemistry, vol. 15, pp. 94–107, 2005.

43. M. R. Detty, S. L. Gibson, and S. J. Wagner, "Current clinical and preclinical photosensitizers for use in photodynamic therapy," Journal of Medicinal Chemistry, vol. 47, no. 16, pp. 3897–3915, 2004.

44. H. Sachdeva, D. Dwivedi, R. R. Bhattacharjee, S. Khaturia, and R. Saroj, "NiO nanoparticles: an efficient catalyst for the multicomponent one-pot synthesis of novel spiro and condensed indole derivatives," Journal of Chemistry, vol. 2013, Article ID 606259, 10 pages, 2013.

45. H. Sachdeva and D. Dwivedi, "Lithium-acetate-mediated biginelli one-pot multicomponent synthesis under solvent-free conditions and cytotoxic activity against the human lung cancer cell line A549 and breast cancer cell line MCF7," The Scientific World Journal, vol. 2012, Article ID 109432, 9 pages, 2012.

46. H. Sachdeva, R. Saroj, S. Khaturia, and D. Dwivedi, "Operationally simple green synthesis of some Schiff bases using grinding chemistry technique and evaluation of antimicrobial activities," Green Processing and Synthesis, vol. 1, pp. 469–477, 2012.

47. H. Sachdeva, D. Dwivedi, and S. Khaturia, "Aqua mediated facile synthesis of 2-(5/7-fluorinated-2-oxoindolin-3-ylidene)-N-(4-substituted phenyl) hydrazine carbothioamides,"Research Journal of Pharmaceutical, Biological and Chemical Science, vol. 2, no. 2, p. 213, 2011.

48. H. Sachdeva, R. Saroj, S. Khaturia, and H. L. Singh, "Comparative studies of lewis acidity of alkyl-tin chlorides in multicomponent biginelli condensation using grindstone chemistry technique," Journal of the Chilean Chemical Society, vol. 57, no. 1, p. 1012, 2012.
49. S. Patil, S. D. Jadhav, and U. P. Patil, "Natural acid catalyzed synthesis of schiff base under solvent-free condition: as a green approach," Archives of Applied Science Research, vol. 4, no. 2, pp. 1074–1078, 2012.
50. S. Patil, S. D. Jadhav, and M. B. Deshmukh, "Natural acid catalyzed synthesis of schiff base under solvent-free condition: as a green approach,"

Chapter 5

TWO-DIRECTIONAL SYNTHESIS AS A TOOL FOR DIVERSITY-ORIENTED SYNTHESIS: SYNTHESIS OF ALKALOID SCAFFOLDS

Kieron M. G. O'Connell[1], Monica Díaz-Gavilán[2], Warren R. J. D. Galloway[1] and David R. Spring[1]

[1]Department of Chemistry, University of Cambridge, Lensfield Rd, Cambridge, CB2 1EW, UK

[2]Departemento de Química Farmacéutica y Orgánica, Facultad de Farmacia, Campus de Cartuja, s.n. 18071 Granada, Spain

ABSTRACT

Two-directional synthesis represents an ideal strategy for the rapid elaboration of simple starting materials and their subsequent transformation into complex molecular architectures. As such, it is becoming recognised as an enabling technology for diversity-oriented synthesis. Herein, we provide a thorough account of our work combining two-directional synthesis with diversity-oriented synthesis, with particular reference to the synthesis of polycyclic alkaloid scaffolds.

INTRODUCTION

Diversity-oriented synthesis (DOS) aims to prepare structurally diverse compound collections in an efficient manner [1-3]. Of the possible "types of diversity" that can be incorporated into a compound collection, the most important, in terms of creating a functionally (biologically) diverse collection, is generally considered to be scaffold (or skeletal) diversity, i.e., the variation of molecular frameworks between compounds [4,5]. Therefore, one of the key challenges in DOS is the development of strategies that allow the efficient generation of a range of complex molecular scaffolds. A large number of approaches towards this goal have been reported, with some of the most effective being based around the "folding-up" of functionalised linear substrates into cyclic molecular architectures [6-8]. The design and synthesis of these linear substrates can, in itself, represent a significant challenge as it is desirable that these compounds are easily accessible in a small number of synthetic steps. Two-directional synthesis [9-12] offers a powerful method for the synthesis of such substrates, because each synthetic transformation has the potential to provide twice as much molecular complexity compared to standard approaches.

We have recently reported a strategy for DOS that combined two-directional synthesis with the use of these folding reaction pathways [13]. In this work, two-directional synthesis was used to rapidly generate a series of linear aminoalkenes, which were then folded into bicyclic and tricyclic scaffolds through Lewis acid mediated cascade processes. The compounds produced in this campaign were reminiscent of naturally occurring alkaloids, such as the Coccinellidae natural products, which are secreted by ladybirds to deter predators [14]. The total synthesis of one of their number, myrrhine, was also achieved by the elaboration of one of the compounds produced. In this article, the work is presented in more detail, alongside additional results from our work combining two-directional synthesis with DOS. Treated together, we believe these works provide useful insights into the potential utility of two-directional synthesis as an enabling technology for DOS. The initial DOS campaign was largely inspired by the pioneering work of Robert Stockman on combining two-directional synthesis with tandem reactions to create complex molecular architectures [15-18]. The key folding step in the DOS was the Lewis acid mediated pairing reaction of a nucleophilic amino group with suitable electrophilic functionality, provided by Michael acceptor α,β-unsaturated ester groups. Two-directional synthesis was used to append these electrophilic groups at two positions around the linear substrates, allowing bicyclisation processes to be instigated. The scaffold diversity between the products then resulted from the different ring sizes that it was possible to form from these linear substrates. (Figure 1 shows an overview of the DOS strategy.)

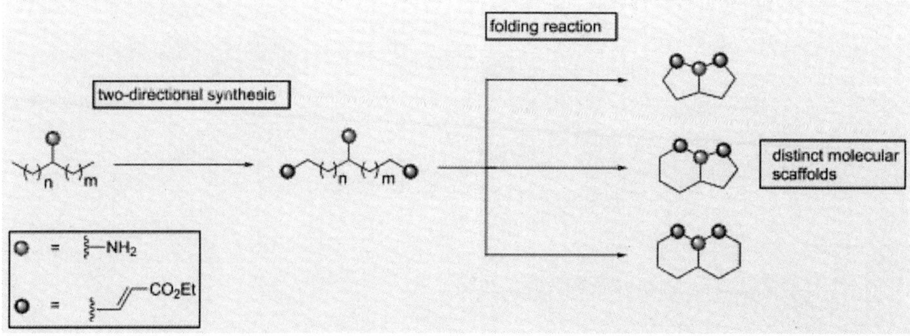

Figure 1. Overview of the DOS strategy.

RESULTS AND DISCUSSION

Synthesis of Linear Precursors

Our initial studies explored the synthesis and reactivity of four N-Boc-aminoalkenes containing α,β-unsaturated ester groups (1–4, Scheme 1), as substrates for intramolecular pairing reactions. The N-Boc protecting group was chosen to give the potential for deprotection to be carried out in tandem with the Lewis acid catalysed cyclisation reactions by intramolecular conjugate addition. Three of these compounds (1–3) were obtained from the corresponding alcohols

[19,20] in three steps: Mitsunobu reaction with NH-Boc-tosylate, followed by tosyl deprotection with magnesium, and finally two-directional cross metathesis with ethyl acrylate to install the desired α,β-unsaturated ester functionality. This sequence provided the desired N-Boc-aminoalkenes in respectable overall yields of 38–56%. Compound 4 was prepared in a four-step sequence from the requisite phenyldialkyl alcohol. Ritter reaction with chloroacetonitrile followed by cleavage of the resulting chloroacetamide with thiourea gave the free amine [21], which was then protected with Boc anhydride. Finally, cross metathesis with ethyl acrylate furnished the desired compound 4 in 24% overall yield.

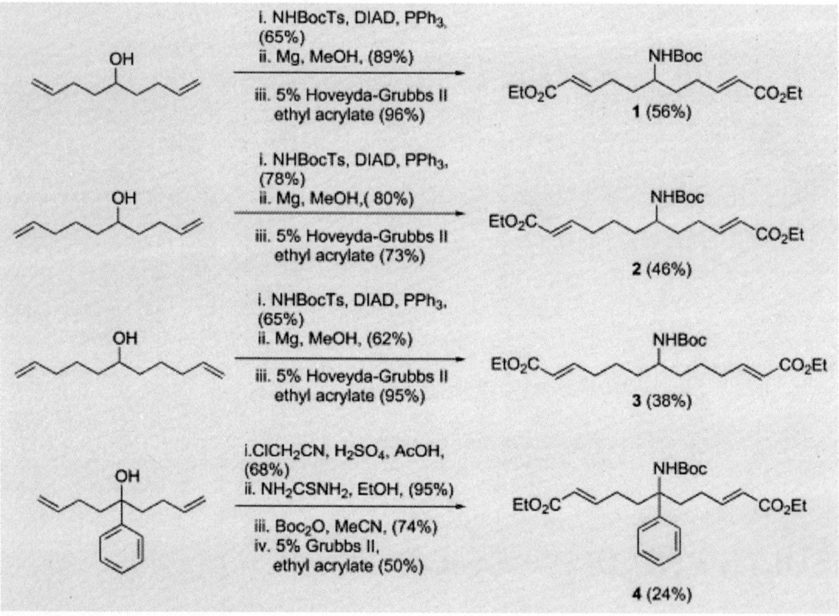

Scheme 1. Synthesis of linear cyclisation precursors 1–4.

Cyclisation Reactions

The first attempts at the tandem Boc-deprotection/bicyclisation of these substrates were performed by using AlCl3 as the Lewis acid (Scheme 2); compounds 1–4 were treated with 1.1 equiv of AlCl3 in dichloromethane at room temperature. These conditions proved effective at promoting bicyclisation for compounds 1, 2 and 4, for which the desired bicyclic products were obtained in 67–85% yield, as a mixture of diastereomers. The cyclisation of 1 gave pyrrolizidine 5 as a mixture of 4,10-trans-7,10-trans (trans-5) (42%) and 4,10-cis-7,10-trans (cis-5) (28%) isomers, which proved to be separable by flash chromatography. The stereochemistry of cis-5 was confirmed by analogy to known 1H and 13C NMR values [22] and by NOESY spectroscopy, which showed enhancements between H-7, H-4 and H-10. In this case, it proved possible to achieve an improved yield of both diastereomers by treating 1 with an excess (163 equiv) of trifluroacetic acid, which furnished trans-5 and cis-5 in 53% and 34% yield, respectively.

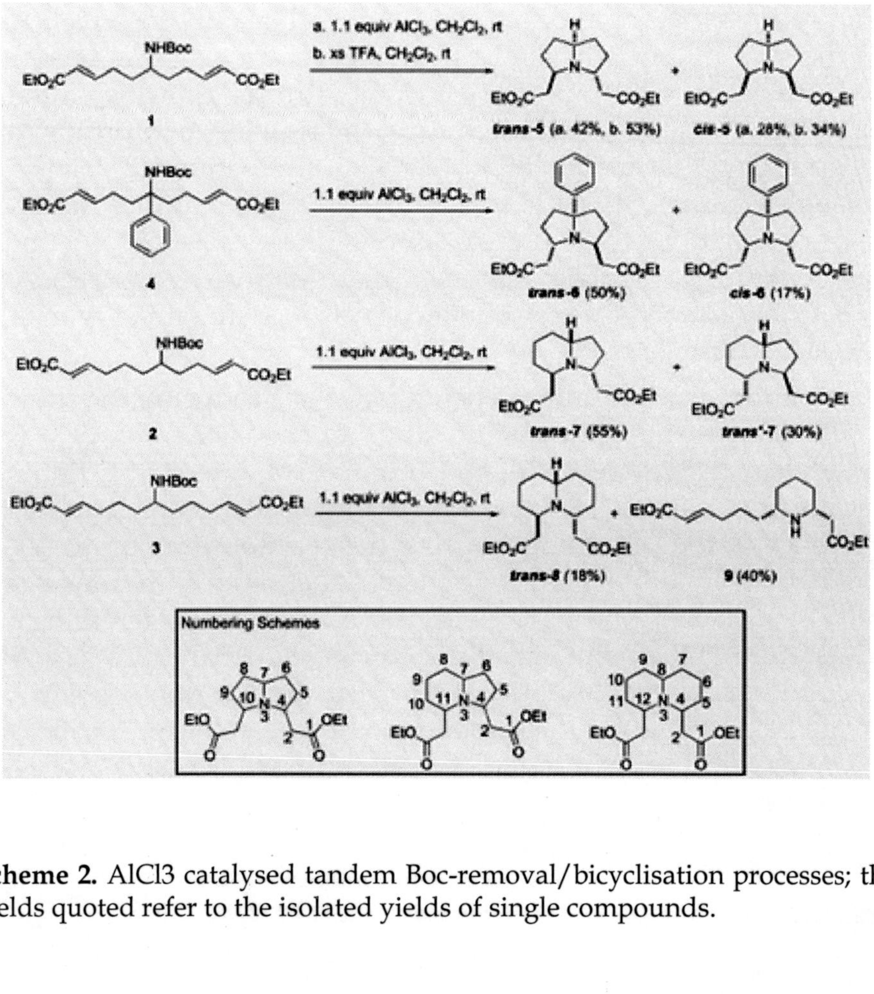

Scheme 2. AlCl3 catalysed tandem Boc-removal/bicyclisation processes; the yields quoted refer to the isolated yields of single compounds.

The corresponding reaction of phenyl-substituted analogue 4 also gave a mixture of two diastereomers; the major diastereomer was 4,10-trans-7,10-cis isomer (trans-6), which was formed in 50% yield, and a 17% yield of the 4,7,10-cis-isomer (cis-6) was also obtained.

Indolizidine 7 was produced in very good overall yield (85%). The unsymmetrical nature of this compound gave rise to the possibility of formation of additional diastereomers when compared to 5 and 6; however, once again only two were formed in any appreciable amount. The compounds obtained were both found to have a trans-fused geometry at the ring junction, as indicated by IR (strong absorbance at 2850 cm−1) and 1H NMR (H-7 chemical shift around 2.4 ppm) spectroscopy [23,24]. The ester-bearing side chains were also found to be trans to each other in both cases, meaning that the two products obtained differ from each other only in which ring has the side chain cis to the ring junction hydrogen. The favoured product was the 4,11-trans-7,11-cis isomer (trans-7) in which the side chain of the 6-membered ring is cis to the ring junction proton; this compound was isolated in 55% yield. The alternative 4,11-trans-4,7-cis isomer (trans′-7) was obtained in 30% yield. Despite the good yields obtained for these three examples, the cyclisation of 3 under these conditions proved disappointing, with 4,12-trans-8,12-cis-quinolizidine (trans-8) only obtained in 18% yield, along with 40% of monocyclic species 9.

In light of the difficulties encountered in producing the desired bicyclic species, an optimisation study of the cyclisation of 3 was undertaken, which resulted in a number of interesting findings that are summarised in Table 1. The initial alterations made little difference to the process; increasing the reaction time up to seven days had essentially no effect on the product ratio, with trans-8 and 9 obtained in 21% and 37% yield, respectively, and increasing the temperature to the point of reflux in dichloromethane also had little effect on the conversion. However, when the reaction solvent was changed to toluene and the reaction mixture heated to reflux, trans-8 and 9 were still formed in similar proportions (24% and 21%), but in this case an additional product was also isolated in 30% yield. This compound was found to be tricyclic compound 10 (Scheme 3a), which was obtained as a single diastereomer with all of the ring junction protons on the same face. This all–cis-stereochemistry was surprising, as so far 8 had only been obtained with the side chains in trans- configuration; however the configuration of 10 was unambiguously confirmed by X-ray crystallography. In some ways the formation of a tricyclic species such as 10 was not altogether surprising, as a similar tricyclic species was generated in Stockman's synthesis of hippodamine [25]. In that work, bicycle trans-8 was transformed into the corresponding tricyclic compound (possessing cis–trans ring junction stereochemistry) by a base-mediated Dieckmann cyclisation (Scheme 3b). It seems in our case that, under the correct conditions, a Dieckmann reaction can be made to occur in tandem with the Boc-deprotection and double-conjugate addition processes. This one-pot, four-step reaction process is extremely interesting for the amount of molecular complexity generated in a single transformation, and also because the all-cis-stereochemistry of 10, which differs from the cis–trans-stereochemistry observed by Stockman for the Dieckman cyclisation of trans-8. For these reasons, further investigations into the process were carried out.

Table 1. Overview of the Lewis acid mediated folding reactions of 3.

solvent	Lewis acid (equiv)	temp.	% yield			
			trans-8	*cis*-8	9	10
DCM[a]	$AlCl_3$ (1.1)	rt	18	—	40	—
DCM[b]	$AlCl_3$ (1.1)	rt	21	—	37	—
toluene[a]	$AlCl_3$ (1.1)	reflux	24	—	21	30
toluene[a]	$AlCl_3$ (3)	reflux	43	—	13	—
toluene[a]	$Sc(OTf)_3$ (3)	reflux	50	—	21	—
toluene[a]	$Sn(OTf)_2$ (0.5)	reflux	—	9	—	72
toluene[a]	$Sn(OTf)_2$ (1.1)	reflux	30	13	—	23
toluene	$Sn(OTf)_2$ (3)	reflux	29	—	10	—

[a]Reaction stirred overnight; [b]reaction stirred for seven days.

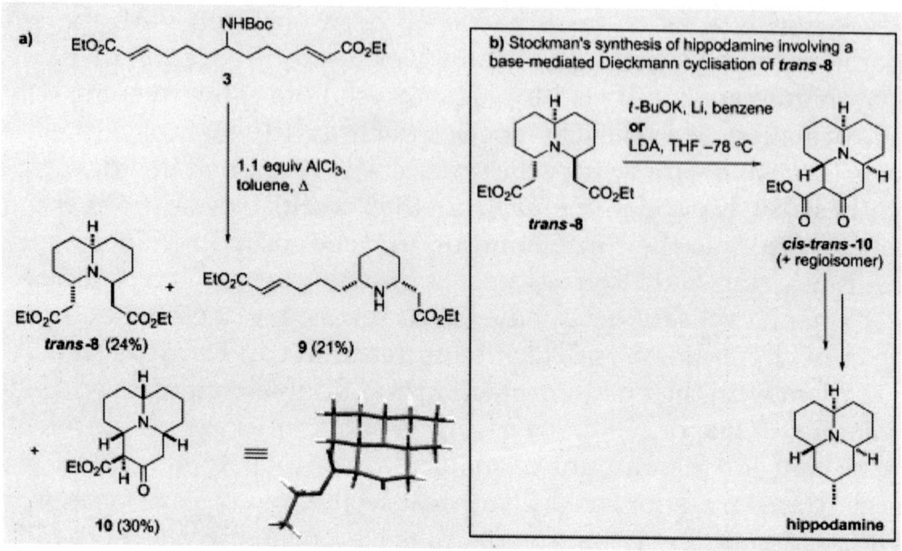

Scheme 3. (a) AlCl3 catalyzed formation of tricyclic alkaloid 10 along with an X-ray crystal structure of 10; (b) base-mediated Dieckmann cyclisation of trans-8 employed by Stockman and co-workers in the synthesis of hippodamine [25].

The amount of AlCl3 was increased to 3 equiv; however, this led to the suppression of tricycle formation in favour of a slight increase in the yield of trans-8 to 43% (a 13% yield of 9 was also obtained). Switching the Lewis acid to Sc(OTf)3 and still using 3 equiv, gave a slight improvement in the yields of the trans-8 to 50% and 9 to 21% but no formation of 10. Switching the Lewis acid again to Sn(OTf)2 resulted in a decrease in the yields of trans-8 to 29% and 9 to 10% and again no formation of 10. The amount of Lewis acid was then reduced to 0.5 equiv, which dramatically altered the course of the reaction. Performing the reaction under reflux in toluene with 0.5 equiv Sn(OTf)2 produced tricycle 10 in 72% isolated yield. Thus both trans-8 and 10 could be accessed in good yields from the same substrate simply by varying the amount and identity of the Lewis acid used.

Interestingly, the catalytic variant of the reaction also produced 9% of the bicyclic cis-8, which had not been isolated from any of the previous reactions. For completeness, one further reaction was performed by using 1.1 equiv of Sn(OTf)2; careful purification of this reaction gave 23% of 10, 30% of trans-8 and 13% of cis-8. The presence of the previously undetected cis-8 in these two final reactions was intriguing and led us to speculate as to whether a different mechanistic pathway could be in operation depending on the amount of Lewis acid used.

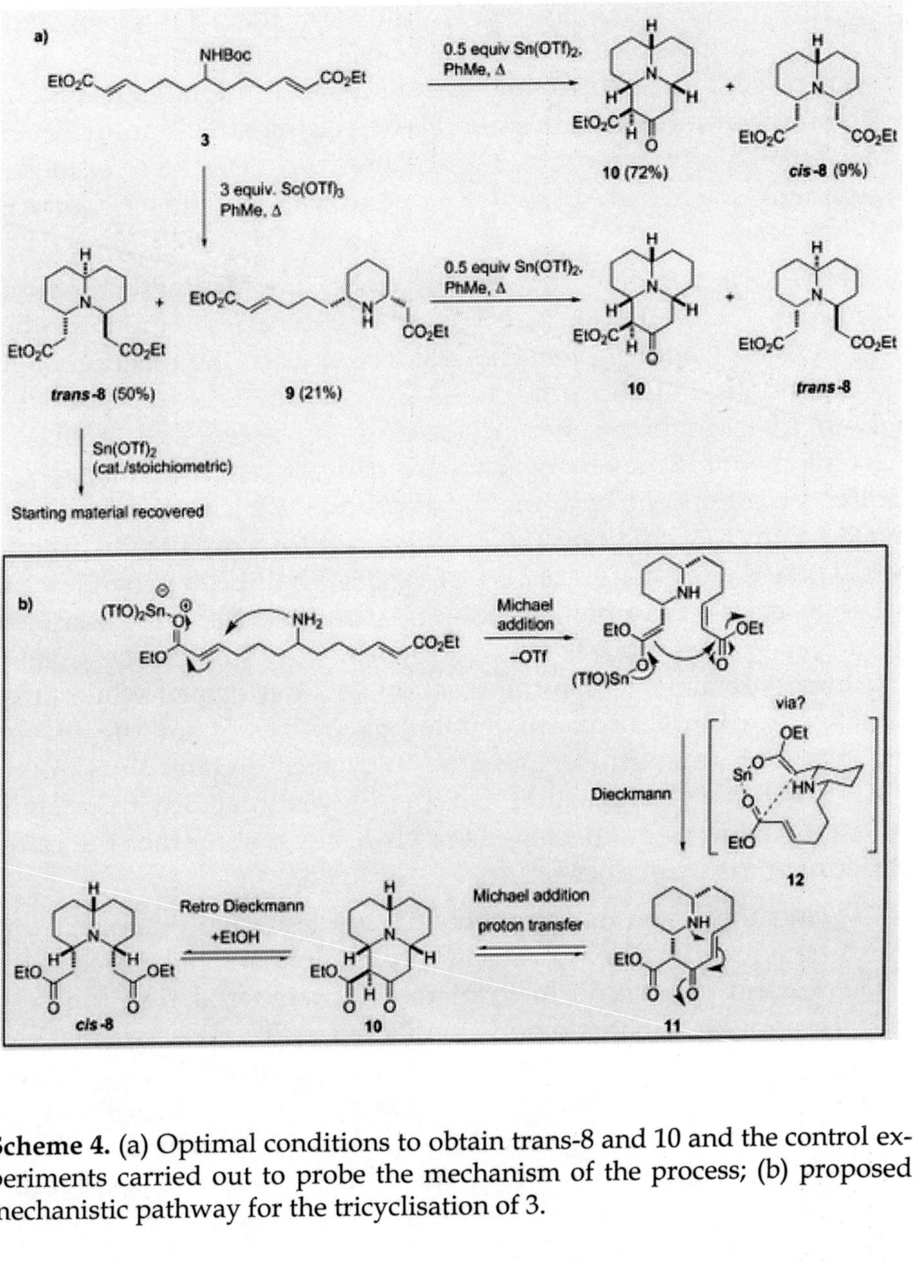

Scheme 4. (a) Optimal conditions to obtain trans-8 and 10 and the control experiments carried out to probe the mechanism of the process; (b) proposed mechanistic pathway for the tricyclisation of 3.

A number of factors led to this mechanistic speculation; principal among them was the fact that cis-8 was never detected in reactions in which 10 was not formed. In all of the earlier experiments the only bicycle detected was trans-8, implying that the double-conjugate addition process heavily favours the formation of this compound. Therefore it was considered that cis-8 could be forming from 10; suggesting, somewhat counter intuitively, that the Dieckmann cyclisation to give 6,10-bridged bicycle 11 could in some cases be favoured over the expected double-conjugate addition. Transannular conjugate addition across the 10-membered ring of 11 would give 10, and a retro-Dieckmann process could then form cis-8. Several control experiments were run in an attempt to validate this hypothesis (Scheme 4a).

Trans-8 was treated with both stoichiometric and catalytic amounts of Sn(OTf)2 and there was no evidence of tricycle formation in either case, with only starting material recovered from the reactions. The direct formation of 10 from trans-8 was not thought to be possible due to their differing stereochemistry; however, an alternative diastereomer of 10 may be expected to form (as observed by Stockman for the corresponding base-mediated process) [25]. Monocycle 9 was treated with catalytic Sn(OTf)2, which led to a mixture of trans-8 and 10. As the formation of 10 from trans-8 had been proven not to occur, it seemed reasonable to assume from this that it is possible to form both species from 9. It is also noteworthy that the best yields of 10 were obtained when the reaction was performed while fitted with a Dean–Stark apparatus containing pieces of sodium to trap the ethanol formed in the Dieckmann cyclisation, and thus inhibit the retro-Dieckmann reaction. Finally, 10 was treated with catalytic Sn(OTf)2, which, as expected, gave cis-8, suggesting that the retro-Dieckmann reaction does occur.

Taking these results into account, we tentatively suggest that the mechanism for the tricyclisation of 3 does indeed proceed via a Dieckmann cyclisation to form the 10-membered ring followed by transannular conjugate addition (Scheme 4b). The fact that this appears to occur when lower amounts of Lewis acid are used (0.5 or 1.1 equiv) but not when three equiv are used indicates that the mechanistic path of the cascade is dependent on the amount of the Lewis acid present. When three equiv of Lewis acid are present (the "stoichiometric" process), it is feasible that during the course of the reaction both of the carbonyl groups are activated simultaneously, allowing the double-conjugate addition to proceed smoothly. However, when fewer equiv are used (the "catalytic" process) there is a relative deficiency of Lewis acid present, and thus, it is less probable that both carbonyl groups can be coordinated to separate Lewis acid molecules simultaneously. If the Dieckmann reaction occurs via a standard 6-membered transition state, it requires both carbonyls to be either coordinated to, or bonded to, a single metal centre, and so this could go some way towards explaining the apparent course of the reaction. Assuming that the Lewis acid remains bonded to the carbonyl of the enol form of the ester group after the first conjugate addition, there may be insufficient Lewis acid free in solution to activate the second ester group separately, leaving it effectively inert to conjugate addition (and Dieckmann cyclisation). However, if the second carbonyl group becomes coordinated to the metal centre that is bonded to the first, the Dieckmann reaction becomes the most favourable process and so occurs preferentially to the second conjugate addition. For these reasons, we cautiously postulate that the Dieckmann reaction may occur via a chelated transition state such as 12, in which both carbonyls are coordinated to a single metal centre.

The data in Table 1 provides further support for the suggestion that different mechanisms can operate depending on the amount of Lewis acid used. This support is provided by the fact that the formation of 10 proceeds best under truly catalytic conditions: when 0.5 equiv Sn(OTf)2 were used, a 72% yield of 10 was obtained compared to 23% when 1.1 equiv were used. In fact, it appears that when 1.1 equiv of Lewis acid are used both mechanisms can occur, as indicated by the formation of both trans-8 and 10 in these reactions, but when 0.5 equiv are used the formation of trans-8 is not observed suggesting that under these conditions the simple double-conjugate addition cannot occur. The identity of the Lewis acid used in the reaction seems not to overtly affect the course of the reaction in terms of the products obtained, as similar product distributions were observed for the reactions using three equiv of AlCl3, Sc(OTf)3 and Sn(OTf)2. The formation of 10 was also not limited to the use of Sn(OTf)2, as a 30% yield of 10 was obtained for the reaction using 1.1 equiv of AlCl3.

Between them, the catalytic and stoichiometric variants of this reaction provide a useful illustration of the use of reagent-based diversification within a predominantly substrate-based strategy. Attempts were then made to apply this reagent-based diversification to the other linear substrates (Scheme 5). Compound 2 was treated with 0.5 equiv of Sn(OTf)2 in acetonitrile, which surprisingly did not lead to the formation of the 6-6-5-tricyclic species, instead producing trans′-7 in 69% yield along with 10% trans-7. While this reaction did not produce any tricyclic species, it was an interesting result, as the selectivity for trans′-7 over trans-7 was the opposite of that observed for the original AlCl3 catalysed process.

Performing the reaction in toluene with 1.1 equiv of Sn(OTf)2 did produce the expected tricyclic species in 50% yield as a mixture of two diastereomers (cis-13 and trans-13) in a 4:1 ratio, proving that the folding of this linear substrate into tricyclic species in one-pot was also possible. Unfortunately, this did not prove to be the case for 1; all attempts to transform 1 into a tricyclic species in one pot, using either stoichiometric or catalytic Lewis acid were unsuccessful. However, it did prove possible to perform the Dieckmann reaction on the bicyclic compounds cis-5 and trans-5 to access the tricyclic architectures. The bicyclic species were treated with LDA at −18 °C, which affected the desired cyclisation in both cases. Strangely, the process proved far more efficient for trans-5, giving trans-14 in 91% yield, compared to 38% for the cis-isomer. In both cases the 1H NMR suggested that these compounds exist as the usually disfavoured enol tautomer.

These reactions, along with the reactions of 1, clearly illustrate the power of this two-directional approach to DOS. Using this methodology it was possible to access five bicyclic and tricyclic scaffolds covering a range of 3D shapes, including pyrrolizidines, indolizidines and quinolizidines, along with 6-6-6 and 5-6-6 azatricyclic species in a single transformation from a small collection of structurally simple linear starting materials. One further tricyclic scaffold (5-5-6) was also accessible in one further synthetic step. The effective introduction of reagent-based diversification into the strategy was extremely satisfying, as by altering the choice of Lewis acid and reaction temperature we were able to adjust the course and selectivity of the reactions to generate different scaffolds and stereochemistry. This combination of reagent and substrate-based approaches for the generation of molecular diversity can afford many interesting possibilities not achievable by either approach alone.

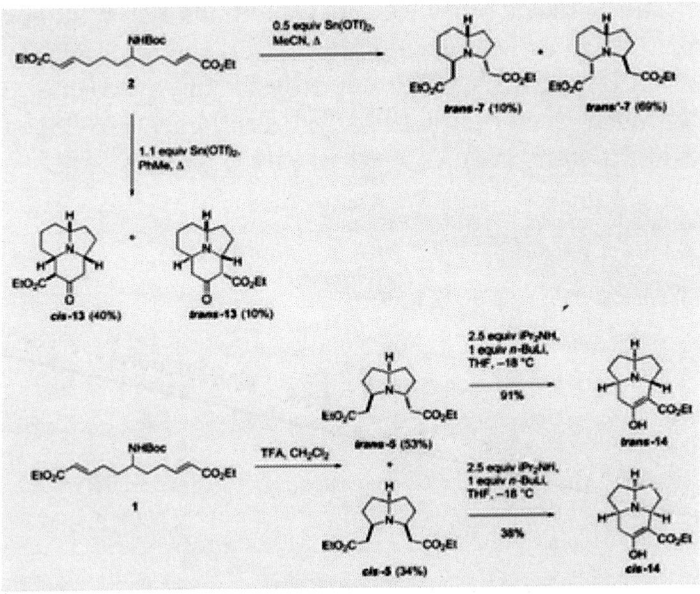

Scheme 5. DOS of 5-5-6 and 5-6-6 tricyclic alkaloids 13 and 14.

Total Synthesis of Myrrhine

Inspired by Stockman's syntheses of the related alkaloids hippodamine and epi-hippodamine [26], tricyclic species 10 was identified as a potential intermediate for the total synthesis of myrrhine, which was then achieved in three steps from 10 (Scheme 6). Compound 10 was treated with Na2CO3 in a mixture of EtOH and H2O under reflux to achieve ester saponification, which was followed by decarboxylation, proceeding smoothly to give the corresponding ketone in 76% yield. The ketone was then transformed to the exocyclic alkene 15 in 61% yield by Wittig reaction with the appropriate phosphonium salt. The final step in the synthesis was a diastereoselective reduction of the double bond with hydrogen gas and Raney-nickel. This was achieved in a moderate 57% yield and with good (~10:1) diastereoselectivity by complexing the nitrogen lone pair with tosic acid, effectively blocking the undesirable face of the tricycle during the course of the reduction. This reduction also produced 5% of the unnatural (or as yet undiscovered) isomer epi-myrrhine. The N-oxide of myrrhine, which has also not been isolated from natural sources, was then synthesised in 96% yield by treating myrrhine with mCPBA. This synthesis of myrrhine compares favourably with previously reported syntheses [26-28], achieving the feat in eight steps and 7% overall yield.

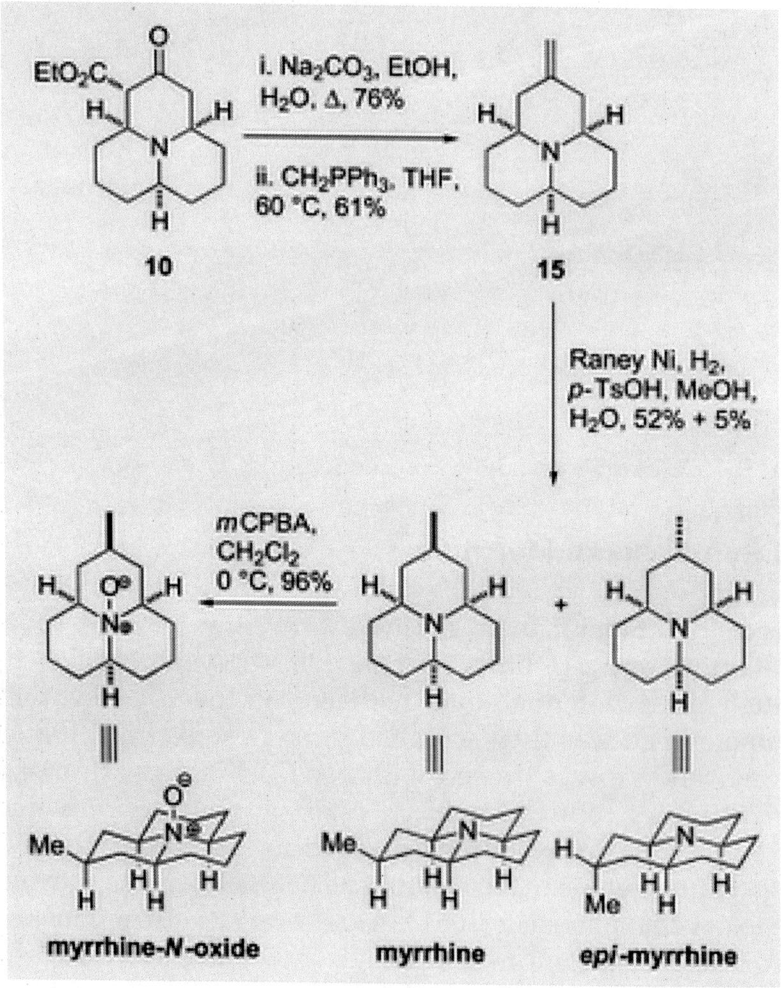

Scheme 6. Total synthesis of myrrhine, epi-myrrhine and myrrhine-N-oxide.

Alternative Starting Materials

The evident efficiency of two-directional synthesis in a DOS context, as exemplified by our synthesis of these alkaloid scaffolds, has led us to continue investigations in this area and to explore the potential utility of this approach for a range of different substrates. Among these substrates, two that stand out as particularly promising are nitromethane and tris(hydroxymethyl)aminomethane (Tris) 16.

Nitromethane is of great interest to us as a potential DOS substrate, as we have a long-standing interest in developing divergent reaction pathways from small and simple starting materials [29,30]. As it represents a one-carbon unit, nitromethane is an ideal substrate for investigation. For Tris, the central quarternary carbon centre is the key point of interest, as we believe that, with the judicious selection of appendages, it should allow us to access a range of bicyclic structures, including examples of fused, bridged and spiro bicycles. Preliminary studies into the utility of these substrates in a DOS context have yielded some promising results.

Our work with nitromethane has so far led to the synthesis of meso-diphenylpyrrolizidine 17, which was achieved in three steps (Scheme 7). Nitromethane was treated with NaOH, and the resulting anion was used to displace the chloride from 3-chloro-1-phenylpropan-1-one giving a 90% yield of the nitroketone. Two-directional synthesis of diketone 18 in this fashion did not prove to be feasible; however, it was achieved in good yield by Michael addition of the nitroketone anion to phenylvinylketone. Subjecting diketone 18 to H2 gas and Raney-nickel then reduced the nitro group and effected the desired double reductive amination to give 17 in 30% yield. It is likely that the scope of this sequence could be extended to include indolizidine and quinolizidine scaffolds, and so provide an alternative route to these frameworks, instead of the double Michael addition strategy.

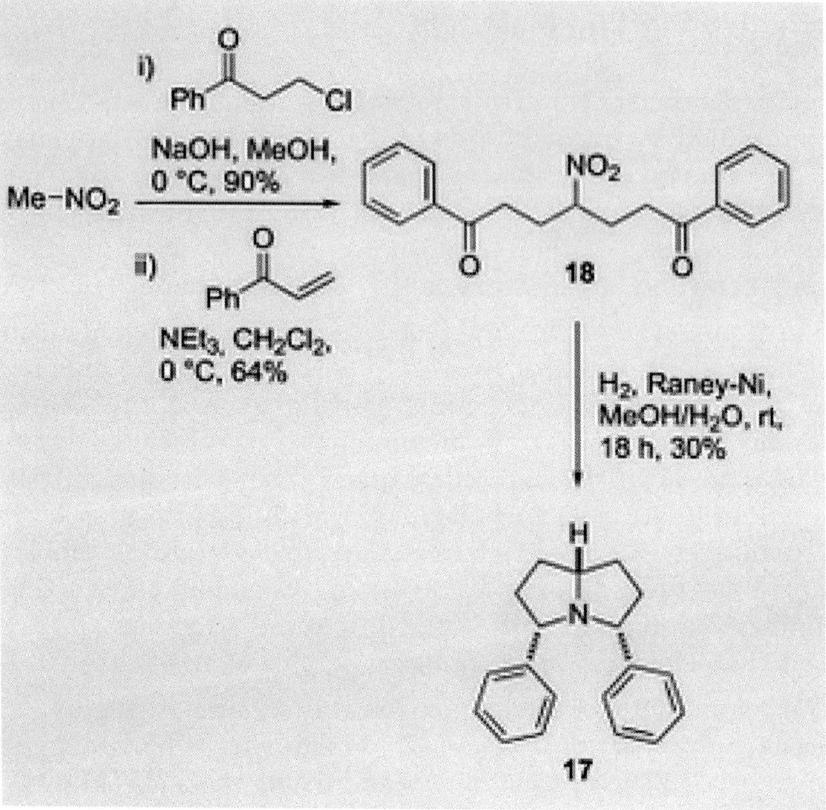

Scheme 7. Use of nitromethane in DOS: Synthesis of meso-diphenylpyrrolidizine 17.

The three hydroxy and single amino groups of Tris give the potential for many variations in substitution; however, our studies have so far focused on allylated derivatives, in particular the triallyl derivative 19 (Scheme 8). The synthesis of 19 was achieved in two steps from Tris: N-Boc protection proceeded in 70% yield, and was followed by alkylation with an excess of allyl bromide to provide the desired triallyl species in 66% yield. Cross metathesis of 19 with ethyl acrylate was then performed. Fortunately, it proved to be possible to achieve some selectivity for different products by varying the catalyst used. Treating 19 with 3% Grubbs II catalyst in neat ethyl acrylate at room temperature gave monoester 20 in 41% yield, whereas performing the reaction with 5% Hoveyda-Grubbs II catalyst gave a 73% yield of triester 21.

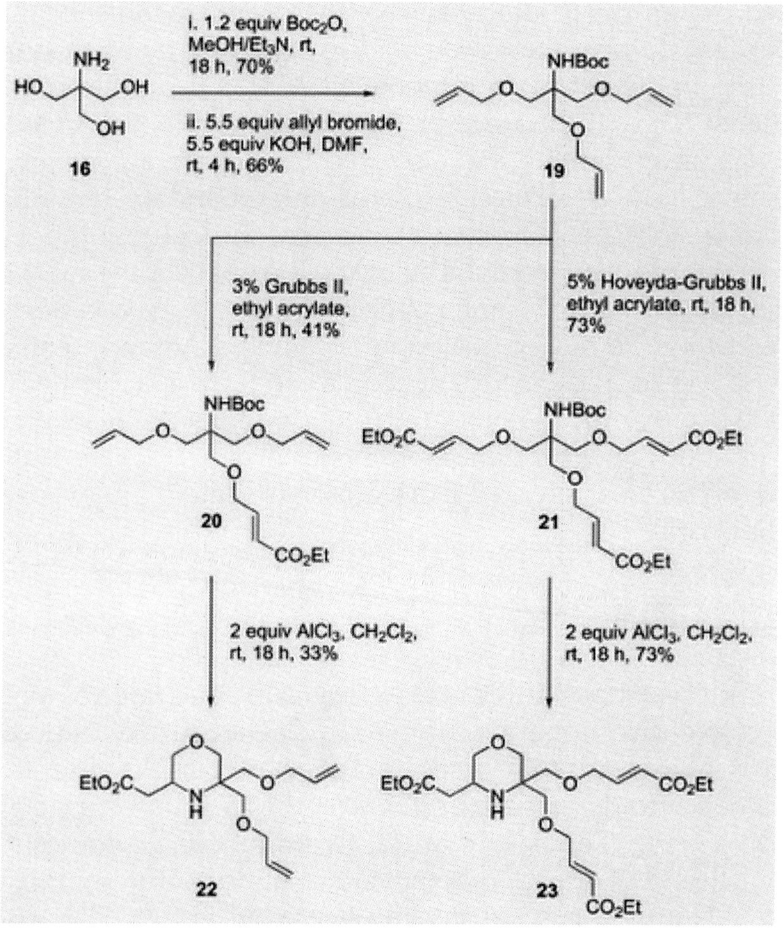

i. 1.2 equiv Boc_2O, MeOH/Et_3N, rt, 18 h, 70%
ii. 5.5 equiv allyl bromide, 5.5 equiv KOH, DMF, rt, 4 h, 66%

3% Grubbs II, ethyl acrylate, rt, 18 h, 41%

5% Hoveyda-Grubbs II, ethyl acrylate, rt, 18 h, 73%

2 equiv $AlCl_3$, CH_2Cl_2, rt, 18 h, 33%

2 equiv $AlCl_3$, CH_2Cl_2, rt, 18 h, 73%

Scheme 8. Use of Tris as a substrate for DOS: Synthesis of decorated morpholines 22 and 23.

These two metathesis products were then subjected to the tandem deprotection-cyclisation conditions, this time by using two equiv of AlCl3 in dichloromethane at room temperature. For monoester 20, only a single conjugate addition was possible, leaving the pendant terminal alkene groups as handles for further reactivity. This cyclisation proceeded to give the expected, functionalised morpholine scaffold 22 in a moderate 33% yield. A similar product was also obtained for 21, with a single conjugate addition process occurring to give 23 in 73% yield and no trace of the bicyclic product detected. A number of attempts were then made to force the reaction to occur, by varying the Lewis acid to Sn(OTf)2 and Sc(OTf)3 and by heating under reflux in toluene, all of which produced varying amounts of 23, with no trace of the bicyclic product detected. This result was disappointing and also, to some degree, surprising given the relative ease of formation of the corresponding quinolizidine compounds. We speculate that the relative difficulty in the reaction is due to the reduced availability of the lone pair of the morpholine nitrogen compared to the piperidine nitrogen (pKa 8.36 versus 11.22) [31], retarding the rate of nucleophilic attack. Another possibility is that the oxygens in the pendant chains can form a stable hydrogen-bonded species with the NH group that inhibits the reaction of the nitrogen with the Michael acceptor ester groups. Studies involving the use of Tris as a potential DOS substrate remain on-going within our laboratories.

CONCLUSION

The work presented in this article serves to illustrate the potential power of two-directional synthesis in DOS. The use of two-directional synthesis allowed us to access a range of bicyclic and tricyclic molecular scaffolds, rapidly and efficiently, by following a common reaction scheme. The nature of the two-directional synthesis lends itself to the formation of bicyclic compounds by the folding up of doubly substituted precursors, and it proved to be a very effective strategy for the synthesis of natural-product-like alkaloid scaffolds. Our work so far in this area has focused mainly on the synthesis of fused bicyclic compounds; however, we hope in the future to be able to apply a two-directional synthesis approach to the DOS of a wide range of molecular scaffolds and structural classes.

ACKNOWLEDGEMENTS

The authors thank UCB Celltech, the ERC, EU, EPSRC, BBSRC, MRC, Wellcome Trust and Frances and Augustus Newman Foundation for funding.

REFERENCES

1. Galloway, W. R. J. D.; Isidro-Llobet, A.; Spring, D. R. Nat. Commun. 2010, 1, No. 80. doi:10.1038/ncomms1081
2. Nielsen, T. E.; Schreiber, S. L. Angew. Chem., Int. Ed. 2008, 47, 48–56. doi:10.1002/anie.200703073
3. Spandl, R. J.; Díaz-Gavilán, M.; O'Connell, K. M. G.; Thomas, G. L.; Spring, D. R. Chem. Rec. 2008, 8, 129–142. doi:10.1002/tcr.20144
4. Galloway, W. R. J. D.; Díaz-Gavilán, M.; Isidro-Llobet, A.; Spring, D. R. Angew. Chem., Int. Ed. 2009, 48, 1194–1196. doi:10.1002/anie.200805452
5. Dow, M.; Fisher, M.; James, T.; Marchetti, F.; Nelson, A. Org. Biomol. Chem. 2012, 10, 17–28. doi:10.1039/c1ob06098h
6. Burke, M. D.; Berger, E. M.; Schreiber, S. L. Science 2003, 302, 613–618. doi:10.1126/science.1089946
7. Morton, D.; Leach, S.; Cordier, C.; Warriner, S.; Nelson, A. Angew. Chem., Int. Ed. 2009, 48, 104–109. doi:10.1002/anie.200804486
8. O'Leary-Steele, C.; Pedersen, P. J.; James, T.; Lanyon-Hogg, T.; Leach, S.; Hayes, J.; Nelson, A. Chem.–Eur. J. 2010, 16, 9563–9571. doi:10.1002/chem.201000707
9. Poss, C. S.; Schreiber, S. L. Acc. Chem. Res. 1994, 27, 9–17. doi:10.1021/ar00037a002
10. Vrettou, M.; Gray, A. A.; Brewer, A. R. E.; Barrett, A. G. M. Tetrahedron 2007, 63, 1487–1536. doi:10.1016/j.tet.2006.09.109
11. Kennedy, A.; Nelson, A.; Perry, A. Beilstein J. Org. Chem. 2005, 1, No. 2. doi:10.1186/1860-5397-1-2
12. Gignoux, C.; Newton, A. F.; Barthelme, A.; Lewis, W.; Alcaraz, M.-L.; Stockman, R. A. Org. Biomol. Chem. 2012, 10, 67–69. doi:10.1039/c1ob06380d
13. Díaz-Gavilán, M.; Galloway, W. R. J. D.; O'Connell, K. M. G.; Hodkingson, J. T.; Spring, D. R. Chem. Commun. 2010, 46, 776–778. doi:10.1039/b917965h
14. Happ, G. M.; Eisner, T. Science 1961, 134, 329–331. doi:10.1126/science.134.3475.329
15. Stockman, R. A. Tetrahedron Lett. 2000, 41, 9163–9165. doi:10.1016/S0040-4039(00)01640-3
16. Stockman, R. A.; Sinclair, A.; Arini, L. G.; Szeto, P.; Hughes, D. L. J. Org. Chem. 2004, 69, 1598–1602. doi:10.1021/jo035639a

17. Rejzek, M.; Stockman, R. A.; van Maarseveen, J. H.; Hughes, D. L. Chem. Commun. 2005, 4661–4662. doi:10.1039/b508969g
18. Robbins, D.; Newton, A. F.; Gignoux, C.; Legeay, J.-C.; Sinclair, A.; Rejzek, M.; Laxon, C. A.; Yalamanchili, S. K.; Lewis, W.; O'Connell, M. A.; Stockman, R. A. Chem. Sci. 2011, 2, 2232–2235. doi:10.1039/c1sc00371b
19. Boyer, F.-D.; Hanna, I. Eur. J. Org. Chem. 2006, 471–482. doi:10.1002/ejoc.200500645
20. Schwartz, B. D.; McErlean, C. S. P.; Fletcher, M. T.; Mazomenos, B. E.; Konstantopoulou, M. A.; Kitching, W.; De Voss, J. J. Org. Lett. 2005, 7, 1173–1176. doi:10.1021/ol050143w
21. Jirgensons, A.; Kauss, V.; Kalvinsh, I.; Gold, M. R. Synthesis 2000, 1709–1712. doi:10.1055/s-2000-8208
22. Scarpi, D.; Occhiato, E. G.; Guarna, A. J. Org. Chem. 1999, 64, 1727-1732. doi:10.1021/jo981946i
23. Crabb, T. A.; Newton, R. F.; Jackson, D. Chem. Rev. 1971, 71, 109–126. doi:10.1021/cr60269a005
24. Rader, C. P.; Young, R. L., Jr.; Aaron, H. S. J. Org. Chem. 1965, 30, 1536–1539. doi:10.1021/jo01016a048
25. Rejzek, M.; Stockman, R. A.; Hughes, D. L. Org. Biomol. Chem. 2005, 3, 73–83. doi:10.1039/b413052a
26. Ayer, W. A.; Dawe, R.; Eisner, R. A.; Furuichi, K. Can. J. Chem. 1976, 54, 473–481. doi:10.1139/v76-064
27. Mueller, R. H.; Thompson, M. E.; DiPardo, R. M. J. Org. Chem. 1984, 49, 2217–2231. doi:10.1021/jo00186a029
28. Gerasyuto, A. I.; Hsung, R. P. J. Org. Chem. 2007, 72, 2476–2484. doi:10.1021/jo062533h
29. Wyatt, E. E.; Fergus, S.; Galloway, W. R. J. D.; Bender, A.; Fox, D. J.; Plowright, A. T.; Jessiman, A. S.; Spring, D. R. Chem. Commun. 2006, 3296–3298. doi:10.1039/b607710b
30. Thomas, G. L.; Spandl, R. J.; Glansdorp, F. G.; Welch, M.; Bender, A.; Cockfield, J.; Lindsay, J. A.; Bryant, C.; Brown, D. F. J.; Loiseleur, O.; Rudyk, H.; Ladlow, M.; Spring, D. R. Angew. Chem., Int. Ed. 2008, 47, 2808–2812. doi:10.1002/anie.200705415
31. Hall, H. K., Jr. J. Am. Chem. Soc. 1957, 79, 5441–5444. doi:10.1021/ja01577a030

Chapter 6

MODULAR SYNTHESIS OF POLYPHENOLIC BENZOFURANS, AND APPLICATION IN THE TOTAL SYNTHESIS OF MALIBATOL A AND SHOREAPHENOL

David Y.-K. Chen *, Qiang Kang and T. Robert Wu

Chemical Synthesis Laboratory at Biopolis, Institute of Chemical and Engineering Sciences (ICES),

Agency for Science, Technology and Research (A*STAR), 11 Biopolis Way, The Helios Block, No. 03-08, 138667, Singapore

INTRODUCTION

Polyphenolic secondary metabolites have attracted growing interest from the scientific community in recent years [1-6]. However, despite their fascinating molecular architectures and diverse biological properties, chemical syntheses of these natural products and/or designed analogues have been scarce [7,8]. With this in mind, and as a continuation of our chemical and biological investigations of polyphenolic natural products [9,10], we set out to develop a general strategy for the synthesis of dimeric, resveratrol-derived benzofurans represented by generic structure 1, as shown in Figure

1. We further demonstrated the developed technology in the total synthesis of malibatol A (2) and shoreaphenol (3), two dimeric resveratrol polyphenolic benzofurans isolated from Hopea malibato and Shorea robusta, respectively [11-13].

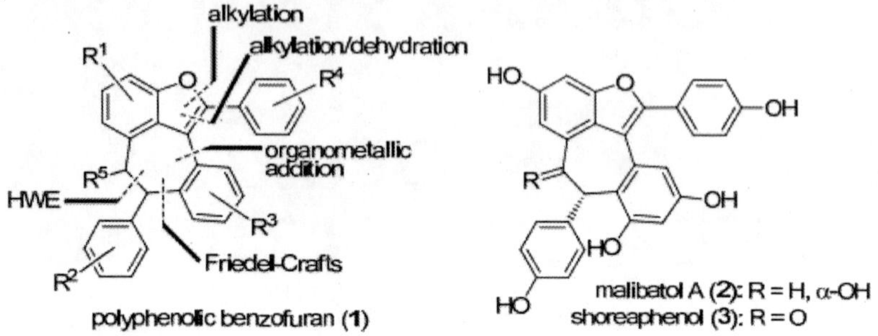

alkylation
alkylation/dehydration
organometallic addition
HWE
Friedel-Crafts
polyphenolic benzofuran (1)
malibatol A (2): R = H, α-OH
shoreaphenol (3): R = O

Figure 1. Generic molecular structure of polyphenolic benzofuran 1 and structures of malibatol A (2) and shoreaphenol (3)

RESULTS AND DISCUSSION

Recognizing the hexacyclic structure represented by 1 containing four substituted phenyl rings, we envisaged a modular approach where each one of the phenyl rings can be installed independently and sequentially. Therefore, as outlined in Scheme 1, the proposed synthesis begin with stilbene aldehyde 4, a building block with two aromatic domains brought together through a Horner-Wadswoth-Emmons (HWE) olefination reaction [14] and a subsequent Vilsmeier formylation [15]. Introduction of a third aromatic domain through the addition of an organometallic aryl species 5 to aldehyde 4, followed by subsequent oxidation (IBX) should give ketone 6. The intermediate benzylic alcohol obtained prior to IBX oxidation has previously been demonstrated by Snyder and co-workers as a versatile intermediate to access a number of resveratrol derived natural products [7,8]. Carbonyl-directed selective demethylation of 6 should lead to phenol 7, setting the stage for the attachment of the final aromatic moiety through an alkylation with benzyl halide 8 or a Mitsunobu reaction [16] with benzyl alcohol 9.

Scheme 1. General, modular strategy for the construction of hexacyclic benzofuran 1

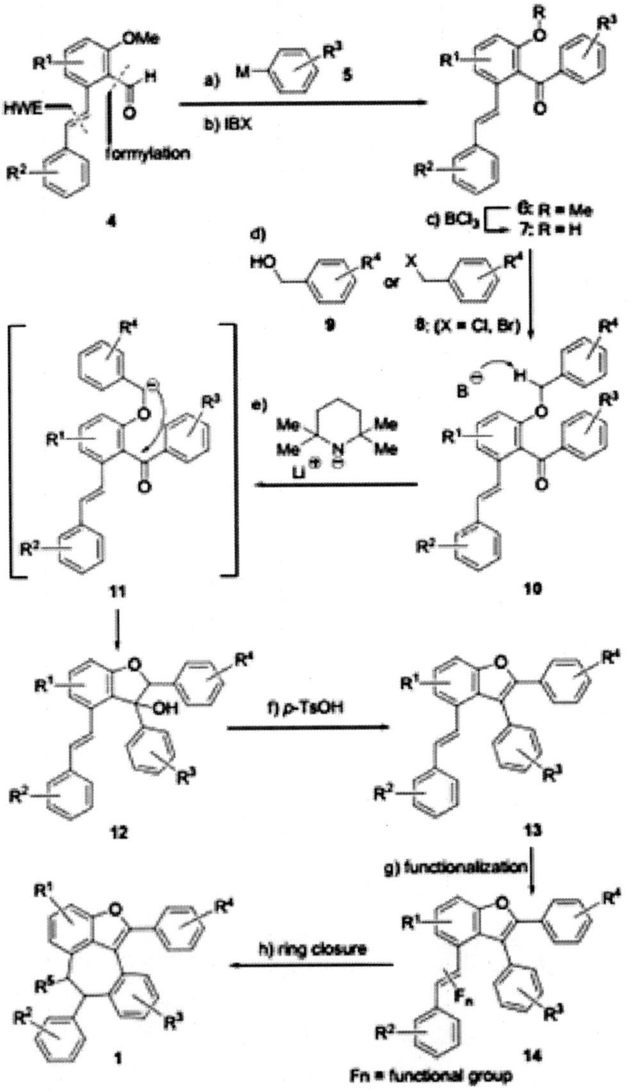

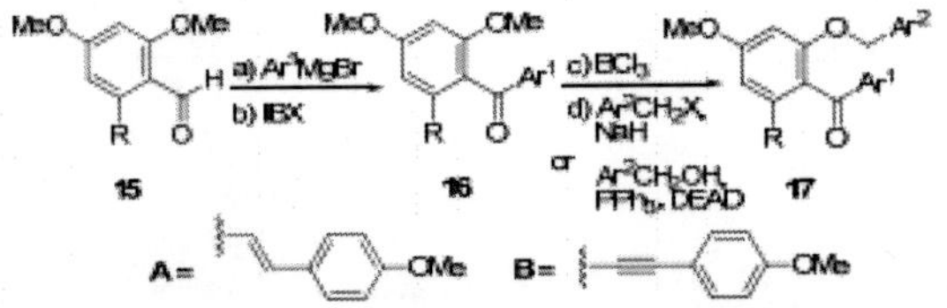

With benzyl ether 10 in hand, the formation of the benzofuran ring is anticipated through its initial benzylic deprotonation (LiTMP), followed by an intramolecular cyclization (11 to 12) and subsequent dehydration (12 to 13, p-TsOH•H2O), to deliver pentacyclic benzofuran 13 [17]. Finally, the olefinic functionality in stilbene 13 should serve as a versatile handle for either direct seven-membered ring formation, or further transformation (14, e.g. epoxidation) leading to functionalized hexacycles 1 upon ring closure With this general strategy in mind, its realization to generate a library of benzofuran polyphenols is illustrated in Tables 1-□. As shown in Table 1, aryl ketones 16 and benzyl ethers 17 were efficiently prepared in 85-90% yield (over the two steps from 15) and 71-95% yield (over the two steps from 16), respectively. Next, benzofuran formation from keto benzyl ethers 17 under the two-step procedure generally proceeded in good yields (71-85% yield, Table 2), apart from the failure of p-bromo substrate to participate in the cyclization (entry 3, Table 2) and the less satisfactory dehydration for the acid sensitive furanyl substrate (entry 5, Table 2).

Table 1. Preparation of ketone 16 and benzyl ether 17.

Entry	R	Ar^1	Ar^2	16 Yield (%)[b]	17 Yield (%)[b]
1(a)	H	C_6H_5	C_6H_5	85%	89%
2(b)	A	3,5-$(MeO)_2C_6H_3$	C_6H_5	88%	95%
3(c)	A	3,5-$(MeO)_2C_6H_3$	4-$(Br)C_6H_4$	88%	92%
4(d)	A	3,5-$(MeO)_2C_6H_3$	4-$(MeO)C_6H_4$	88%	90%
5(e)[a]	A	3,5-$(MeO)_2C_6H_3$	2-furyl	88%	71%
6(f)	A	C_6H_5	4-$(MeO)C_6H_4$	85%	91%
7(g)	A	3,4,5-$(MeO)_3C_6H_2$	4-$(MeO)C_6H_4$	90%	95%
8(h)	A	3,4-$(MeO)_2C_6H_3$	4-$(MeO)C_6H_4$	87%	90%
9(i)	B	3,5-$(MeO)_2C_6H_3$	4-$(MeO)C_6H_4$	86%	87%

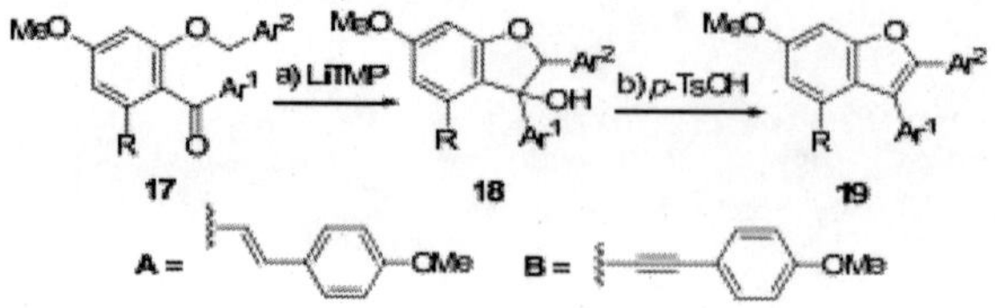

Table 2. Preparation of benzofuran 19

Entry	R	Ar^1	Ar^2	19 Yield (%)[a]
1(a)	H	C_6H_5	C_6H_5	83
2(b)	A	3,5-$(MeO)_2C_6H_3$	C_6H_5	71
3(c)	A	3,5-$(MeO)_2C_6H_3$	4-(Br)C_6H_4	0
4(d)	A	3,5-$(MeO)_2C_6H_3$	4-(MeO)C_6H_4	87
5(e)	A	3,5-$(MeO)_2C_6H_3$	2-furyl	38
6(f)	A	C_6H_5	4-(MeO)C_6H_4	80
7(g)	A	3,4,5-$(MeO)_3C_6H_2$	4-(MeO)C_6H_4	85
8(h)	A	3,4-$(MeO)_2C_6H_3$	4-(MeO)C_6H_4	81
9(i)	B	3,5-$(MeO)_2C_6H_3$	4-(MeO)C_6H_4	85

Finally, closure of the seven-membered ring was carried out under acidic conditions (p-TsOH•H_2O) to give cyclized compound 20 in high yields (90-95% yield, entries 1, 2, 5-7, Table 3). The incompatibility of the furanyl functionality under the acidic conditions was once again observed (entry 3, Table 3), and the electronically less favoured substrate 19d failed to participate in the Friedel-Crafts type cyclization (entry 4, Table 3).

MeO Ar2 R R1 19 a) p-TsOH MeO Ar2 R1 MeO 20 A= OMe B= OMe

Table 3. Friedel□Crafts type cyclization of benzofurans 20

Entry	R	R^1	Ar^2	20 Yield (%)[a]
1(a)	A	3,5-$(MeO)_2$	C_6H_5	95
2(b)	A	3,5-$(MeO)_2$	4-(MeO)C_6H_4	90
3(c)	A	3,5-$(MeO)_2$	2-furyl	0
4(d)	A	H	4-(MeO)C_6H_4	0
5(e)	A	3,4,5-$(MeO)_3$	4-(MeO)C_6H_4	92
6(f)	A	3,4-$(MeO)_2$	4-(MeO)C_6H_4	90
7(g)	B	3,4-$(MeO)_2$	4-(MeO)C_6H_4	95

In addition, we demonstrated a one-pot procedure to prepare hexacyclic benzofuran 20b directly from keto benzyl ether stilbene 17d (Scheme 2). This highly efficient, cascade process involving deprotonation-cyclization (LiTMP), dehydration and Friedel-Crafts ring-closure (*p*-TsOH) illustrated the utility of the developed methodology in the synthesis of highly functionalized, polycyclic polyphenols, a useful structural class for both chemical and biological investigations.

Scheme 2. One-pot preparation of hexacyclic benzofuran 20b

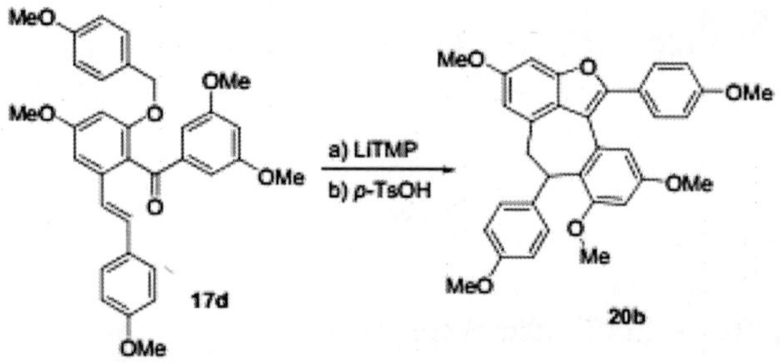

Next, the developed methodology was applied to the total synthesis of malibatol A (2) [18] and shoreaphenol (3), as shown in Scheme 3 [19]. In this instance, with pentacyclic benzofuran 19d in hand, construction of the oxygen-substituted, seven-membered ring in the malibatol A (2) and shoreaphenol (3) framework called for an intramolecular Friedel-Crafts type epoxide-opening process.

Thus, epoxidation of stilbene 19d under the bromohydrin protocol (NBS, NaOH), followed by treatment of the resulting epoxide (21) with BBr3 resulted the concomitant cyclization and global demethylation as a one-pot process, presumably through the intermediacy of 22, giving racemic malibatol A (2) as a single diastereoisomer in 20% yield. Oxidation of malibatol A (2) in the presence of PDC then afforded shoreaphenol (3), despite the modest yield of 46%. Both malibatol A (2) and shoreaphenol (3) exhibited spectroscopic data (1

H- and 13C-NMR) and mass spectrometry data matching those reported for the natural substances [11-13]

Scheme 3. Total synthesis of malibatol A (2) and shoreaphenol (3).

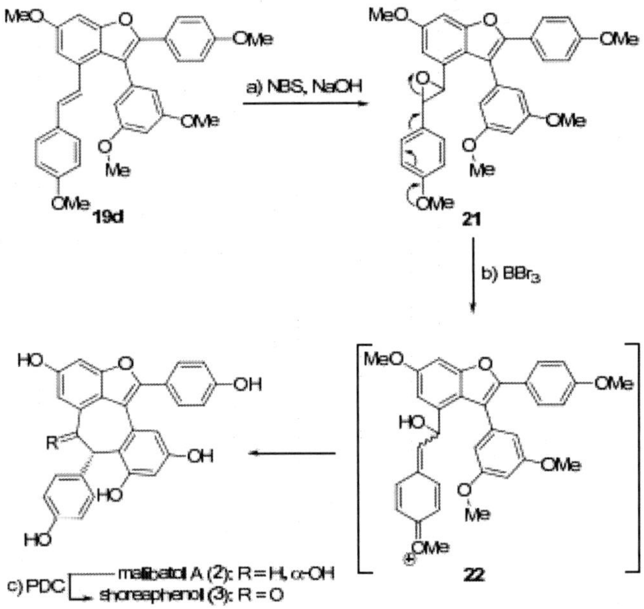

Experimental

GENERAL

All reactions were carried out under a nitrogen or argon atmosphere with dry solvents under anhydrous conditions, unless otherwise noted. Dry tetrahydrofuran (THF) and methylene chloride (CH2C12) were obtained by passing commercially available pre-dried, oxygen-free formulations through activated alumina columns. Methanol (MeOH), N,N'-dimethylformamide (DMF), dimethyl sulfoxide (DMSO) and benzene were purchased in anhydrous form and used without further purification. Acetone, water, ethyl acetate (EtOAc), diethyl ether (Et2O), methylene chloride (CH2C12), and hexanes were purchased at the highest commercial quality and used without further purification, unless otherwise stated. Reagents were purchased at the highest mercial quality and used without further purification, unless otherwise stated. Yields refer to chromatographically and spectroscopically (1H-NMR) homogeneous materials, unless otherwise stated. Reactions were monitored by thin-layers chromatography (TLC) carried out on 0.25 mm E. Merck silica gel plates (60F-254) using UV light as visualizing agent and an ethanolic solution of ammonium molybdate and anisaldehyde and heat as developing agents. E. Merck silica gel (60, particle size 0.040-0.063 mm) was used for flash column chromatography. 1H and 13C-NMR spectra were recorded at 600 and 150 MHz, respectively, on a Bruker AV-600 instrument and calibrated using residual undeuterated solvent as an internal reference.

The following abbreviations were used to explain the multiplicities: s = singlet, d = doublet, t = triplet, q = quartet, quint = quintet, m = multiplet, pent = pentet, hex = hexet, br = broad. IR spectra were recorded on a Perkin-Elmer Spectrum One FTIR spectrometer with diamond ATR accessory. Melting points (m.p.) are uncorrected and were recorded on a Buchi B-540 melting point apparatus. High resolution mass spectra (HRMS) were recorded on an Agilent ESI TOF (time of flight) mass spectrometer at 3500 V emitter voltage.

General procedure A (Preparation of diaryl ketones 16, Table 1)

To a solution of aldehyde 15 (2.0 mmol) in THF (20 mL) at 0 °C was added the appropriate Grignard reagent (0.5 M in THF, 3.0 mmol). The resulting mixture was stirred for 0.5 h before it was quenched with NH4Cl (5 mL, sat. aq.). The layers were separated and the aqueous layer was extracted with EtOAc (3 × 10 mL). The combined organic layers were washed with brine (10 mL), dried

(Na_2SO_4) and concentrated in vacuo to afford the crude benzyl alcohol, which was used directly without further purification. To the solution of crude benzyl alcohol (obtained as above) in DMSO (5 mL) at 23 °C was added IBX (1.15 g, 4.1 mmol) in one portion. The resulting mixture was stirred for 2 h before it was quenched with Na2S2O3 (5 mL, sat. aq.). The layers were separated and the aqueous layer was extracted with Et2O (3 × 30 mL). The combined organic layers were washed with brine (30 mL), dried (Na2SO4) and concentrated in vacuo. Flash column chromatography (silica gel) afforded diaryl ketone 16. Using this general procedure the following compounds were prepared: (2,4-Dimethoxyphenyl)

(phenyl)methanone (16a). From 2,4-dimethoxybenzaldehyde and phenyl magnesium bromide. Flash column chromatography (silica gel, hexanes-EtOAc 4:1) afforded ketone 16a (412 mg, 85%) as a pale yellow foam. All physical properties of this compound were identical to those reported in literature [20]. column chromatography (silica gel, hexanes-EtOAc 2:1) afforded ketone 16b (764 mg, 88%) as a pale yellow foam. All physical properties of this compound were identical to those reported in literature [8].

(E)-(2,4-Dimethoxy-6-(4-methoxystyryl)phenyl)(phenyl)methanone (16f). From (E)-2,4-dimethoxy-6-(4-methoxystyryl)benzaldehyde and phenylmagnesium bromide. Flash column chromatography (silica gel, hexanes-EtOAc 2:1) afforded ketone 16f (636 mg, 85%) as a pale yellow foam. 16f: Rf = 0.45 (silica gel, hexanes-EtOAc 2:1); IR (film) vmax 2938, 1661, 1595, 1510, 1253, 1161, 1078, 920, 830, 721 cm-1; 1H-NMR (CD3CN): □ = 7.80-7.78 (m, 2 H), 7.58-7.55 (m, 1 H), 7.46-7.43 (m, 2 H), 7.25 (d, J = 9.0 Hz, 2 H), 7.13 (d, J = 16.2 Hz, 1 H), 6.97 (d, J = 1.8 Hz, 1 H), 6.81 (d, J = 9.0 Hz, 2 H), 6.70 (d, J = 16.2 Hz, 1 H), 6.56 (d, J = 1.8 Hz, 1 H), 3.89 (s, 3 H), 3.71 (s, 3 H), 3.63 ppm (s, 3 H); 13C-NMR (CD3OD): □ = 197.2, 161.6, 159.7, 158.3, 138.3, 137.4, 133.5, 131.2, 129.4, 129.1, 128.7, 127.9, 122.4, 121.0, 114.1, 101.4, 97.7, 55.5, 55.3, 54.9 ppm; HRMS (ESI): calcd for C24H22O4Na+ [M + Na+] 397.1410, found 397.1406.

(E)-(2,4-Dimethoxy-6-(4-methoxystyryl)phenyl)(3,4,5-trimethoxyphenyl)methanone (16g). From (E)-

2,4-dimethoxy-6-(4-methoxystyryl)benzaldehyde and 3,4,5-trimethoxyphenylmagnesium bromide. Flash column chromatography (silica gel, hexanes-EtOAc 2:1) afforded ketone 16g (836 mg, 90%) as a pale yellow foam. 16g: Rf = 0.25 (silica gel, hexanes-EtOAc 2:1); IR (film) vmax 2938, 1661, 1578, 1511, 1413, 1327, 1156, 1126, 834 cm-1; 1H-NMR (CD3CN): = 7.28 (d, J = 8.4 Hz, 2 H), 7.13 (d, J = 16.2 Hz, 1 H), 7.08 (s, 2 H), 6.95 (d, J = 2.4 Hz, 1 H), 6.84 (d, J = 8.4 Hz, 2 H), 6.67 (d, J = 16.2 Hz, 1 H), 6.56 (d, J = 2.4 Hz, 1 H), 3.89 (s, 3 H), 3.77 (s, 3 H), 3.75 (s, 6 H), 3.74 (s, 3 H), 3.68 ppm (s, 3 H); 13C-NMR (CD3OD): □ = 195.8, 161.5, 159.7, 158.2, 153.3, 142.7, 137.5, 133.6, 131.1, 129.4, 127.9, 122.6, 120.8, 114.1, 106.6, 101.4, 97.7, 60.0, 55.7, 55.5, 55.3, 54.9 ppm; HRMS (ESI): calcd for C27H28O7Na+ [M + Na+] 487.1727, found 487.1712.

(E)-(2,4-dimethoxy-6-(4-methoxystyryl)phenyl) (3,4-dimethoxyphenyl)methanone (16h). From (E)-2,4-dimethoxy-6-(4-methoxystyryl) benzaldehyde and 3,4-dimethoxy phenylmagnesium bromide. Flash column chromatography (silica gel, hexanes-EtOAc 2:1) afforded ketone 16h (755 mg, 87%) as a pale yellow foam. 16h: Rf = 0.20 (silica gel, hexanes-EtOAc 2:1); IR (film) vmax 2937, 1739, 1653, 1595, 1511, 1267, 1158, 835 cm-1; 1

H-NMR (CD3CN): □ = 7.50 (d, J = 1.8 Hz, 1 H), 7.26 (d, J = 9.0 Hz, 2 H), 7.20 (dd, J = 8.4, 2.4 Hz, 1 H), 7.13 (d, J = 16.2 Hz, 1 H), 6.94 (d, J = 2.4 Hz, 1 H), 6.90 (d, J = 8.4 Hz, 1 H), 6.83 (d, J = 9.0 Hz, 2 H), 6.65 (d, J = 16.2 Hz, 1 H), 6.55 (d, J = 1.8 Hz, 1 H), 3.89 (s, 3 H), 3.84 (s, 3 H), 3.83 (s, 3 H), 3.74 (s, 3 H), 3.67 ppm (s, 3 H); 13C-NMR (CD3CN): □ = 195.5, 161.3, 159.7, 158.0, 153.8, 149.2, 137.1, 131.3, 130.9, 129.4, 127.8, 125.1, 122.5, 121.3, 114.1, 110.5, 110.1, 101.1, 97.7, 55.5, 55.5, 55.3, 55.2, 54.9 ppm; HRMS (ESI): calcd for C26H26O6Na+ [M + Na+] 457.1621, found 457.1610.

(2,4-Dimethoxy-6-((4-methoxyphenyl)ethynyl)phenyl) (3,5dimethoxyphenyl)methanone (16i). From 2,4-dimethoxy-6-[(4-methoxyphenyl)ethynyl]benzaldehyde and 3,5-dimethoxy phenylmagnesium bromide. Flash column chromatography (silica gel, hexanes-EtOAc 2:1) afforded ketone 16i (743 mg, 86%) as a pale yellow foam. 16i: Rf = 0.28 (silica gel, hexanes-EtOAc 2:1); IR (film) vmax 2938, 1671, 1590, 1569, 1510, 1247, 1153, 1065, 832 cm-1; 1H-NMR (CD3CN): □ = 7.10 (d, J = 9.0 Hz, 2 H), 6.92.(d, J = 2.4 Hz, 2 H), 6.84 (d, J = 9.0 Hz, 2 H), 6.73 (d, J = 2.4 Hz, 1 H), 6.72 (t, J = 2.4 Hz, 1 H), 6.66 (d, J = 2.4 Hz, 1 H), 3.86 (s, 3 H), 3.77 (s, 6 H), 3.75 (s, 3 H), 3.72 ppm (s, 3 H); 13C-NMR (CD3CN): □ = 195.7, 162.3, 162.0, 161.0, 158.8, 140.7, 133.6, 125.0, 123.2, 115.0, 114.8, 108.6, 107.7, 106.1, 100.3, 94.2, 86.4, 56.5, 56.3, 56.2, 55.9 ppm; HRMS (ESI): calcd for C26H24O6Na+ [M + Na+] 455.1465, found 455.1467.

GENERAL PROCEDURE B (PREPARATION OF BENZYL ETHERS 17, TABLE 1)

To a solution of diaryl ketone 16 (1.0 mmol) in CH2Cl2 (10 mL) at 0 °C was added BCl3 (1.0 M in CH2Cl2, 1.5 mL, 1.5 mmol) dropwise. The resulting mixture was stirred for 1 h before it was quenched with NH4Cl (10 mL, sat. aq.). The layers were separated and the aqueous layer was extracted with CH_2C_{12} (3 × 10 mL). The combined organic layers were washed with brine (20 mL), dried (Na2SO4) and concentrated in vacuo to afford the crude phenol, which was used directly without further purification. To a solution of the crude phenol (obtained as above) in DMF (5 mL) at 0 °C was added NaH (80 mg, 60% wt/wt in mineral oil, 2.0 mmol). The resulting mixture was stirred for 0.5 h before benzyl bromide (or chloride) (1.4 mmol) was added. The reaction mixture was warmed to 23 °C and the progress was monitored by TLC analysis. Upon completion of the reaction (<4 h for most cases), the reaction mixture was quenched with NH4Cl (20 mL, sat. aq.). The layers were separated and the aqueous layer was extracted with Et2O (3 × 20 mL). The combined organic layers were washed with brine (30 mL), dried (Na2SO4) and concentrated in vacuo. Flash column chromatography (silica gel) afforded the desired benzyl ether 17. Using the described general procedure the following substances were prepared:

(2-(Benzyloxy)-4-methoxyphenyl)(phenyl)methanone (17a). From ketone 16a and benzyl bromide. Flash column chromatography (silica gel, hexanes-EtOAc 4:1) afforded benzyl ether 17a (283 mg, 89%) as a yellow foam. 17a: Rf = 0.35 (silica gel, hexanes-EtOAc 4:1); IR (film) vmax 1651, 1601, 1579, 1501, 1446, 1272, 1166, 1120, 737, 697 cm-1; 1H-NMR (CD3CN): □ = 7.73-7.72 (m, 2 H), 7.57-7.55 (m, 1 H), 7.46-7.41 (m, 3 H), 7.21-7.16 (m, 3 H), 6.93-6.92 (m, 2 H), 6.69 (d, J = 2.4 Hz, 1 H), 6.64 (dd, J = 8.4, 2.4 Hz, 1 H), 4.97 (s, 2 H), 3.84 ppm (s, 3 H); 13C-NMR (CD3CN): □ = 195.6, 163.4, 158.3, 139.2, 136.5, 132.5, 131.6, 129.2, 128.3, 128.2, 127.6, 126.9, 121.7, 105.6, 99.7, 69.8, 55.4 ppm; HRMS (ESI): calcd for C21H18O3Na+ [M + Na+] 341.1148, found 341.1156

(E)-(2-(Benzyloxy)-4-methoxy-6-(4-methoxystyryl)phenyl)(3,5 dimethoxyphenyl)methanone (17b). From ketone 16b and benzyl bromide. Flash column chromatography (silica gel, hexanes-EtOAc 2:1) afforded benzyl ether 17b (485 mg, 95%) as a yellow foam. 17b: Rf = 0.45 (silica gel, hexanes-EtOAc 2:1); IR (film) vmax 1667, 1594, 1511, 1301, 1204, 1156, 1065, 829 cm-1; 1H-NMR (CD3CN): □ = 7.31 (d, J = 9.0 Hz, 2 H), 7.24-7.23 (m, 3 H), 7.14 (d, J = 16.2 Hz, 1 H), 7.04-7.02 (m, 2 H), 6.96 (d, J = 2.4 Hz, 1 H), 6.88 (d, J = 2.4 Hz, 2 H), 6.86 (d, J = 9.0 Hz, 2 H), 6.73-6.70 (m, 2 H), 6.59 (d, J = 1.8 Hz, 1 H), 4.99 (s, 2 H), 3.88 (s, 3 H), 3.76 (s, 3 H), 3.75 ppm (s, 6 H); 13C-NMR (CD3CN): □ = 196.8, 161.5, 161.1, 159.8, 157.2, 140.8, 137.7, 136.6, 131.3, 129.4,

128.3, 127.9, 127.8, 127.2, 122.4, 121.3, 114.1, 106.7, 105.1, 101.7, 99.0, 70.0, 55.3, 55.3, 54.9 ppm; HRMS (ESI): calcd for C32H30O6Na+ [M + Na+] 533.1934, found 533.1951.

(E)-(2-((4-Bromobenzyl)oxy)-4-methoxy-6-(4-methoxystyryl) phenyl)(3,5 dimethoxyphenyl)methanone (17c). From ketone 16b and 4-bromobenzyl bromide. Flash column chromatography (silica gel, hexanes-EtOAc 2:1) afforded benzyl ether 17c (542 mg, 92%) as a yellow foam. 17c: Rf = 0.41 (silica gel, hexanes-EtOAc 2:1); IR (film) vmax 2837, 1666, 1593, 1510, 1300, 1156, 1066, 806 cm-1; 1H- NMR (CD3CN): □ = 7.37 (d, J = 8.4 Hz, 2 H), 7.31 (d, J = 9.0 Hz, 2 H), 7.15 (d, J = 16.2 Hz, 1 H), 6.97 (d, J = 2.4 Hz, 1 H), 6.93 (d, J = 8.4 Hz, 2 H), 6.86 (d, J = 9.0 Hz, 2 H), 6.84 (s, 2 H), 6.72 (d, J = 16.2 Hz, 1 H), 6.70 (t, J = 2.4 Hz, 1 H), 6.58 (d, J = 2.4 Hz, 1 H), 4.95 (s, 2 H), 3.88 (s, 3 H), 3.75 (s, 3 H), 3.74 ppm (s, 6 H); 13C-NMR (CD3CN): □ = 196.8, 161.5, 161.1, 159.8, 157.1, 140.9, 137.9, 135.9, 131.3, 131.3, 129.4, 129.1, 127.9, 122.3, 121.3, 121.1, 114.1, 106.6, 105.1, 101.9, 99.0, 69.3, 55.3, 55.3, 54.9 ppm; HRMS (ESI): calcd for C32H29BrO6Na+ [M + Na+] 611.1039, found 611.1033. (E)-(3,5-Dimethoxyphenyl)(4-methoxy-2-((4-methoxybenzyl)oxy)-6-(4 methoxystyryl)phenyl)-methan one (17d). From ketone 16b and 4-methoxybenzyl chloride. Flash column chromatography (silica gel, hexanes-EtOAc 4:1) afforded benzyl ether 17d (486 mg, 90%) as a pale yellow solid. 17d: Rf = 0.40 (silica gel, hexanes-EtOAc 2:1); m.p. = 118-119 °C (hexanes-EtOAc); IR (film) vmax 2970, 1738, 1594, 1512, 1352, 1302, 1249, 1204, 1156, 1065, 834 cm-1; 1H-NMR (CDCl3): □ = 7.31 (d, J = 9.0 Hz, 2 H), 7.03 (d, J = 15.0 Hz, 1 H), 6.98 (d, J = 2.4 Hz, 2 H), 6.92 (d, J = 9.0 Hz, 2 H), 6.86 (d, J = 2.4 Hz, 1 H), 6.85 (d, J = 14.4 Hz, 1 H), 6.82 (d, J = 9.0 Hz, 2 H), 6.74 (d, J = 9.0 Hz, 2 H), 6.66 (t, J = 2.4 Hz, 1 H), 6.45 (d, J = 2.4 Hz, 1 H), 4.88 (s, 2 H), 3.87 (s, 3 H), 3.77 (s, 6 H), 3.76 (s, 3 H), 3.75 ppm (s, 3 H); 13C-NMR (CDCl3): □ = 197.3, 161.1, 160.6, 159.3, 158.9, 157.3, 140.9, 137.8, 130.8, 129.4, 128.4, 128.2, 127.9, 122.8, 121.6, 113.8, 113.4, 106.9, 105.3, 101.2, 98.9, 69.8, 55.4, 55.3, 55.1, 55.0 ppm; HRMS (ESI): calcd for C33H32O7Na+[M + Na+] 563.2040, found 563.2037

(E)-(3,5-Dimethoxyphenyl)(2-(furan-2-ylmethoxy)-4-methoxy-6-(4 methoxystyryl)phenyl)methanone (17e). To a solution of phenol 16b (420 mg, 1.0 mmol) in THF (10 mL) at 23 °C was added PPh3 (786 mg, 3.0 mmol). The resulting mixture was cooled to 0 °C before a solution of DEAD (522 mg, 3.0 mmol) and furfuryl alcohol (294 mg, 3.0 mmol) in THF (2 mL) were added. The resulting mixture was warmed to 23 °C and stirred for 12 h before it was quenched with NH4Cl (5 mL, sat. aq.). The layers were separated and the aqueous layer was extracted with EtOAc (3 × 10 mL). The combined organic layers were washed with brine (10 mL), dried (Na2SO4) and concentrated in vacuo. Flash column chromatography (silica gel, hexanes-EtOAc 4:1) afforded furanyl ether 17e (355 mg, 71%) as a yellow oil. 17e: Rf = 0.72 (silica gel, benzene-EtOAc 8:1); IR (film) vmax 2937, 1667, 1592, 1510, 1456, 1300, 1155, 1063, 819 cm-1; 1H-NMR (CD3CN): □ = 7.38 (d, J = 1.2 Hz, 1 H), 7.28 (d, J = 9.0 Hz, 2 H), 7.13 (d, J = 16.2 Hz, 1 H), 6.97 (d, J = 2.4 Hz, 1 H), 6.85 (d, J = 9.0 Hz, 2 H), 6.82 (d, J = 2.4 Hz, 2 H), 6.69 (t, J = 2.4 Hz, 1 H), 6.68 (d, J = 2.4 Hz, 1 H), 6.65 (t, J = 16.2 Hz, 1 H), 6.33-6.32 (m, 1 H), 6.28 (d, J = 3.6 Hz, 1 H), 4.95 (s, 2 H), 3.90 (s, 3 H), 3.75 (s, 3 H), 3.74 ppm (s, 6 H); 13C-NMR (CD3CN): □ = 197.4, 162.2, 161.9, 160.6, 157.6, 150.7, 144.2, 141.3, 138.5, 132.2, 130.2, 128.8, 123.1, 122.3, 115.0, 111.3, 111.1, 107.6, 106.0, 103.0, 100.1, 63.5, 56.2, 56.1, 55.8 ppm; HRMS (ESI): calcd for C30H28O7Na+ [M + Na+] 523.1727, found 523.1725. 2:1) afforded benzyl ether 17f (437 mg, 91%) as a yellow foam. 17f: Rf = 0.50 (silica gel, hexanes EtOAc 2:1); IR (film) vmax 1661, 1595, 1511, 1249, 1163, 1033, 827, 721 cm-1; 1H-NMR (CD3CN): □ = 7.77-7.75 (m, 2 H), 7.62-7.59 (m, 1 H), 7.47 (t, J = 7.8 Hz, 2 H), 7.28 (d, J = 9.0 Hz, 2 H), 7.14 (d, J = 16.2 Hz, 1 H), 6.97 (d, J = 1.8 Hz, 1 H), 6.89 (d, J = 8.4 Hz, 2 H), 6.83 (d, J = 9.0 Hz, 2 H), 6.74 (d, J = 8.4 Hz, 2 H), 6.72 (d, J = 16.2 Hz, 1 H), 6.61 (d, J = 1.8 Hz, 1 H), 4.89 (s, 2 H), 3.89 (s, 3 H), 3.74 (s, 3 H), 3.71 ppm (s, 3 H); 13C-NMR (CD3CN): □ = 198.5, 162.8, 161.0, 160.6, 158.7, 139.9, 139.0, 134.7, 132.5, 130.7, 130.4, 130.3, 130.0, 129.7, 129.2, 123.7, 122.8, 115.4, 114.9, 103.0, 100.4, 71.2, 56.6, 56.2, 56.1 ppm; HRMS (ESI): calcd for C31H28O5Na+ [M + Na+] 503.1829, found 503.1817.

(E)-(4-Methoxy-2-((4-methoxybenzyl)oxy)-6-(4-methoxystyryl) phenyl)(3,4,5 trimethoxyphenyl) methan one (17g). From ketone 16g and 4-methoxybenzyl chloride. Flash column chromatography (silica gel, hexanes-EtOAc 2:1) afforded benzyl ether 17g (542 mg, 95%) as a yellow foam. 17g: Rf = 0.30 (silica gel, hexanes-EtOAc 2:1); IR (film) vmax 1655, 1595, 1512, 1413, 1327, 1249, 1157, 1126, 832 cm-1; 1H-NMR (d6-acetone): □ 7.35 (d, J = 9.0 Hz, 2 H), 7.22 (d, J = 15.6 Hz, 1 H), 7.08 (s, 2 H), 7.05 (d, J = 2.4 Hz, 1 H), 6.97 (d, J = 8.4 Hz, 2 H), 6.88 (d, J = 8.4 Hz, 2 H), 6.85 (d, J = 15.6 Hz, 1 H), 6.77 (d, J = 8.4 Hz, 2 H), 6.68 (d, J = 2.4 Hz, 1 H), 4.99 (s, 2 H), 3.91 (s, 3 H), 3.81 (s, 3

H), 3.77 (s, 9 H), 3.74 ppm (s, 3 H); 13C-NMR (d6-acetone): □ = 195.3, 161.4, 159.8, 159.3, 157.5, 153.4, 142.8, 137.7, 134.4, 130.8, 129.6, 128.7, 128.6, 127.9, 122.7, 121.6, 114.0, 113.4, 106.6, 101.4, 99.0, 69.6, 59.8, 55.6, 54.9, 54.6, 54.5 ppm; HRMS (ESI): calcd for C34H34O8Na+ [M + Na+] 593.2145, found 593.2137.

(E)-(3,4-Dimethoxyphenyl)(4-methoxy-2-((4-methoxybenzyl) oxy)-6-(4 methoxystyryl)phenyl)methan one (17h). From ketone 16h and 4-methoxybenzyl chloride. Flash column chromatography (silica gel, hexanes-EtOAc 2:1) afforded benzyl ether 17h (486 mg, 90%) as a yellow foam. 17h: Rf = 0.25 (silica gel, hexanes-EtOAc 2:1); IR (film) vmax 1652, 1594, 1510, 1265, 1249, 1159, 1121, 1024, 833 cm-1; 1H-NMR (CD3CN): □ 7.45 (d, J = 1.8 Hz, 1 H), 7.27 (d, J = 9.0 Hz, 2 H), 7.20 (dd, J = 8.4, 1.8 Hz, 1 H), 7.13 (d, J = 16.2 Hz, 1 H), 6.98 (d, J = 8.4 Hz, 2 H), 6.95 (d, J = 2.4 Hz, 1 H), 6.89 (d, J = 8.4 Hz, 1 H), 6.83 (d, J = 9.0 Hz, 2 H), 6.76 (d, J = 9.0 Hz, 2 H), 6.70 (d, J = 16.2 Hz, 1 H), 6.60 (d, J = 2.4 Hz, 1 H), 4.91 (s, 2 H), 3.87 (s, 3 H), 3.83 (s, 3 H), 3.80 (s, 3 H), 3.73 (s, 3 H), 3.72 ppm (s, 3 H); 13C NMR (CD3CN): □ = 195.7, 161.2, 159.7, 159.3, 157.1, 153.8, 149.2, 137.3, 131.6, 130.9, 129.4, 129.0, 128.5, 127.8, 124.9, 122.5, 121.9, 114.1, 113.6, 110.5, 110.2, 101.4, 99.2, 69.9, 55.6, 55.3, 55.3, 54.9, 54.8 ppm; HRMS (ESI): calcd for C33H32O7Na+ [M + Na+] 563.2040, found 563.2057.

(3,5-Dimethoxyphenyl)(4-methoxy-2-((4-methoxybenzyl)oxy)-6-((4-methoxyphenyl)ethynyl)phenyl) methanone (17i). From ketone 16i and 4-methoxybenzyl chloride. Flash column chromatography (silica gel, hexanes-EtOAc 2:1) afforded benzyl ether 17i (468 mg, 87%) as a yellow foam. 17i: Rf = 0.27 (silica gel, hexanes-EtOAc 2:1); IR (film) vmax 2937, 1671, 1591, 1511, 1300, 1248, 1155,

1063, 832 cm-1; 1H -NMR (CD3CN): □ = 7.12 (d, J = 8.4 Hz, 2 H), 7.07 (d, J = 9.0 Hz, 2 H), 6.88 (d, J = 2.4 Hz, 2 H), 6.84 (d, J = 9.0 Hz, 2 H), 6.81 (d, J = 8.4 Hz, 2 H), 6.74 (d, J = 2.4 Hz, 1 H), 6.72 (t, J = 2.4 Hz, 1 H), 6.71 (d, J = 2.4 Hz, 1 H), 4.97 (s, 2 H), 3.85 (s, 3 H), 3.77 (s, 6 H), 3.76 (s, 3 H), 3.74 ppm (s, 3 H); 13C-NMR (CD3CN): □ = 195.8, 162.2, 162.0, 161.0, 160.3, 157.8, 140.9, 133.6, 130.0, 129.1, 125.5, 123.4, 115.0, 114.8, 114.5, 108.9, 107.6, 106.0, 101.8, 94.1, 86.4, 70.9, 56.3, 56.2, 55.9, 55.7 ppm; HRMS (ESI): calcd for C33H30O7Na+ [M + Na+] 561.1883, found 561.1883.

General procedure C (Preparation of benzofurans 19, Table 2)

To a solution of benzyl ether 17 (0.2 mmol) in THF (2 mL) at 0

°C was added LiTMP (0.5 M in THF, 2 mL, 1.0 mmol). The resulting mixture was stirred at 0 °C and the progress was monitored by TLC analysis. Upon completion of the reaction (~ 2 h for most cases), the reaction mixture was quenched with NH4Cl (5 mL, sat. aq.). The layers were separated and the aqueous layer was extracted with EtOAc (3 × 5 mL). The combined organic layers were washed with brine (10 mL), dried (Na_2SO_4) and concentrated in vacuo to afford crude tertiary alcohol 18, which was used directly without further purification. To a solution of the crude tertiary alcohol 18 (obtained as above) in CH_2C_{12} (3 mL) at 23 °C was added p-TsOH•H2O (38 mg, 0.2 mmol). The resulting mixture was stirred for 1 h before it was quenched with NaHCO3 (3 mL, sat. aq.). The layers were separated and the aqueous layer was extracted with CH2Cl2 (3 × 5 mL). The combined organic layers were washed with brine (5 mL), dried (Na2SO4) and concentrated in vacuo. Flash column chromatography (silica gel) afforded the desired benzofuran 19. Using this general procedure the following compounds were prepared:

6-Methoxy-2,3-diphenylbenzofuran (19a). From benzyl ether 17a. Flash column chromatography (silica gel, hexanes-EtOAc 4:1) afforded benzofuran 19a (50 mg, 83%) as a pale yellow oil. All physical properties of this compound were identical to those reported in literature [21].

(E)-3-(3,5-Dimethoxyphenyl)-6-methoxy-4-(4-methoxystyryl)-2-phenylbenzofuran (19b). From benzyl ether 17b. Flash column chromatography (silica gel, hexanes-EtOAc 4:1) afforded benzofuran 19b (70 mg, 71%) as a yellow oil. 19b: Rf = 0.42 (silica gel, hexanes-EtOAc 2:1); IR (film) vmax 1601, 1510, 1420, 1249, 1204, 1143, 1064, 1033, 808, 693 cm-1; 1H-NMR (CD3CN): □ = 7.58-7.57 (m, 2 H), 7.32-7.25 (m, 3 H), 7.15 (d, J = 2.4 Hz, 1 H), 7.06 (d, J = 2.4 Hz, 1 H), 7.01 (d, J = 16.2 Hz, 1 H), 6.99 (d, J = 8.4 Hz, 2 H), 6.83 (d, J = 9.0 Hz, 2 H), 6.79 (d, J = 16.2 Hz, 1 H), 6.72 (t, J = 2.4 Hz, 1 H),

6.65 (d, J = 2.4 Hz, 2 H), 3.88 (s, 3 H), 3.77 (s, 3H), 3.74 ppm (s, 6 H); 13C-NMR (CD3CN):□ = 163.1, 160.8, 159.9, 156.3, 150.6, 138.0, 133.6, 131.9, 131.2, 130.1, 129.8, 129.3, 128.8, 127.1, 123.4, 122.8, 119.3, 115.3, 109.9, 108.0, 101.0, 96.1, 56.8, 56.6, 56.3 ppm; HRMS (ESI): calcd for C32H28O5Na+ [M + Na+] 515.1829, found 515.1837.

(E)-3-(3,5-Dimethoxyphenyl)-6-methoxy-2-(4-methoxyphenyl)-4-(4-methoxystyryl)benzofuran (19d). From benzyl ether 17d. Flash column chromatography (silica gel, hexanes:-EtOAc 2:1) afforded benzofuran 19d (91 mg, 87%) as a yellow solid. 19d: Rf = 0.52 (silica gel, hexanes-EtOAc 2:1); m.p. = 61-62 °C (hexanes-EtOAc); IR (film) vmax 2936, 1603, 1509, 1250, 1153, 1143, 1033, 833, 808 cm-1; 1H-NMR (CD3CN): □ = 7.43 (d, J = 8.4 Hz, 2 H), 7.07 (d, J = 2.4 Hz, 1 H), 6.95-6.89 (m, 4 H), 6.78-6.75 (m, 5 H), 6.68 (t, J = 2.4 Hz, 1 H), 6.59 (d, J = 2.4 Hz, 2 H), 3.82 (s, 3 H), 3.73 (s, 3 H), 3.70 (s, 6 H), 3.69 ppm (s, 3 H); 13C-NMR (CD3CN): □ = 161.6, 159.4, 159.3, 158.1, 154.8, 149.4, 137.0, 131.8, 130.0, 128.4, 127.4, 127.3, 123.1, 122.4, 121.7, 116.2, 113.9, 113.8, 108.7, 106.4, 99.5, 94.7, 55.3, 55.2, 54.9, 54.8 ppm; HRMS (ESI): calcd for C33H30O6Na+ [M + Na+] 545.1934, found 545.1951.

(E)-3-(3,5-Dimethoxyphenyl)-2-(furan-2-yl)-6-methoxy-4-(4-methoxystyryl)benzofuran (19e). From furanyl ether 17e. Flash column chromatography (silica gel, hexanes-EtOAc 2:1) afforded benzofuran 19e (36 mg, 38%) as a yellow oil. 19e: Rf = 0.40 (silica gel, hexanes-EtOAc 2:1); IR (film) vmax 2936, 1600, 1510, 1421, 1250, 1152, 1064, 819 cm-1; 1H-NMR (CD3CN): □ = 7.50 (d, J = 1.2 Hz, 1 H), 7.16 (d, J = 2.4 Hz, 1 H), 7.06 (d, J = 2.4 Hz, 1 H), 7.02 (d, J = 15.6 Hz, 1 H), 7.01 (d, J = 8.4 Hz, 2 H), 6.84-6.82 (m, 3 H), 6.70 (t, J = 2.4 Hz, 1 H), 6.62 (d, J = 2.4 Hz, 2 H), 6.46 (q, J = 1.8 Hz, 1 H), 6.38 (d, J = 3.0 Hz, 1 H), 3.88 (s, 3 H), 3.77 (s, 3 H), 3.76 ppm (s, 6 H); 13C-NMR (CD3CN): □ = 162.2, 160.4, 159.5, 156.1, 146.2, 143.9, 143.4, 136.3, 133.1, 130.7, 129.8, 128.3, 123.0, 121.5, 114.9, 112.5, 109.4, 107.7, 100.6, 95.8, 56.4, 56.1, 55.8 ppm; HRMS (ESI): calcd for C30H26O6Na+ [M + Na+] 505.1621, found 505.1603.

(E)-6-Methoxy-2-(4-methoxyphenyl)-4-(4-methoxystyryl)-3-phenylbenzofuran (19f). According to General Procedure C using benzyl ether 17f. Flash column chromatography (silica gel, hexanes-EtOAc 4:1) afforded benzofuran 19f (74 mg, 80%) as a yellow oil. 19f: Rf = 0.60 (silica gel, hexanes-EtOAc 2:1); IR (film) vmax 1605, 1511, 1251, 1175, 1144, 1033, 833, 702 cm-1; 1H-NMR (d6-acetone): □ = 7.62-7.59 (m, 3 H), 7.52-7.50 (m, 2 H), 7.45 (d, J = 9.0 Hz, 2 H), 7.18 (d, J = 2.4 Hz, 1 H), 7.09 (d, J = 2.4 Hz, 1 H), 7.03 (d, J = 16.2 Hz, 1 H), 6.96 (d, J = 8.4 Hz, 2 H), 6.86 (d, J = 9.0 Hz, 2 H), 6.80 (d, J = 8.4 Hz, 2 H), 6.73 (d, J = 16.2 Hz, 1 H), 3.91 (s, 3 H), 3.77 ppm (s, 6 H); 13C-NMR (d6-acetone): □ = 159.5, 159.5, 158.4, 155.0, 149.7, 135.1, 131.9, 130.7, 129.9, 129.3, 128.6, 128.0, 127.6, 127.3, 123.2, 121.9, 121.8, 116.4, 113.8, 113.8, 106.5, 94.7, 55.2, 54.7, 54.6 ppm; HRMS (ESI): calcd for C31H26O4Na+ [M + Na+] 485.1723, found 485.1713.

(E)-6-Methoxy-2-(4-methoxyphenyl)-4-(4-methoxystyryl)-3-(3,4,5-trimethoxyphenyl)benzofuran (19g).

According to General Procedure C using benzyl ether 17g. Flash column chromatography (silica gel, hexanes:EtOAc 2:1) afforded benzofuran 19g (94 mg, 85%) as a yellow oil. 19g: Rf = 0.34 (silica gel, hexanes:EtOAc 2:1); IR (film) vmax 2936, 1604, 1511, 1409, 1250, 1126, 1033, 838 cm-1; 1H-NMR (CD3CN): □ = 7.50 (d, J = 9.0 Hz, 2 H), 7.14 (d, J = 1.8 Hz, 1 H), 7.05-6.99 (m, 4 H), 6.86 (d, J = 9.0 Hz, 2 H), 6.81 (d, J = 16.2 Hz, 1 H), 6.78 (d, J = 9.0 Hz, 2 H), 6.75 (s, 2 H), 3.88 (s, 3 H), 3.86 (s, 3 H), 3.76 (s, 3 H), 3.75 (s, 3 H), 3.68 ppm (s, 6 H); 13C-NMR (CD3CN): □ = 159.5, 159.4, 158.2, 154.8, 154.0, 149.7, 137.9, 132.0, 130.1, 130.0, 128.8, 127.4, 127.3, 123.2, 122.3, 121.7, 116.4, 114.0, 113.9, 107.9, 106.6, 94.8, 60.3, 55.9, 55.4, 54.9, 54.9 ppm; HRMS (ESI): calcd for C34H32O7Na+ [M + Na+] 575.2040, found 575.2048. (E)-3-(3,4-Dimethoxyphenyl)-6-methoxy-2-(4-methoxyphenyl)-4-(4 methoxystyryl)benzofuran (19h). From benzyl ether 17h. Flash column chromatography (silica gel, hexanes-EtOAc 2:1) afforded benzofuran 19h (85 mg, 81%) as a yellow oil. 19h: Rf = 0.35 (silica gel, hexanes-EtOAc 2:1); IR (film)vmax 2923, 1604, 1509, 1247, 1138, 1026, 833, 734 cm-1; 1H-NMR (CD3CN): □ = 7.49 (d, J = 9.0 Hz, 2 H), 7.12 (d, J = 1.8 Hz, 1 H), 7.08 (d, J = 8.4 Hz, 1 H), 7.03 (d, J = 2.4 Hz, 1 H), 7.02 (d, J = 2.4 Hz, 1 H), 6.99-6.96 (m, 4 H), 6.85 (d, J = 9.0 Hz, 2 H), 6.79 (d, J = 9.0 Hz, 2 H), 6.73 (d, J = 16.2 Hz, 1 H), 3.92 (s, 3 H), 3.88 (s, 3 H), 3.77 (s, 3 H), 3.76 (s, 3 H), 3.67 ppm (s, 3 H); 13C-NMR (CD3CN): □ = 159.4, 159.4, 158.2, 154.8, 149.9, 149.8, 149.3, 132.0, 130.0, 128.5, 127.5, 127.3, 126.9, 123.3, 122.9, 122.3, 122.0, 116.3, 114.2, 113.9, 113.9, 112.2, 106.4, 94.8, 55.6, 55.5, 55.4, 54.9 ppm; HRMS (ESI): calcd for C33H30O6Na+ [M + Na+] 545.1934, found 545.1946.

3-(3,5-Dimethoxyphenyl)-6-methoxy-2-(4-methoxyphenyl)-4-((4 methoxyphenyl)ethynyl)benzofuran (19i). From benzyl ether 17i. Flash column chromatography (silica gel, hexanes-EtOAc 4:1) afforded benzofuran 19i (88 mg, 85%) as a yellow oil. 19i: Rf = 0.70 (silica gel, hexanes-EtOAc 2:1); IR (film) vmax 2935, 1604, 1510, 1485, 1248, 1152, 1035, 831 cm-1; 1H-NMR (CD3CN): □ = 7.47 (d, J = 9.0 Hz, 2 H), 7.16 (d, J = 2.4 Hz, 1 H), 7.02 (d, J = 8.4 Hz, 2 H), 6.99 (d, J = 2.4 Hz, 1 H), 6.87 (d, J = 9.0 Hz, 2 H), 6.85 (d, J = 8.4 Hz, 2 H), 6.64 (d, J = 2.4 Hz, 2 H), 6.56 (t, J = 2.4 Hz, 1 H), 3.87 (s, 3 H), 3.79 (s, 3 H), 3.77 (s, 3 H), 3.70 ppm (s, 6 H); 13C-NMR (CD3CN): □ = 161.8,

160.7, 160.6, 158.5, 155.4, 151.5, 135.8, 133.7, 128.6, 123.9, 123.7, 117.3, 116.4, 116.0, 115.5, 114.8, 114.7, 109.9, 100.4, 97.6, 94.6, 86.1, 56.5, 55.9, 55.8 ppm; HRMS (ESI): calcd for C33H28O6Na+ [M + Na+] 543.1778, found 543.1773

General procedure D (Preparation of hexacyclic benzofurans 20, Table 3)

To a solution of benzofuran 19 (0.04 mmol) in CH2Cl2 (6 mL) at 23 °C was added p-TsOH•H2O (22.8 mg, 0.12 mmol). The resulting mixture was heated to 40 °C and stirred for 8 hours before it was quenched with NaHCO3 (3 mL, sat. aq.). The layers were separated and the aqueous layer was extracted with CH2Cl2 (3 × 5 mL). The combined organic layers were washed with brine (5 mL), dried (Na_2SO_4) and concentrated in vacuo. Flash column chromatography (silica gel) afforded the desired hexacyclic benzofuran 20. The following compounds were prepared via this general procedure:

1,3,8-Trimethoxy-11-(4-methoxyphenyl)-5-phenyl-10,11 dihydrobenzo[6,7]cyclohepta[1,2,3-cd]benzo furan (20a). From benzofuran 19b. Flash column chromatography (silica gel, hexanes-EtOAc 4:1) afforded hexacyclic benzofuran 20a (18.7 mg, 95%) as a yellow oil. 20a: Rf = 0.57 (silica gel, hexanes-EtOAc 2:1); IR (film) vmax 2933, 1598, 1509, 1461, 1248, 1144, 1065, 853, 696 cm-1; 1H-NMR (CD3CN): □ = 7.66-7.64 (m, 2 H), 7.47-7.41 (m, 3 H), 7.01 (d, J = 8.4 Hz, 2 H), 6.78 (d, J = 1.8 Hz, 1 H), 6.71 (d, J = 1.8 Hz, 1 H), 6.59 (d, J = 8.4 Hz, 2 H), 6.58 (d, J = 2.4 Hz, 1 H), 6.53 (d, J = 2.4 Hz, 1 H), 5.44 (d, J = 5.4 Hz, 1 H), 3.81 (s, 3 H), 3.78 (s, 3 H), 3.72 (dd, J = 16.2, 5.4 Hz, 1 H), 3.58 (s, 3 H), 3.44 (s, 3 H), 3.37 ppm (d, J = 16.2 Hz, 1 H); 13C-NMR (CD3CN): □ = 158.6, 158.3, 158.1, 157.2, 154.2, 150.6, 135.4, 134.1, 133.9, 131.7, 128.9, 128.7, 128.7, 128.3, 123.9, 119.8, 112.9, 112.6, 107.2, 97.8, 92.7, 55.8, 55.2, 54.5, 54.4, 36.7, 36.6 ppm; HRMS (ESI): calcd for C32H28O5Na+ [M + Na+] 515.1829, found 515.1835.

1,3,8-Trimethoxy-5,11-bis(4-methoxyphenyl)-10,11-dihydrobenzo[6,7]cyclohepta[1,2,3-cd]benzofuran (20b). From benzofuran 19d. Flash column chromatography (silica gel, hexanes-EtOAc 4:1) afforded hexacyclic benzofuran 20b (18.8 mg, 90%) as a yellow oil. 20b: Rf = 0.49 (silica gel, hexanes-EtOAc 2:1); IR (film) vmax 2933, 1599, 1508, 1460, 1248, 1143, 1067, 1032, 834 cm-1; 1H-NMR

(CD3CN): □ = 7.57 (d, J = 9.0 Hz, 2 H), 7.00-6.98 (m, 4 H), 6.75 (d, J = 1.8 Hz, 1 H), 6.69 (d, J = 1.8 Hz, 1 H), 6.61 (d, J = 2.4 Hz, 1 H), 6.58 (d, J = 9.0 Hz, 2 H), 6.51 (d, J = 2.4 Hz, 1 H), 5.43 (d, J = 6.0 Hz, 1 H), 3.83 (s, 3 H), 3.80 (s, 3 H), 3.76 (s, 3 H), 3.69 (dd, J = 16.2, 6.0 Hz, 1 H), 3.57 (s, 3 H), 3.47 (s, 3 H), 3.35 ppm (d, J = 16.2 Hz, 1 H); 13C-NMR (CD3CN): □ = 160.2, 158.6, 158.3, 157.8, 157.2, 154.0, 150.7, 135.2, 134.2, 134.1, 130.2, 128.3, 123.9, 123.8, 119.9, 116.1, 114.1, 112.9, 112.3, 106.9, 97.6, 92.6, 55.8, 55.2, 55.1, 54.5, 54.4, 36.7, 36.6 ppm; HRMS (ESI): calcd for C33H30O6Na+ [M + Na+] 545.1934, found 545.1948.

5-(Furan-2-yl)-1,3,8-trimethoxy-11-(4-methoxyphenyl)-10,11 dihydrobenzo[6,7]cyclohepta[1,2,3-cd] benzofuran (20c). From benzofuran 19e. The desired product 20c was not obtained in this reaction. 8-Methoxy-5,11-bis(4-methoxyphenyl)-10,11-dihydrobenzo[6,7]cyclohepta[1,2,3-cd]benzofuran (20d). From benzofuran 19f. The desired product 20d was not obtained in this reaction. 1,2,3,8-Tetramethoxy-5,11-bis(4-methoxyphenyl)-10,11-dihydrobenzo[6,7]cyclohepta[1,2,3-cd]benzo furan (20e). From benzofuran 19g. Flash column chromatography (silica gel, hexanes-EtOAc 4:1) afforded hexacyclic benzofuran 20e (20.3 mg, 92%) as a yellow oil. 20e: Rf = 0.48 (silica gel, hexanes EtOAc 2:1); IR (film) vmax 2933, 1611, 1508, 1316, 1248, 1143, 1037, 835 cm-1; 1H-NMR (CD3CN): □ = 7.60 (d, J = 8.4 Hz, 2 H), 7.02-7.00 (m, 4 H), 6.89 (s, 1 H), 6.75 (s, 1 H), 6.69 (s, 1 H), 6.58 (d, J = 8.4 Hz, 2 H), 5.31 (d, J = 5.4 Hz, 1 H), 3.84 (s, 3 H), 3.83 (s, 3 H), 3.76 (s, 3 H), 3.74 (dd, J = 16.2, 5.4 Hz, 1 H), 3.73 (s, 3 H), 3.40 (d, J = 16.2 Hz, 1 H), 3.39 ppm (s, 3 H); 13C-NMR (CD3CN): □ = 160.2, 157.9, 157.2, 154.0, 152.2, 151.2, 150.1, 141.4, 134.9, 134.1, 130.1, 129.4, 128.3, 128.0, 123.9, 119.9, 115.8, 114.1, 112.9, 112.4, 110.2, 92.7, 61.2, 60.2, 55.2, 55.1, 54.8, 54.5, 38.0, 36.8 ppm; HRMS (ESI): calcd for C33H30O6Na+ [M + Na+] 545.1934, found 545.1948

2,3,8-Trimethoxy-5,11-bis(4-methoxyphenyl)-10,11-dihydrobenzo[6,7]cyclohepta[1,2,3-cd]benzofuran (20f). From benzofuran 19h. Flash column chromatography (silica gel, hexanes-EtOAc 2:1) afforded hexacyclic benzofuran 20f (18.8 mg, 90%) as a yellow oil. 20f: Rf = 0.35 (silica gel, hexanes-EtOAc 2:1); IR (film) vmax 2934, 1610, 1511, 1499, 1251, 1177, 1143, 1035, 834, 793 cm-1; 1H-NMR (CD3CN): □ = 8.07 (d, J = 9.0 Hz, 2 H), 7.23 (s, 1 H), 7.08 (d, J = 9.0 Hz, 2 H), 6.86 (d, J = 8.4 Hz, 2 H), 6.70 (d, J = 8.4 Hz, 2 H), 6.64 (d, J = 2.4 Hz, 1 H), 6.57 (s, 1 H), 6.21 (d, J = 2.4 Hz, 1 H), 4.63 (dd, J =

6.6, 2.4 Hz, 1 H), 3.90 (s, 3 H), 3.74 (s, 3 H), 3.69 (s, 3 H), 3.66 (s, 3 H), 3.65 (s, 3 H), 3.62 (dd, J = 14.4, 2.4 Hz, 1 H), 3.15 ppm (dd, J = 14.4, 6.6 Hz, 1 H); 13C-NMR (CD3CN): □ = 193.6, 165.3, 165.0, 162.1, 158.8, 153.5, 150.5, 148.4, 139.9, 139.6, 136.3, 132.9, 131.8, 130.4, 127.8, 122.6, 114.9, 114.6, 114.1, 113.4, 112.7, 107.6, 56.3, 56.3, 56.2, 56.1, 55.6, 49.7, 41.4 ppm; HRMS (ESI): calcd for C33H30O6Na+ [M + Na+] 545.1934, found 545.1943.

1,3,8-Trimethoxy-5,11-bis(4-methoxyphenyl)benzo[6,7] cyclohepta[1,2,3-cd]benzofuran (20g). From benzofuran 19i. Flash column chromatography (silica gel, hexanes-EtOAc 4:1) afforded hexacyclic benzofuran 20g (19.8 mg, 95%) as a yellow oil. 20g: Rf = 0.70 (silica gel, hexanes-EtOAc 2:1); IR (film) vmax 2919, 1738, 1606, 1505, 1462, 1365, 1247, 1199, 833 cm-1; 1H-NMR (CD3CN): □ = 7.82 (d, J = 9.0 Hz, 2 H), 7.21 (d, J = 9.0 Hz, 2 H), 7.08 (d, J = 9.0 Hz, 2 H), 6.86 (d, J = 9.0 Hz, 2 H), 6.78 (d, J = 1.8 Hz, 1 H), 6.64 (d, J = 1.8 Hz, 1 H), 6.62 (d, J = 2.4 Hz, 1 H), 6.59 (s, 1 H), 6.29 (d, J = 2.4 Hz, 1 H), 3.86 (s, 3 H), 3.80 (s, 3 H), 3.78 (s, 3 H), 3.43 (s, 3 H), 3.17 ppm (s, 3 H); 13C-NMR (CD3CN): □ = 161.4, 161.4, 161.3, 159.8, 158.8, 155.0, 151.9, 141.9, 141.7, 138.2, 133.8, 132.4, 131.0, 127.4, 127.1, 125.0, 120.6, 116.4, 115.1, 114.1, 111.9, 106.2, 100.3, 94.6, 56.2, 56.0, 56.0, 55.7, 55.3 ppm; HRMS (ESI): calcd for C33H30O6Na+ [M + Na+] 543.1784, found 543.1745.

ONE-POT PREPARATION OF HEXACYCLIC BENZOFURAN 20B

To a solution of benzyl ketone 17d (100 mg, 0.185 mmol) in THF (3 mL) at 0 °C was added LiTMP (0.5 M in THF, 1.9 mL, 0.93 mmol). The resulting mixture was stirred at 0 °C for 30 min before it was quenched with NH4Cl (5 mL, sat. aq.). The layers were separated and the aqueous layer was extracted with EtOAc (3 × 5 mL). The combined organic layers were washed with brine (5 mL), dried (Na2SO4) and concentrated in vacuo to afford crude alcohol, which was used directly without further purification. To a solution of the crude alcohol (obtained as above) in CH2Cl2 (3 mL) at 23 °C was added p-TsOH•H2O (105 mg, 0.56 mmol). The resulting mixture was heated to 45 °C and stirred for 8 h before it was quenched with NaHCO3 (3 mL, sat. aq.). The layers were separated and the aqueous layer was extracted with CH2C12 (3 × 5 mL). The combined organic layers were washed with brine (5 mL), dried (Na2SO4) and concentrated in vacuo. Flash column chromatography (silica gel, hexanes-EtOAc 2:1) afforded the desired hexacyclic benzofuran 20b (77 mg, 80%) as a yellow oil

(E)-(3,5-Dimethoxyphenyl)(2-hydroxy-4-methoxy-6-(4-methoxystyryl)phenyl)methanone (16b'). To a solution of ketone 16b (13.0 g, 30 mmol) in CH2C12 (100 mL) at 0 °C was added BC13 (1 M in CH2C12, 45 mL, 45 mmol) dropwise. The resulting mixture was stirred for 1 h before it was quenched with NaHCO3 (50 mL, sat. aq.). The layers were separated and the aqueous layer was extracted with CH2C12 (3 × 50 mL). The combined organic layers were washed with brine (100 mL), dried (Na2SO4) and concentrated in vacuo. Flash column chromatography (silica gel, hexanes-EtOAc-CH_2C_{12} 4:1:1) afforded phenol 16b' (12 g, 95%) as a yellow solid. 16b': Rf = 0.45 (silica gel, hexanes-EtOAc 2:1); m.p. = 139-140 °C (hexanes-EtOAc); IR (film) vmax 2939, 1600, 1511, 1457, 1254, 1204, 1157, 1064, 840, 808 cm-1; 1H-NMR (CDC13): □ = 11.5 (br s, 1 H), 6.89 (d, J = 8.4 Hz, 2 H), 6.74 (d, J = 8.4 Hz, 2 H), 6.72 (d, J = 2.4 Hz, 2 H), 6.67 (d, J = 2.4 Hz, 1 H), 6.65 (d, J = 16.2 Hz, 1 H), 6.48 (t, J = 2.4 Hz, 1 H), 6.47 (d, J = 2.4 Hz, 1 H), 6.46 (d, J = 15.6 Hz, 1 H), 3.88 (s, 3 H), 3.78 (s, 3 H), 3.68 ppm (s, 6 H); 13C-NMR (CDC13): □ = 200.1, 164.7, 164.6, 160.6, 159.4, 142.7, 142.7, 130.0, 129.5, 127.7, 126.9, 113.8, 113.3, 106.9, 106.3, 104.4, 99.9, 55.6, 55.2 ppm; HRMS (ESI): calcd for C25H24O6Na+ [M + Na+] 443.1465, found 443.1454.

3-(3,5-Dimethoxyphenyl)-6-methoxy-2-(4-methoxyphenyl)-4-(3-(4 methoxyphenyl)oxiran-2-yl)benzo furan (21). To a solution of benzofuran 19d (4.20 g, 8.04 mmol) in DMSO (50 mL) and water (10 mL) at 0 °C was added NBS (1.57 g, 8.84mmol) in one portion. The resulting mixture was stirred for 0.5 h before it was quenching with Na2S2O3 (50 mL, sat. aq.). The layers were separated and the aqueous layer was extracted with Et2O (3 × 100 mL). The combined organic layers were washed with brine

(100 mL), dried (Na2SO4) and concentrated in vacuo to afford the crude bromohydrin, which was use directly without further purification. To a solution of crude bromohydrin (obtained as above) in Et2O (100 mL) at 23 °C were added NaOH (4 M, aq., 30 mL) and PhEt3NCl (1.83 g, 8.04 mmol). The resulting mixture was stirred

for 2 h before the layers were separated, and the aqueous layer was extracted with Et2O (3 × 100 mL). The combined organic layers were washed with brine (100 mL), dried (Na2SO4) and concentrated in vacuo. Flash column chromatography (silica gel,hexanes-EtOAc 2:1) afforded epoxide 21 (3.25 g, 75%, over the two steps) as a yellow solid. 21: Rf = 0.55 (silica gel, hexanes-EtOAc 2:1); m.p. = 147-148 °C (hexanes/CH2C12); IR (film) vmax 2939, 1738, 1611, 1587, 1512, 1204, 1154, 1033, 832 cm-1; 1H-NMR (CDC13): □ = 7.47 (d, J = 9.0 Hz, 2 H), 7.02 (d, J = 1.8 Hz, 1 H), 6.89 (d, J = 8.4 Hz, 2 H), 6.86 (d, J = 1.8 Hz, 1 H), 6.80 (d, J = 8.4 Hz, 2 H), 6.79 (d, J = 9.0 Hz, 2 H), 6.54 (br, 1 H), 6.31 (br, 1 H), 6.18 (t, J = 1.8 Hz, 1 H), 3.89 (s, 3 H), 3.83 (s, 3 H), 3.79 (d, J = 2.4 Hz, 1 H), 3.77 (s, 3 H), 3.74 (br, 3 H), 3.53 (d, J = 2.4 Hz, 1 H), 3.30 ppm (br, 3 H); 13C-NMR (CDC13): □ = 161.0, 159.7, 159.2, 158.2, 154.2, 149.7, 136.0, 131.5, 128.5, 127.2, 126.9, 123.2, 122.5, 115.4, 113.8, 113.6, 107.7 (br), 105.9, 99.7, 95.3, 63.1, 59.1, 55.8, 55.3, 55.2, 54.7 (br) ppm; HRMS (ESI): calcd for C33H30O7Na+ [M + Na+] 561.1883, found 561.1898.

Malibatol A (2): To a solution of epoxide 21 (100 mg, 0.19 mmol) in CH2C12 (30 mL) at -78 °C was added BBr3 (1.0 M in CH2C12, 2.28 mL, 2.28 mmol). The resulting mixture was warmed to 23 °C and stirred for 2 h before it was quenched with NaHCO3 (10 mL, sat. aq.). The layers were separated and the aqueous layer was extracted with EtOAc (3 × 20 mL). The combined organic layers were washed with brine (20 mL), dried (Na2SO4) and concentrated in vacuo. Flash column chromatography (silica gel, CH2C12-MeOH 9:1) afforded malibatol (2, 17.8 mg, 20%) as a tan oil. Compound 2: Rf = 0.23 (silica gel, CH2C12-MeOH 9:1); IR (film) vmax 3323, 2918, 1612, 1510, 1433, 1366, 1231, 1139, 833 cm-1; 1

H-NMR (CD3OD): □ = 7.45 (d, J = 8.6 Hz, 2 H), 7.02 (d, J = 8.6 Hz, 2 H), 7.01 (d, J = 2.4 Hz, 1 H), 6.80 (d, J = 8.6 Hz, 2 H), 6.57 (dd, J = 2.4, 1.2 Hz, 1 H), 6.51 (d, J = 2.4 Hz, 1 H),

6.33 (d, J = 9.0 Hz, 2 H), 6.30 (d, J = 2.4 Hz, 1 H), 5.46 (brs, 1 H), 5.28 ppm (m, 1 H); 13C-NMR (CD3OD): □ = 159.1, 157.4, 156.7, 156.2, 155.3, 155.1, 151.2, 139.6, 135.8, 133.4, 130.9, 130.6, 124.6, 121.2, 119.0, 117.3, 116.4, 114.7, 109.9, 109.7, 102.1, 95.9, 74.8, 48.9 ppm; HRMS (ESI): calcd for C28H20O7Na+ [M + Na+] 491.1101, found 491.1092

Shoreaphenol (3): To a solution of malibatol A (2) (5 mg, 10.7 μmol) in THF (1 mL) at □□ °C was added PDC (4.8 mg, 12.8 μmol).

The resulting mixture was stirred for 1 h before it was quenched with Na2S2O3 (1 mL, sat. aq.). The layers were separated and the aqueous layer was extracted with EtOAc (3 × 2 mL). The combined organic layers were washed with brine (3 mL), dried (Na2SO4) and concentrated in vacuo. Flash column chromatography (silica gel, CH2Cl2-MeOH 5:1) afforded shoreaphenol (3, 2.3 mg, 46%) as a yellow oil. Compound 3: Rf = 0.26 (silica gel, CH2Cl2-MeOH 9:1); IR (film) vmax 3339, 1738, 1612, 1366, 1216, 829 cm-1; 1H-NMR (d6-acetone): □ = 7.70 (d, J = 8.4 Hz, 2 H), 7.33 (d, J = 2.4 Hz, 1 H), 7.04 (d, J = 1.8 Hz, 1 H), 6.98 (d, J = 8.4 Hz, 2 H), 6.85 (d, J = 7.8 Hz, 2 H), 6.70 (d, J = 2.4 Hz, 1 H), 6.57 (d, J = 2.4 Hz, 1 H), 6.55 (d, J = 8.4 Hz, 2 H), 6.12 (brs, 1 H), 5.28 ppm (m, 1 H); 13C-NMR (d6-acetone): □ = 196.3, 159.5, 158.3, 157.7, 156.4, 156.1, 154.9, 153.3, 135.2, 131.1, 131.0, 130.6, 128.5, 123.1, 122.4, 116.6, 116.4, 115.6, 114.0, 112.0, 109.0, 103.0, 102.4, 56.1 ppm; HRMS (ESI): calcd for C28H18O7Na+ [M + Na+] 489.0950, found 489.0955.

CONCLUSIONS

In conclusion, a modular and efficient entry to the dimeric resveratrol derived polyphenolic benzofurans has been developed, and applied to the total synthesis of malibatol A (2) and shoreaphenol (3). In view of the largely untapped potential of the polyphenolic secondary metabolites, the synthetic methodology described herein should find wide application in the chemical and biological investigations of this fascinating class of compounds.

REFERENCES AND NOTES

1. Yang, C.S.; Lambert, J.D.; Sang, S. Antioxidative and anti-carcinogenic activities of tea polyphenols. Arch. Toxicol. 2009, 83, 11-21.
2. Bonfili, L.; Cecarini, V.; Amici, M.; Cuccioloni, M.; Angeletti, M.; Keller, J.N.; Eleuteri, A.M. Natural polyphenols as proteasome modulators and their role as anti-cancer compounds. FEBS J. 2008, 275, 5512-5526.
3. Lafay, S.; Gil-Izquierdo, A. Bioavailability of phenolic acids. Phytochem. Rev. 2008, 7, 301-311.
4. Habauzit, V.; Horcajada, M.-N. Phenolic phytochemicals and bone. Phytochem. Rev. 2008, 7, 313-344.

5. Halliwell, B. Are polyphenols antioxidants or pro-oxidants? What do we learn from cell culture and in vivo studies? Arch. Biochem. Biophys. 2008, 476, 107-112.

6. Korkina, L.G.; Pastore, S.; De Luca, C.; Kostyuk, V.A. Metabolism of plant polyphenols in the skin: Beneficial versus deleterious effects. Curr. Drug Metab. 2008, 9, 710-729.

7. Snyder, S.A.; Breazzano, S.P.; Ross, A.G.; Lin, Y.; Zografos, A.L. Total synthesis of diverse carbogenic complexity within the resveratrol class from a common building block. J. Am. Chem. Soc. 2009, 131, 1753-1765.

8. Snyder, S.A.; Zografos, L.; Lin, Y. Total synthesis of resveratrol-based natural products: A chemoselective solution. Angew. Chem. Int. Ed. 2007, 46, 8186-8191.

9. Nicolaou, K.C.; Wu, R.T.; Kang, Q.; Chen, D.Y.K. Total synthesis of hopeahainol A and hopeanol. Angew. Chem. Int. Ed. 2009, 48, 3340-3343.

10. Nicolaou K.C.; Kang Q.; Wu. R.T.; Lim C.S.; Chen D.Y.K. Total synthesis and biological evaluation of the resveratrol-derived polyphenol natural products hopeanol and hopeahainol A. J. Am. Chem. Soc. 2010, 132, 7540-7548.

11. Dai, J.-R.; Hallock, Y.F.; Cardellina, J.H., II; Boyd, M.R. HIV-Inhibitory and cytotoxic oligostilbenes from the leaves of Hopea malibato. J. Nat. Prod. 1998, 61, 351-353.

12. Saraswathy, A.; Purushothaman, K.K.; Patra, A.; Dey, A.K.; Kundu, A.B. Shoreaphenol, a polyphenol from Shorea Robusta. Phytochemisry 1992, 31, 2561-2562.

13. Tanaka, T.; Ito, T.; Ido, Y.; Nakaya, K.; Iinuma, M.; Chelladurai, V. Hopeafuran and a C-glucosyl resveratrol isolated from stem wood of hopea utilis. Chem. Pharm. Bull. 2001, 49, 785-787.

14. Boutagy, J.; Thomas, R. Olefin synthesis with organic phosphonate carbanions. Chem. Rev. 1974, 74, 87-99.

15. Jones, G.; Stanforth, S.P. The vilsmeier reaction of non-aromatic compounds. Org. React. 1997, 49, 1-330.

16. Hughes, D.L. The mitsunobu reaction. Org. React. 1992, 42, 335-656.

17. Kraus, G.A.; Gupta, V. A new synthetic strategy for the synthesis of bioactive stilbene dimmers. A direct synthesis of amurensin. Tetrahedron Lett. 2009, 50, 7180.

18. Kraus, G.A.; Kim, I. Synthetic approach to malibatol A. Org. Lett. 2003, 5, 1191-1192.

19. Kim, I.; Choi, J. A versatile approach to oligostilbenoid natural products - synthesis of permethylated analogues of viniferifuran, malibatol A, and shoreaphenol. Org. Biomol. Chem. 2009, 7, 2788-2795.

20. Hajipour, A.R.; Zarei, A.; Khazdooz, L.; Ruoho, A.E. Simple and efficient procedure for the Friedel-Crafts acylation of aromatic compounds with carboxylic acids in the presence of P2O5/Al2O3 under heterogeneous conditions. Synth. Commun. 2009, 39, 2702-2722.

21. Brady, W.T.; Giang, Y.S.F.; Marchand, A.P.; Wu, A.H. Intramolecular [2 + 2] cycloadditions of ketenes to carbonyl groups. A novel synthesis of substituted benzofurans. J. Org. Chem. 1987, 52, 3457–3461.

Chapter 7

GOLD-CATALYZED CYCLIZATIONS OF ALKYNOL-BASED COMPOUNDS: SYNTHESIS OF NATURAL PRODUCTS AND DERIVATIVES

Benito Alcaide[1,*], Pedro Almendros[2,*] and José M. Alonso[1]

[1]Grupo de Lactamas y Heterociclos Bioactivos, Departamento de Química Orgánica I, Unidad Asociada al CSIC, Facultad de Química, Universidad Complutense de Madrid, 28040-Madrid, Spain

[2]Instituto de Química Orgánica General, CSIC, Juan de la Cierva 3, 28006-Madrid, Spain

ABSTRACT

The last decade has witnessed dramatic growth in the number of reactions catalyzed by gold complexes because of their powerful soft Lewis acid nature. In particular, the gold-catalyzed activation of propargylic compounds has progressively emerged in recent years. Some of these gold-catalyzed reactions in alkynes have been optimized and show significant utility in organic synthesis. Thus, apart from significant methodology work, in the meantime gold-catalyzed cyclizations in alkynol derivatives have become an efficient tool in total synthesis. However, there is a lack of specific review articles covering the joined importance of both gold salts and alkynol-based compounds for the synthesis of natural products and derivatives. The aim of this Review is to survey the chemistry of alkynol derivatives under gold-catalyzed cyclization conditions and its utility in total synthesis, concentrating on the advances that have been made in the last decade and in particular in the last quinquennium.

INTRODUCTION

Organic synthesis has as one of its major points of interest the study of naturally occurring substances, and this remains both a source of information and an intellectual challenge. Thus, a crucial target for organic chemists is to find the appropriate reaction conditions, allowing functional group compatibility and providing high efficiency and atom economy. During the last years, gold-catalyzed cycloisomerization of alkynol-based systems has emerged as a useful tool in this area, allowing the synthesis of different structures such as furans, dihydrofurans, pyrans, furan ones or ketals, among many other heterocyclic systems and naturally occurring structures [1-3].

This overview focuses on the most recent achievements in gold-catalyzed cycloisomerization reactions, for the synthesis of natural products and related compounds. In particular, carbon-carbon and carbon-heteroatom cyclization processes will be considered, paying special attention to reports from the last five years.

CYCLOISOMERIZATION PROCESSES INVOLVING CARBON-CARBON BOND FORMATION

Gold-catalyzed cycloisomerization reactions involving C−C bond formation have recently emerged as an effective methodology to build hydrocarbon rings. Four, five and six membered cyclic structures, as well as medium sized rings are accessible in good yields and under interesting mild reaction conditions using gold salts and gold complexes. Fused bicyclic compounds can also be produced, leading therefore to an attractive series of natural occurring skeletons. Benzofurans represent a recurring motif among natural products. Particularly, 2-substituted and 2,7-disubstituted benzofurans and their derivatives are known to show many different biological activities such as antineoplastic, antiviral, antioxidative or anti-inflammatory properties. Although many routes for the preparation of 2-substituted systems have been developed [4,5], 2,7-substituted benzofurans remain almost unexplored (Figure 1) [6-8].

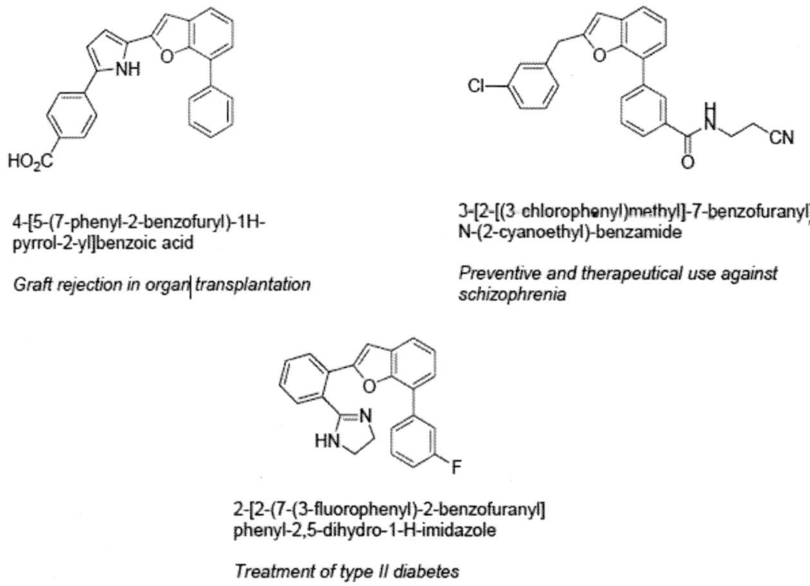

4-[5-(7-phenyl-2-benzofuryl)-1H-pyrrol-2-yl]benzoic acid

Graft rejection in organ transplantation

3-[2-[(3 chlorophenyl)methyl]-7-benzofuranyl]-N-(2-cyanoethyl)-benzamide

Preventive and therapeutical use against schizophrenia

2-[2-(7-(3-fluorophenyl)-2-benzofuranyl] phenyl-2,5-dihydro-1-H-imidazole

Treatment of type II diabetes

Figure 1 7-Aryl benzofuran structure core in different bioactive compounds.

Hashmi et al. have recently proposed an efficient route leading to 7-aryl benzo[b]furans 2 through a gold-catalyzed rearrangement of 3-silyloxy-1,5-enynes [9]. The considerable effort that went into this work, involving a first catalyst screening for substrate 1a and finding the optimal conditions for the dual catalyst system [IPrAuCl]/$AgNTf_2$ is noteworthy. Thus, an easy methodology was performed, using mild conditions, open-air systems and remarkable short reaction times, providing an interesting family of different substituted 7-aryl benzofurans (Scheme 1).

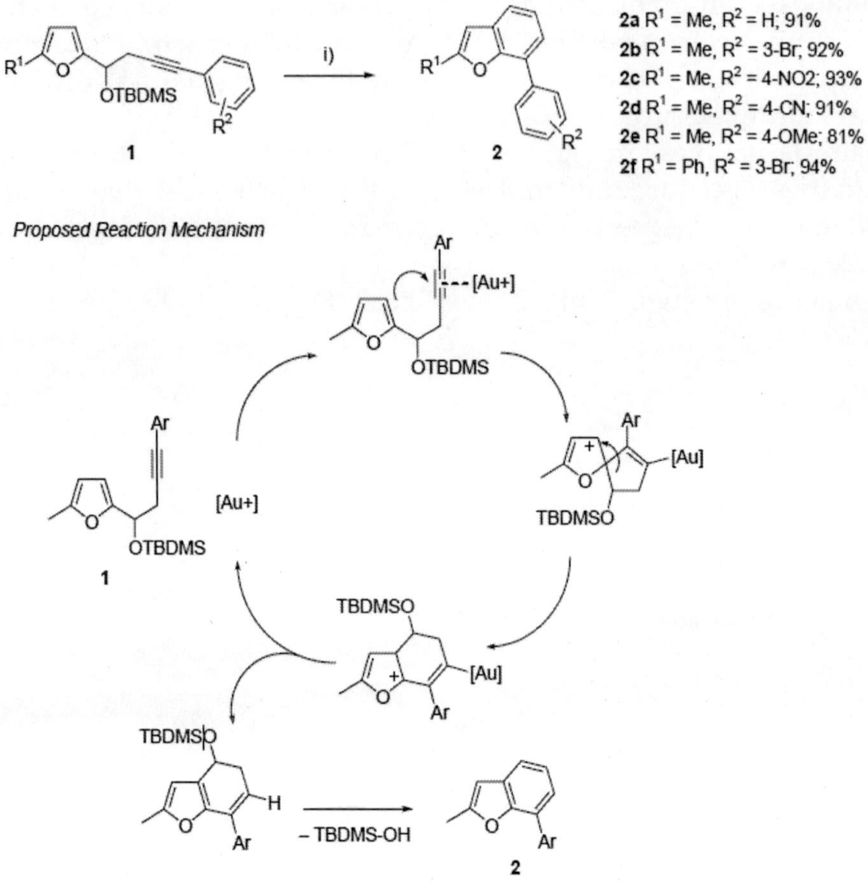

2a R^1 = Me, R^2 = H; 91%
2b R^1 = Me, R^2 = 3-Br; 92%
2c R^1 = Me, R^2 = 4-NO2; 93%
2d R^1 = Me, R^2 = 4-CN; 91%
2e R^1 = Me, R^2 = 4-OMe; 81%
2f R^1 = Ph, R^2 = 3-Br; 94%

Proposed Reaction Mechanism

Scheme 1 Gold(I)-catalyzed rearrangement of 3-silyloxy-1,5-enynes.

Approximately a quarter of biologically active known compounds come from fungi, and among their wide range of properties, the antibiotic activity has attracted much interest [10,11]. Guanacastepene A (Figure 2) is extremely active against methicillin-resistant strains of *Staphylococcus aureus* and vancomicyn-resistant *E. faecalis*, two drug-resistant common pathogens which have generated major concern [12-14].

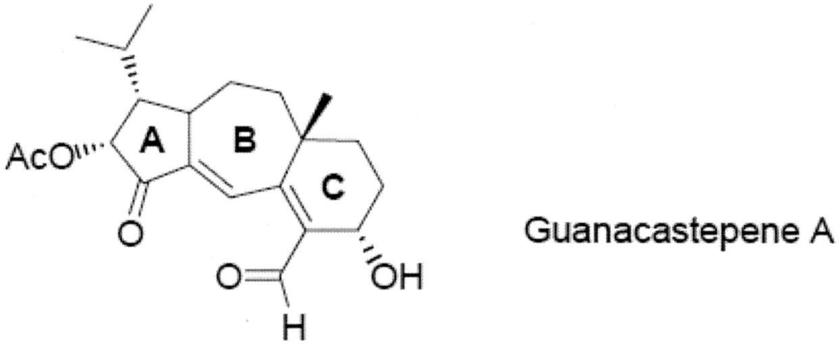

AcO A B C O O H OH Guanacastepene A

Figure 2 Rings A, B and C in natural terpene guanacastepene A.

It has been stated an approach to ring A of guanacastepene, by an unusual gold(I)-catalyzed cycloisomerization of alkynol-based 1,5-enynes [15]. According to the proposed retrosynthesis, most of the functionalities present in the natural terpene would be early introduced, while the presence of the cyclopropyl fused ring could allow the further generation of ring B (Scheme 2).

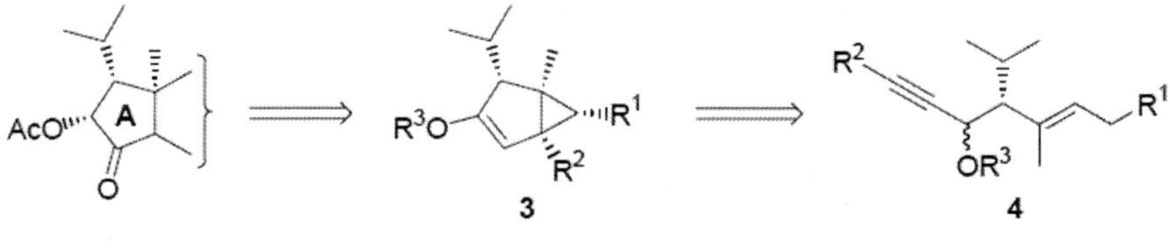

AcO A O ⟹ R^3O R^1 R^2 3 ⟹ R^2 OR^3 R^1 4

Scheme 2 Retrosynthesis of guanacastepene ring A.

Many 1,5-enynes were tested, and an unexpected pattern of reactivity depending on the substituents in substrates 4 was found. Thus, the desired bicyclo[3.1.0] system 3 was obtained only when the reaction was performed with the syn-enynes 4a and 4b, yielding 3a and 3b with good conversions and notable diastereoselectivity. Anti-isomers, or any stereochemical change on the starting 1,5-enynes, resulted in the opposite diastereoselectivity (systems 5), or a dramatic change on the course of the reaction, leading to alkylidene-cyclopentenes 6, cyclohexadienes 7, or α,β-unsaturated aldehydes 8

(Scheme 3). Gold-catalyzed isomerization has been also employed in the search of an appropriate route to (−)-thujopsanone, a derivative of the natural terpene (−)-thujopsene, widely employed in cosmetics [16]. Although the first aim of the authors remained unachieved, and the obtained compound 9 did not exhibit the appreciated properties of the initial target [17], the chemistry developed merits further consideration (Scheme 4) [18]. Thus, it was observed that enynol 10 provided the unexpected ether 11 in the presence of different gold catalysts, in amounts similar to those produced by some other metal salts such as copper or platinum complexes. Interestingly, when the corresponding acetate derivative 12 reacted in the presence of AuCl3, a tandem cycloisomerization/[1,2]-acyl shift took place, leading to adduct 13, precursor of the previously mentioned adduct 9, and a close system to (−)-thujopsanone. Moreover, when the process was tested in the presence of (tBuXPhos)AuNTf2 as catalyst, an unprecedented rearrangement/cycloaddition leading to the tricyclic system 14 was reported (Scheme 5).

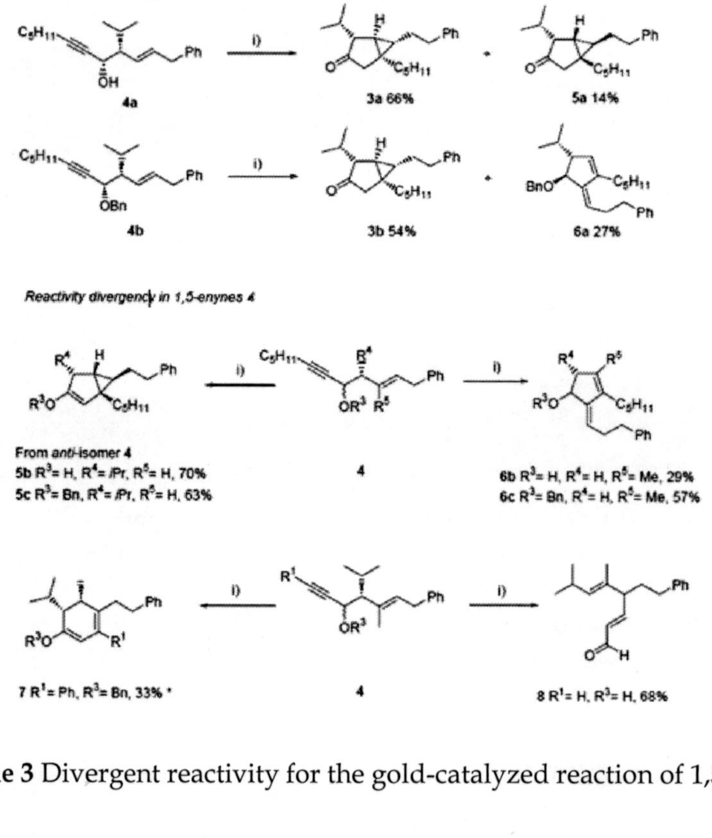

Scheme 3 Divergent reactivity for the gold-catalyzed reaction of 1,5-enynes.

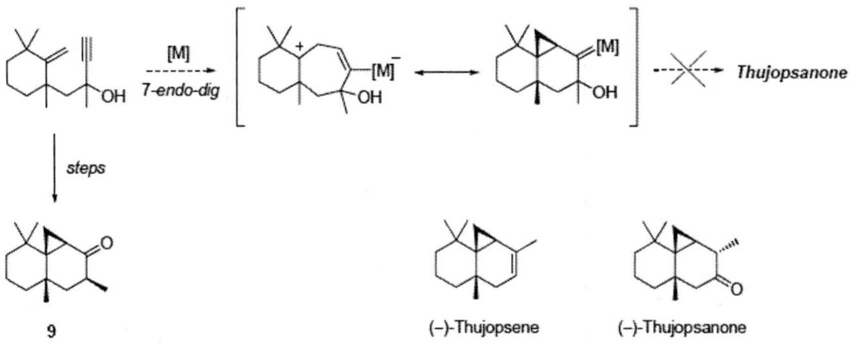

[M]
7-endo-dig
Thujopsanone
steps
9
(–)-Thujopsene
(–)-Thujopsanone

Scheme 4 Projected synthetic route to terpene (−)-thujopsanone.

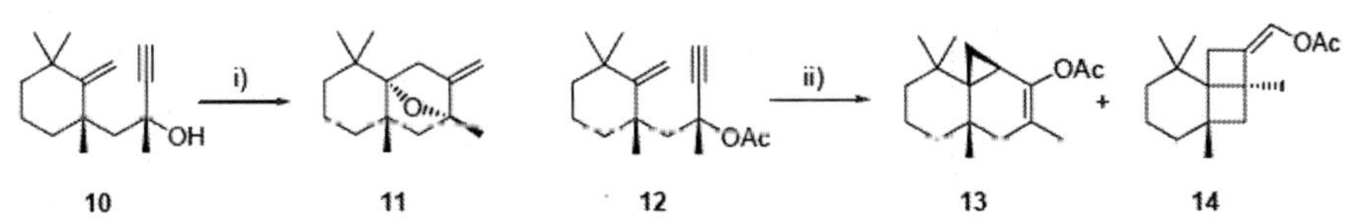

Catalyst	Yield **11**
$[(PPh_3)AuCl]/AgBF_4$	87%
$[(XPhos)AuNTf_2]$	98%
$AuCl_3$	76%
$PtCl_2$	51%
$[Cu(CH_3CN)_4]BF_4$	49%

Catalyst	Yield **13/14**
$[(PPh_3)AuNTf_2]$	60% / 0%
$[(XPhos)AuNTf_2]$	18% / 0%
$AuCl_3$	78% / 0%
$PtCl_2$	43% / 0%
$[Cu(CH_3CN)_4]BF_4$	42% / 0%
$[(tBuXPhos)AuNTf_2]$	9% / 43%

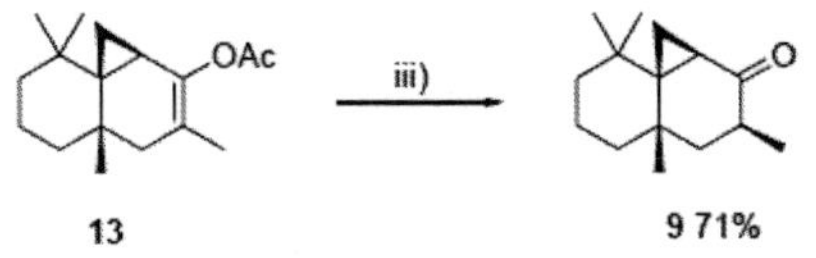

13 → iii) → 9 71%

Scheme 5 Cycloisomerization of enynol 10 and [1,2]-acyl shift rearrangement of acetate **12**, respectively.

Gold-catalyzed cycloisomerization methodology has also been applied to the construction of medium sized rings. Allocolchicinoids, presenting a seven membered ring, are structures related to (−)-colchicine, a natural product with important antimitotic activity (Figure 3). Many of these derivatives also show this kind of mitosis arrest, by inhibiting tubulin polymerization [19-21]. *N*-acetylcolchinol **15**, for instance, is described to bind to tubulin more strongly than colchicine itself. Thus, many reports have appeared describing the synthesis of these structures [22-28].

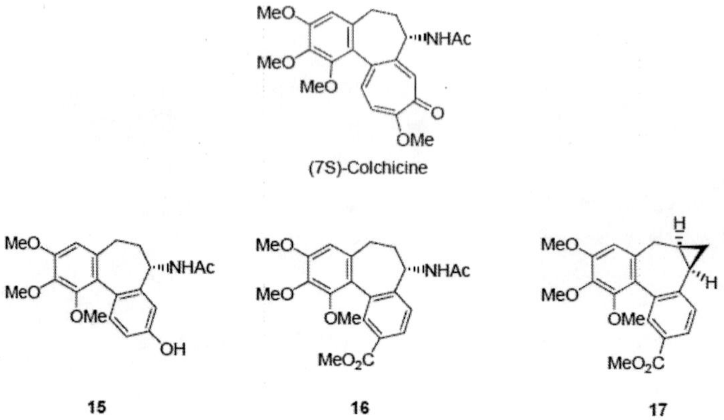

Figure 3 Colchicine and allocolchicinoids systems

Hanna et al. reported the synthesis of derivative 17 [29]. In the proposed sequence, the seven membered ring is formed by a gold(I)-catalyzed 1,2-O-acyl shift, followed by a cyclopropanation step which leads to the fused three member ring (Scheme 6). Thus, gold-catalyzed cyclization of alkynol-based systems has also been stated in this work as a useful tool to create medium sized rings, through an easy methodology providing high yields under mild reaction conditions.

Indole systems are ubiquitous in Nature, appearing in many different alkaloid families. Their wide range of biological activities, and their intriguing chemistry, makes these compounds a target of special interest, and a recurring topic in many studies [30-41]. For instance, the first enantioselective approach to (−)-mersicarpine, an alkaloid isolated from Kopsia plants and exhibiting an unusual tetracyclic structure has been reported. The proposed retrosynthetic analysis included the reaction of an alkynol-based intermediate in the presence of a gold salt, although only the alkyne functional group showed reactivity under these conditions, preserving the hydroxylic group for a further oxidation [42].

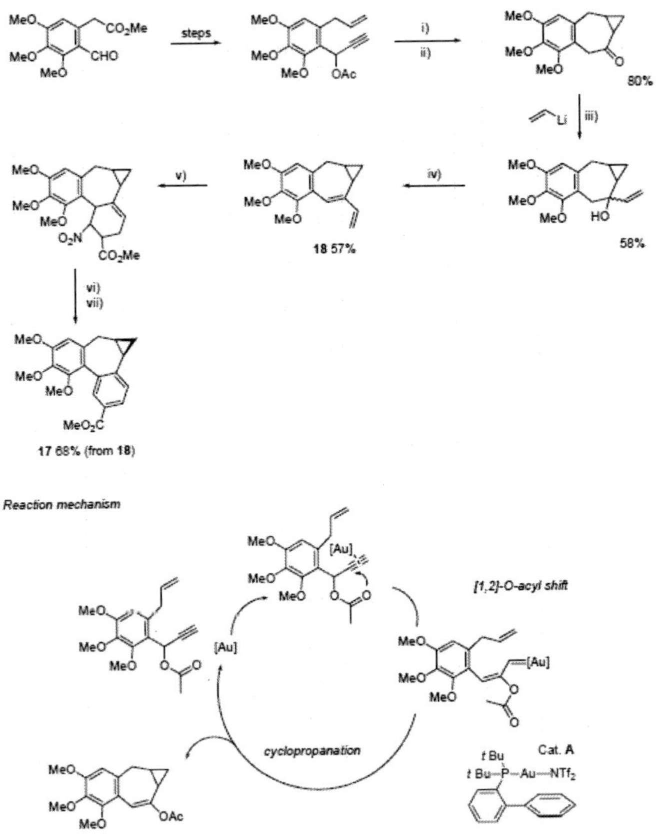

Scheme 6 Synthesis of allocolchicinoid 17 and proposed mechanism for the gold-catalyzed cycloisomerization step.

More interestingly, the reactivity of alkynol-based systems as formal organic synthons has been also explored in the indole chemistry. It has been established the synthesis of the non-natural skeleton 2,3-indoline-fused cyclobutane through a cascade process, including both C−C and C−O bond formation catalyzed by the same gold salt [43]. On the other hand, Echavarren *et al.* described in an exhaustive report about inter- and intramolecular gold-catalyzed reaction of alkynes and indoles some examples starting from alkynols and alkynol-based systems. Carbazole-like systems and related structures were therefore achieved (Scheme 7) [44].

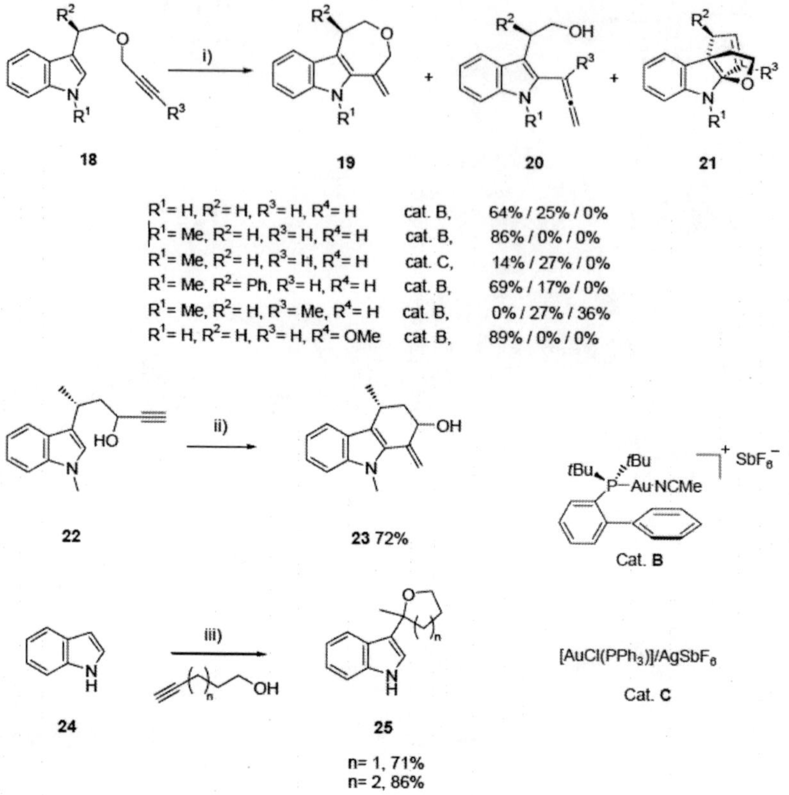

18 19 20 21

R^1= H, R^2= H, R^3= H, R^4= H	cat. B,	64% / 25% / 0%
R^1= Me, R^2= H, R^3= H, R^4= H	cat. B,	86% / 0% / 0%
R^1= Me, R^2= H, R^3= H, R^4= H	cat. C,	14% / 27% / 0%
R^1= Me, R^2= Ph, R^3= H, R^4= H	cat. B,	69% / 17% / 0%
R^1= Me, R^2= H, R^3= Me, R^4= H	cat. B,	0% / 27% / 36%
R^1= H, R^2= H, R^3= H, R^4= OMe	cat. B,	89% / 0% / 0%

22 23 72%

Cat. B

24 25

n= 1, 71%
n= 2, 86%

$[AuCl(PPh_3)]/AgSbF_6$

Cat. C

Scheme 7 Alkynol-based reactivity in indole chemistry.

Inspired by the results of the Echevarren group, Liu et al. described the synthesis of dihydrocyclohepta[b]indoles 26 from (Z)-enynols 27 and indole, through an interesting domino sequence including a first gold(0)-catalyzed Friedel-Craft reaction, followed by a hydroarylation step [45]. The resulting products are of considerable interest, as much as they form the key subunits of several alkaloids, like ambiguine, silicine, caulerpin or caulersin. The reported work includes the optimization of the process, by testing different gold salts and solvents, leading to high reaction conversions through mild conditions (Schemes 8 and 9)

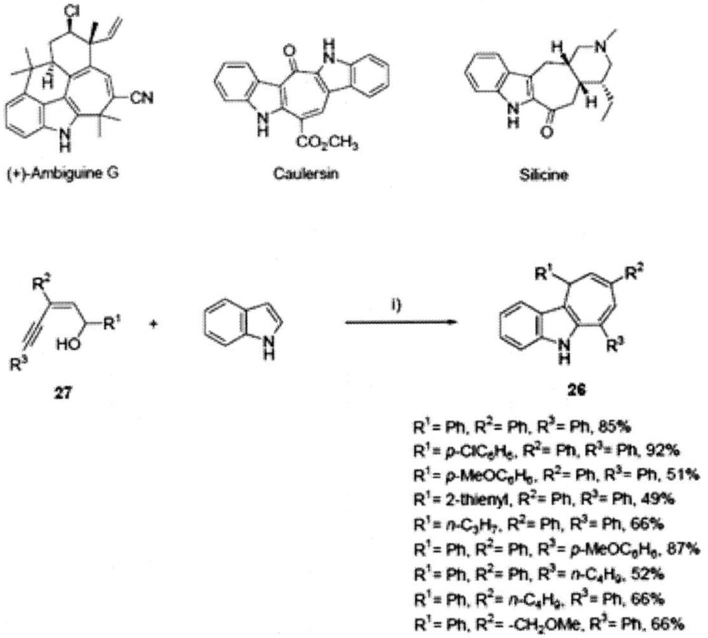

(+)-Ambiguine G   Caulersin   Silicine

27   26

R^1= Ph, R^2= Ph, R^3= Ph, 85%
R^1= p-ClC_6H_4, R^2= Ph, R^3= Ph, 92%
R^1= p-$MeOC_6H_4$, R^2= Ph, R^3= Ph, 51%
R^1= 2-thienyl, R^2= Ph, R^3= Ph, 49%
R^1= n-C_3H_7, R^2= Ph, R^3= Ph, 66%
R^1= Ph, R^2= Ph, R^3= p-$MeOC_6H_4$, 87%
R^1= Ph, R^2= Ph, R^3= n-C_4H_9, 52%
R^1= Ph, R^2= n-C_4H_9, R^3= Ph, 66%
R^1= Ph, R^2= -CH_2OMe, R^3= Ph, 66%

Scheme 8 Synthesis of dihydrocyclohepta[b]indoles, and related natural structures.

CYCLOISOMERIZATION PROCESSES INVOLVING CARBON-HETEROATOM BOND FORMATION

Heterocyclic natural occurring motifs such as furans, pyrans or spiroketals can be easily achieved through heterocyclization processes performed on alkynol-based systems. Gold promoted methodologies provide a convenient route to these structures, allowing mild reaction conditions and high yields. Total synthesis and the preparation of related derivatives have been recently described using both C−N and C−O bond formation.

Cycloisomerization on Alkynol-Based Systems

Chromones are natural heterocycles showing a wide range of biological properties. Thus, many strategies like iodocyclizations

[46], metal-catalyzed cycloadditions [47], or *O*-arylation processes [48] have appeared for the synthesis of these oxacyclic systems. Gold catalyzed cycloisomerization of alkynol based structures 28 have been also stated for the generation of chromones 29 [49]. Interestingly, reaction proceeded with a further migration of group R1, leading to highly functionalized skeletons. Unluckily, only moderate yields were achieved (Scheme 10), inasmuch as isomerization processes competed with the expected Au-based cycloisomerization.

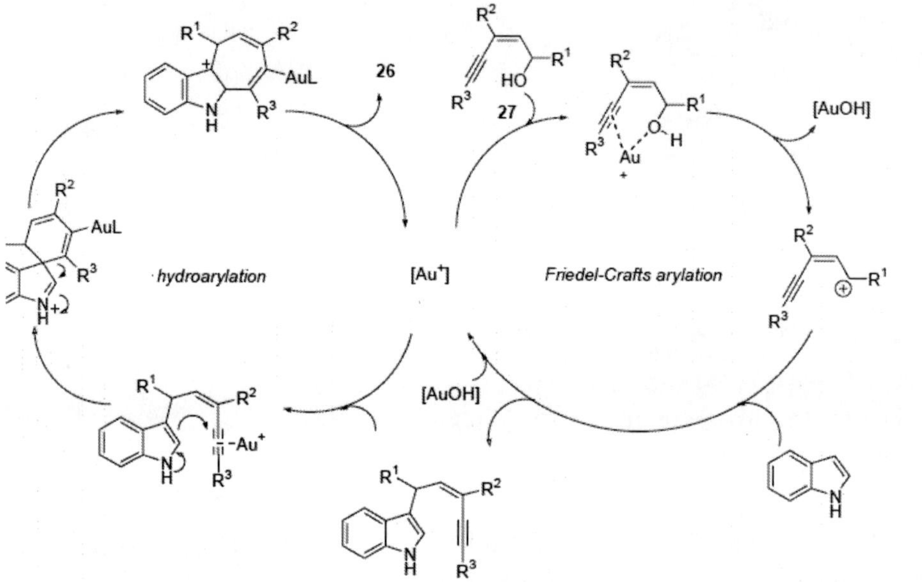

Scheme 9 Proposed reaction mechanism for the tandem gold catalyzed-Friedel-Crafts arylation/hydroarylation process.

A similar approach has been developed for the synthesis of aurone skeletons [50], natural flavonoids, by an easy three step sequence. Aurones exhibit several biological properties [51-55], and its importance had led to several groups to develop convenient synthetic routes [56-62]. Among them, gold-catalyzed oxycyclization provided the best results, as milder reaction conditions and excellent selectivities, avoiding the formation of flavones as byproducts, were achieved (Scheme 11) [63]. In this case, high yields and complete regioselectivity were obtained.

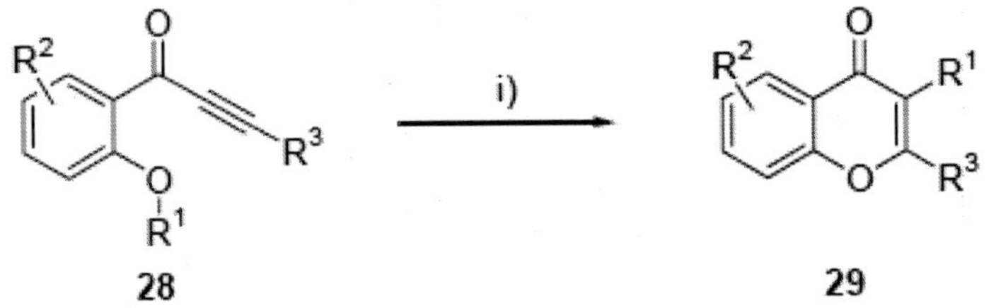

28 i) → **29**

R^1 = Bn, R^2 = 3-MeO, R^3 = *n*Pr, 45%
R^1 = Bn, R^2 = H, R^3 = *n*Pr, 31%
R^1 = Bn, R^2 = 3,5-Br, R^3 = *n*Pr, 40%
R^1 = Bn, R^2 = 3-Allyl, R^3 = *n*Pr, 38%
R^1 = Bn, R^2 = H, R^3 = 4-$MeOC_6H_6$, 25%
R^1 = 4-Cl-Bn, R^2 = H, R^3 = *n*Pr, 40%
R^1 = Allyl, R^2 = H, R^3 = Ph, 38%

Scheme 10 Synthesis of chromones by gold-catalyzed cycloisomerization.

i) (21-94%) ii) iii) (64-99%) *Aurone family*

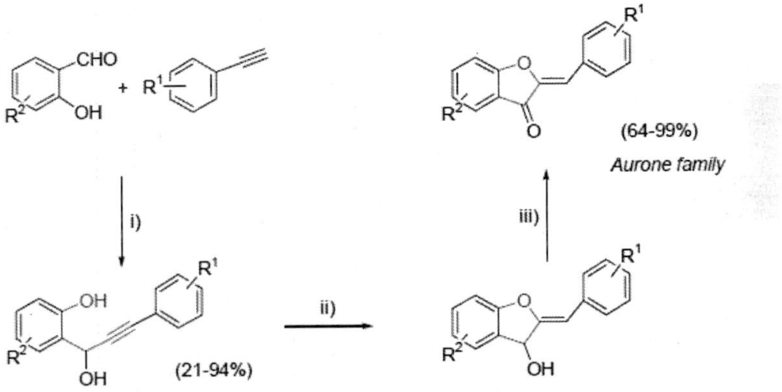

R^1 = H, R^2 = H, 84%
R^1 = H, R^2 = 4-NO_2, 86%
R^1 = H, R^2 = 4-Br, 79%
R^1 = 4-Cl, R^2 = 4-Br, 70%
R^1 = 4-Cl, R^2 = H, 65%
R^1 = 4-OMe, R^2 = H, 70%
R^1 = 3,4-di(OMe), R^2 = 2,5-di(OMe), 83%

Scheme 11 Synthesis of aurone skeleton by gold-catalyzed cycloisomerization.

Moreover, the present methodology was used for the structural revision of two natural products, (Z)-4′-chloroaurone 30 [64], and (Z)-2′-hydroxyaurone 32 [65], proving that the assumed structures were not the correct ones. Thus, flavonoid systems 30 and 32 could be prepared by the above three step strategy which revealed that their spectral data did not match with the previously reported data of the natural isolated ones. Therefore, the isocumarin 31 and the flavone 33 were prepared and probed as the real structures for these natural products (Figure 4).

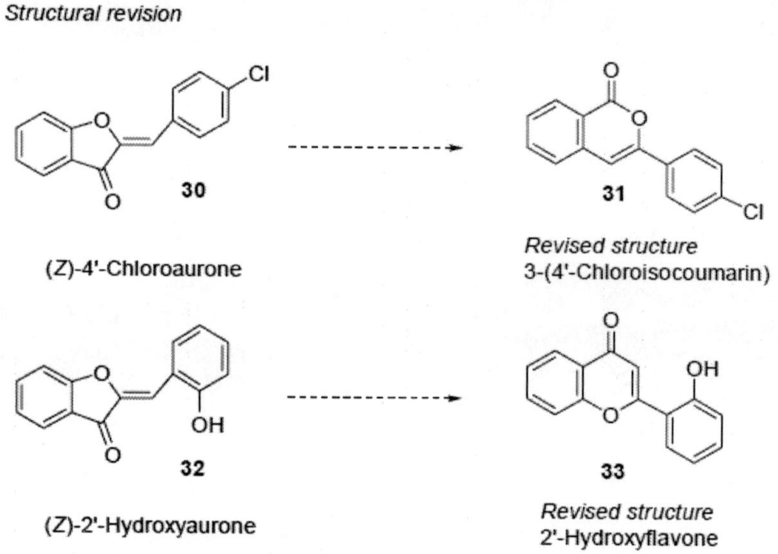

Structural revision

30 — (*Z*)-4'-Chloroaurone → 31 — *Revised structure* 3-(4'-Chloroisocoumarin)

32 — (*Z*)-2'-Hydroxyaurone → 33 — *Revised structure* 2'-Hydroxyflavone

Figure 4 New assignation of structures 31 and 33 by comparison with the prepared by the gold catalysis aurone systems 30 and 32.

Trost et al. recently completed the total synthesis of bryostatin 16 [66,67], a structurally complex macrolide which exhibits a wide range of biological activities [68-71]. Focusing on the proposed 26 step sequence (in the longest linear path, and 39 steps as the total), the gold-catalyzed 6-endo-dig oxycyclization of alkynol 34 to generate the inner dihydropyran cycle D in macrocyclic precursor 35 in 65% yield deserves special attention (Scheme 12).

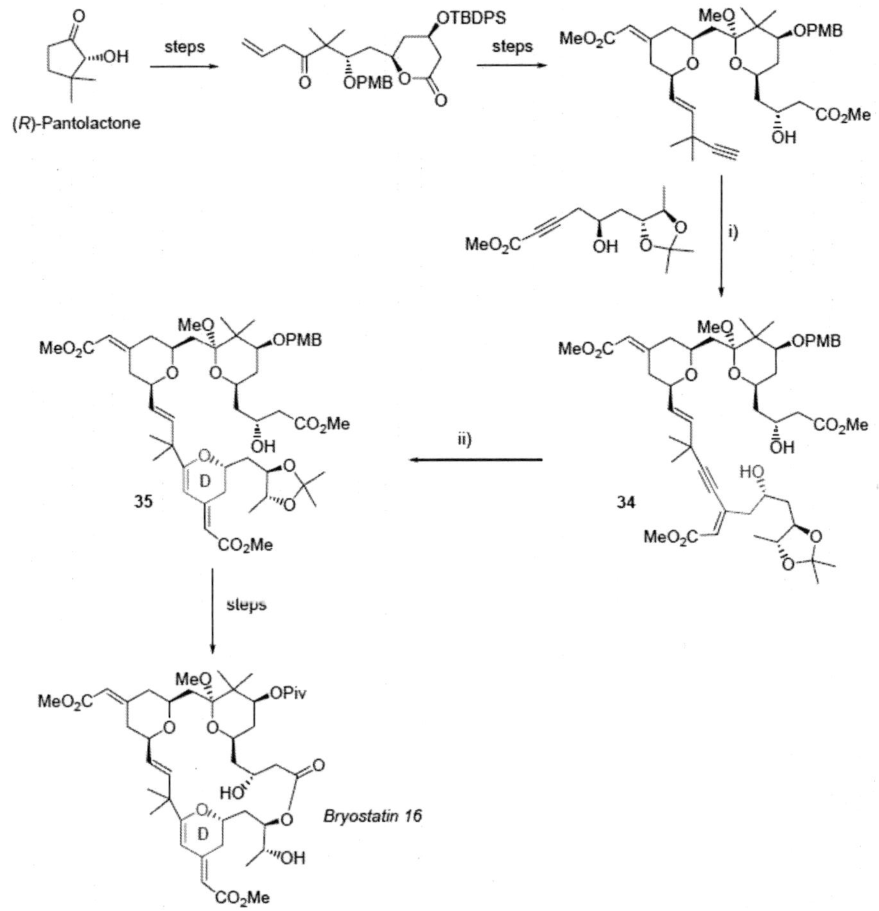

Scheme 12 Gold-based synthesis of dihydropyran ring D in bryostatin total synthesis.

(+)-Cephalostatin 1 is another complex macrolide with interesting biological activity. It has been reported to be a promising anticancer agent for the *p16* tumor suppressor gene, exhibiting high activity and high selectivity between cancer cells and normal cells [72,73]. Because of the small amounts of cephalostatin available from its natural marine sources, a synthetic approach has emerged as the sole viable tool to provide enough material for biological testing [74-78]. On the other hand, the structural complexity of cephalostatin makes this macrocycle an interesting target to develop new skills in organic synthesis.

Fortner et al. have recently described a total synthesis of cephalostatin, involving the construction of both its eastern and western fragments and their further coupling [79]. Along the high quality chemistry developed for this synthesis, we would like to focus on the dihydrofuran ring E construction on compound 36. Thus, gold-catalyzed cycloisomerization emerge again as a useful methodology to convert alkynol systems in oxacyclic skeletons, crucial and recurring motifs for total synthesis. Moreover, the efficiency of gold catalysis to promote a 5-endo-dig process with an 88% conversion, on what is a hindered internal alkyne 37, deserves special consideration (Scheme 13).

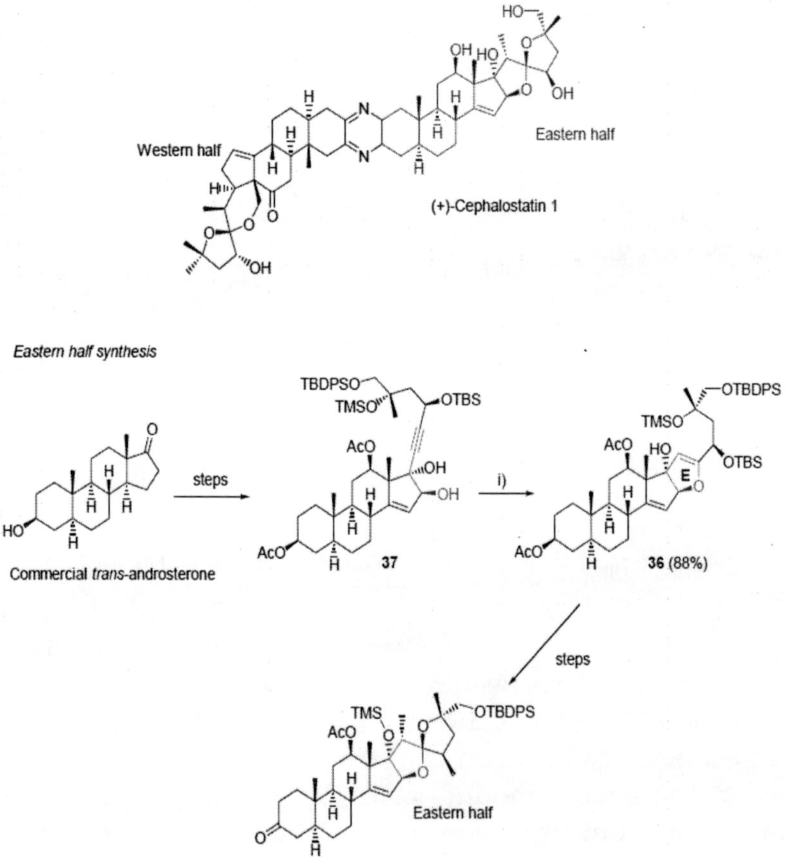

Scheme 13 Synthesis of ring E on the eastern fragment of cephalostatin.

Other natural occurring motifs such as oxazoles and isoxazoles have also been assembled through gold-catalyzed cycloisomerization. Thus, it has been recently established a general method for the synthesis of highly functionalized isoxazoles from alkynyl oxime ethers [80], or an intermolecular alkyne oxidation leading to 2,5-disubstituted oxazoles [81]. Nevertheless, while gold-based alkyne-oxygen cycloisomerization has recently become a hot topic in organic synthesis, only a few examples for alkyne-nitrogen coupling have been described [82-90]. Regarding the synthesis of natural products and derivatives, Chan *et al.* have recently described the synthesis of highly substituted indole skeletons [29], from readily available 2-tosylamino-phenylprop-1-yn-3-ols **38** [91]. The reported work shows a versatile approach to these natural occurring motifs, and develops a fascinating study concerning the chemical reactivity of these substrates under gold-catalyzed conditions. Thus, starting in every case from a 5-*exo*-dig cycloaddition which led to vinyl gold species **39**, different reaction pathways were observed depending on the substituent group R1. It was stated that when R^1 = aryl, reaction proceeded through a Friedel-Craft process, giving indenyl-fused indoles 40. On the other hand, changing to R^1 = H, a protodeauration/1,3-allylic alcohol isomerization took place, leading to indoles 41. The presence of a nucleophile in the reaction media gave place mainly to systems 42, and for $R^1 = CHR^2R^3$, a more facile protodeauration and dehydratation step delivered systems 43 (Scheme 14).

Another example of gold-based C−N cyclization on alkynol systems for the total synthesis of (+)-andrachcinidine (**44**) has been established [92]. This natural alkaloid receives its name from its natural source, the beetle *Andrachne aspera*, and it has been shown to be an interesting chemical defense agent [93]. The proposed reaction sequence started with commercial ketal **45**, which yielded after six steps the nitrogen-containing alkynol **46**. Gold-catalyzed cyclization of **46** provided the piperidine system **47** as a single diastereomer in 89% isolated yield. The reaction mechanism is proposed to follow a first gold-based alkyne hydration providing ketone **48**. Methoxy group cleavage would then generate the corresponding α,β-unsaturated system, which could undergo nucleophile addition building the expected 6-membered heterocycle (Scheme 15). It is noteworthy that no competition between nitrogen and oxygen attack was found, which would led to the less favoured 8-membered heterocycle.

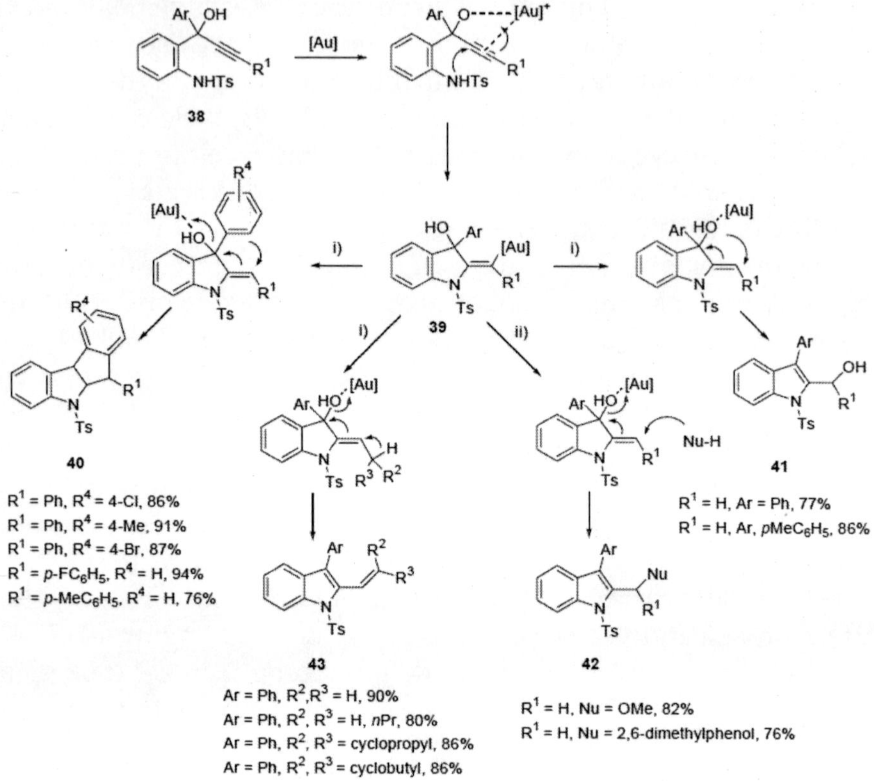

40

R^1 = Ph, R^4 = 4-Cl, 86%
R^1 = Ph, R^4 = 4-Me, 91%
R^1 = Ph, R^4 = 4-Br, 87%
R^1 = p-FC_6H_5, R^4 = H, 94%
R^1 = p-MeC_6H_5, R^4 = H, 76%

41

R^1 = H, Ar = Ph, 77%
R^1 = H, Ar, $p$$MeC_6H_5$, 86%

43

Ar = Ph, R^2,R^3 = H, 90%
Ar = Ph, R^2, R^3 = H, nPr, 80%
Ar = Ph, R^2, R^3 = cyclopropyl, 86%
Ar = Ph, R^2, R^3 = cyclobutyl, 86%

42

R^1 = H, Nu = OMe, 82%
R^1 = H, Nu = 2,6-dimethylphenol, 76%

Scheme 14 Indole synthesis from gold-catalyzed cycloisomerization of 2-tosylaminophenylprop- 1-yn-3-ols

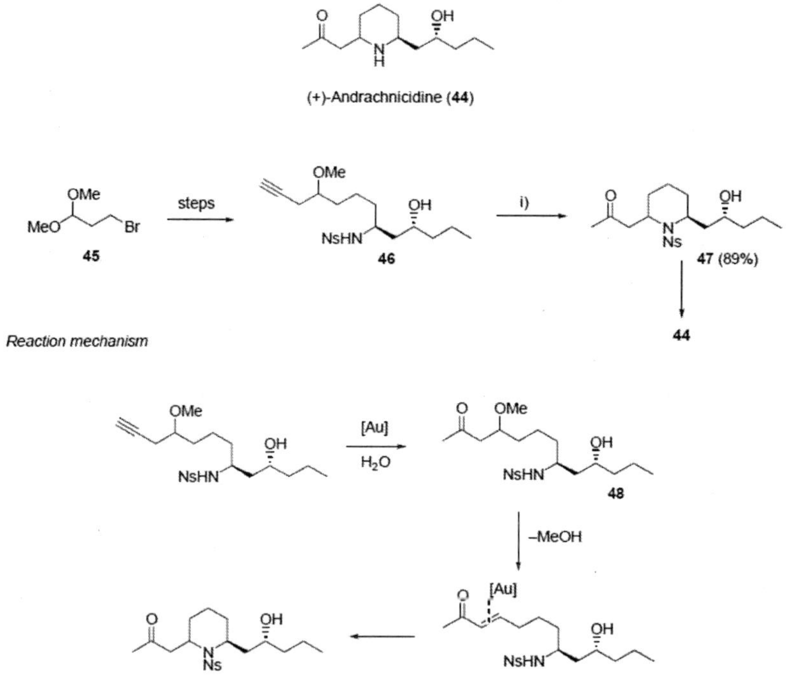

Scheme 15 Synthesis of (+)-and rachcinidine

Cycloisomerization on Alkynediol-Based Systems

Ketals are important key structures, and crucial targets in organic synthesis [94-98]. Fused, bicyclic and spiroketals are recurring motifs in natural compounds, and their preparation is a key step in many total syntheses. In particular, spiroketals represent a structural feature of many biomedically relevant natural and non-natural systems [99-102]. Several methods have been developed for the synthesis of spiroketals, the most common being perhaps the cyclocondensation of ketone diols [103,104].

Nevertheless, gold catalyzed cycloisomerization on alkynediols has emerged as an efficient strategy to build complex ketal systems in just one step, offering specific advantages. For example, Au-catalyzed cycloisomerization of alkynediols are more exotermic, atom economical, and more compatible than ketones under a number of several reaction conditions. Thus, many groups have recently incorporated the present methodology for the synthesis of several natural compounds and derivatives [105-108]. Li et al. have described the preparation of the bisbenzannelated spiroketal core of rubromycins [109].

These natural occurring structures exhibit different biological activities, such as inhibition of DNA polymerase, inhibition of the reverse transcriptase of HIV I, or inhibition of DNA helicase [110-113]. Scheme 16 shows the basic structure motif shared by natural isolated compounds like γ-rubromycin, purpuromycin, or heliquinomycin. According to the described work, easily prepared alkynediols 49 underwent cycloisomerization in the presence of gold catalysis to yield spiroketals 50 with moderate yields, but mainly together with notable amounts of the corresponding benzofuran 51.

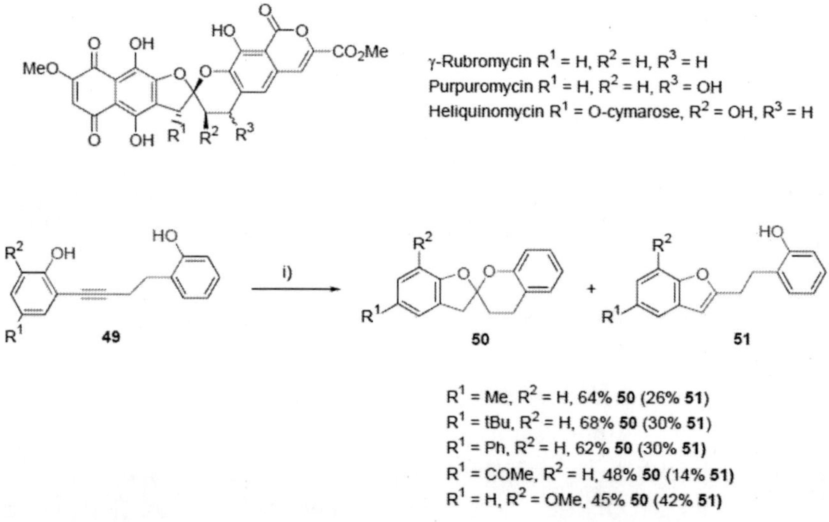

γ-Rubromycin R^1 = H, R^2 = H, R^3 = H
Purpuromycin R^1 = H, R^2 = H, R^3 = OH
Heliquinomycin R^1 = O-cymarose, R^2 = OH, R^3 = H

R^1 = Me, R^2 = H, 64% **50** (26% **51**)
R^1 = tBu, R^2 = H, 68% **50** (30% **51**)
R^1 = Ph, R^2 = H, 62% **50** (30% **51**)
R^1 = COMe, R^2 = H, 48% **50** (14% **51**)
R^1 = H, R^2 = OMe, 45% **50** (42% **51**)

Scheme 16 Synthesis of spiroketal motif of rubromycins

A more effective spiroketalization process was found for the synthesis of cephalosporolides. Concretely, cephalosporolide H 52 is a natural spiroketal isolated from the culture broth of the marine fungus Penicillium sp. This compound presents anti-inflammatory properties by virtue of its inhibitory activity against 3α-hydroxysteroid dehydrogenase [114,115]. Dudley et al. developed a method for cephalosporolide total synthesis based on gold-catalyzed spiroketal generation [116,117]. Starting from pantolactone 53, alkynediol-based system 54 was obtained after a nine step sequence. Gold treatment of 54 yielded the desired structure 55, with an excellent 88% yield. The main inconvenient of the proposed strategy lied on the obtention of 55 as a 1:1 mixture of spiroketal epimers, although further treatment upon zinc chloride chelation provided the expected isomer in 20:1 dr (Scheme 17).

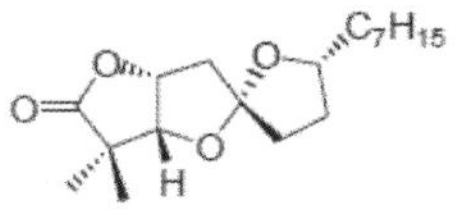

Cephalosporolide H (**52**)

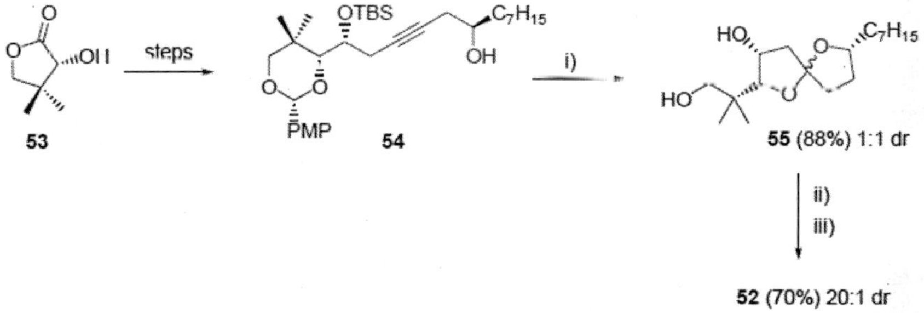

53 steps **54** i) **55** (88%) 1:1 dr ii) iii) **52** (70%) 20:1 dr

Scheme 17 Cephalosporolide H; structure and proposed synthesis

Azaspiracid 56 belongs to a family of marine toxins, responsible for human poisoning and diverse chronic effects on liver, pancreas and thymus [118,119]. Its complete structure has been widely studied [120], and several methods for its synthesis have been reported [121,122]. Forsyth et al. have reported the synthesis of the F–I azaspiracid fragment 57 [123]. In particular, we would like to focus on the construction of F and G rings by a one step gold-catalyzed spiroketalization. Alkynediol-based system 58 was obtained by coupling of subunits 59 and 60, prepared from simple precursors. Treatment of 58 with AuCl provided the desired structure 57 with a high 75% yield as a sole isomer. The reaction mechanism is proposed to follow an initial syn addition of the C6 hydroxy group and the π-activated gold-alkyne complex to build ring F. Protodeauration and protonation of the resultant enol ether at C11 would promote the attack of methoxy oxygen to C10, generating therefore ring G (Scheme 18).

Okadaic acid 61 is a complex natural structure isolated from marine sponges [124,125]. Its biological activities [126-128], together with its attractive chemical structure have attracted much interest among organic chemists. In particular, the presence of several spiroketal motifs in this structure makes it a real challenge from the retrosynthetical point of view. An efficient synthesis of the C15-C38 fragment has been reported, based on the high activity and selectivity of AuCl for the synthesis of spiroketals 62 and 63, starting from alkynediols 64 and 66 respectively [129] (Scheme 19).

Bridged-bicyclic ketals have been also produced through gold-catalyzed cycloisomerization of alkynediols. Based on platensimycin structure, a natural inhibitor of microbial fatty acid biosynthesis [130-132], Corey et al. reported the total synthesis of the near-structural mimic 68 [133]. This new structure presents evidence in the literature suggesting excellent antimicrobial properties [134,135]. Thus, easily achieved alkynediol 69 reacted under gold(III) catalysis delivering ketone 70, which contains the tricyclic core of 68, with an excellent 85% yield and >98% ee (Scheme 20). The route to the desired target is completed in just nine steps, providing a facile and quick methodology to the mentioned bioactive structure.

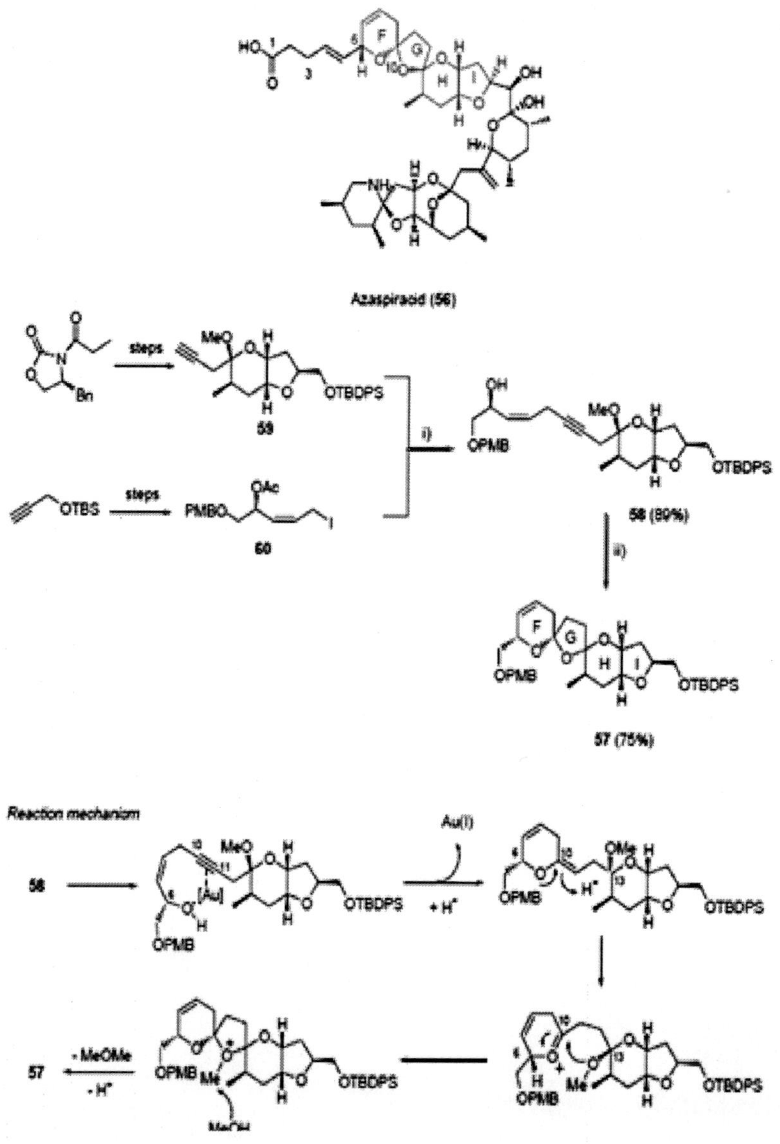

Scheme 18 Synthesis of rings F and G in azaspiracid and reaction mechanism

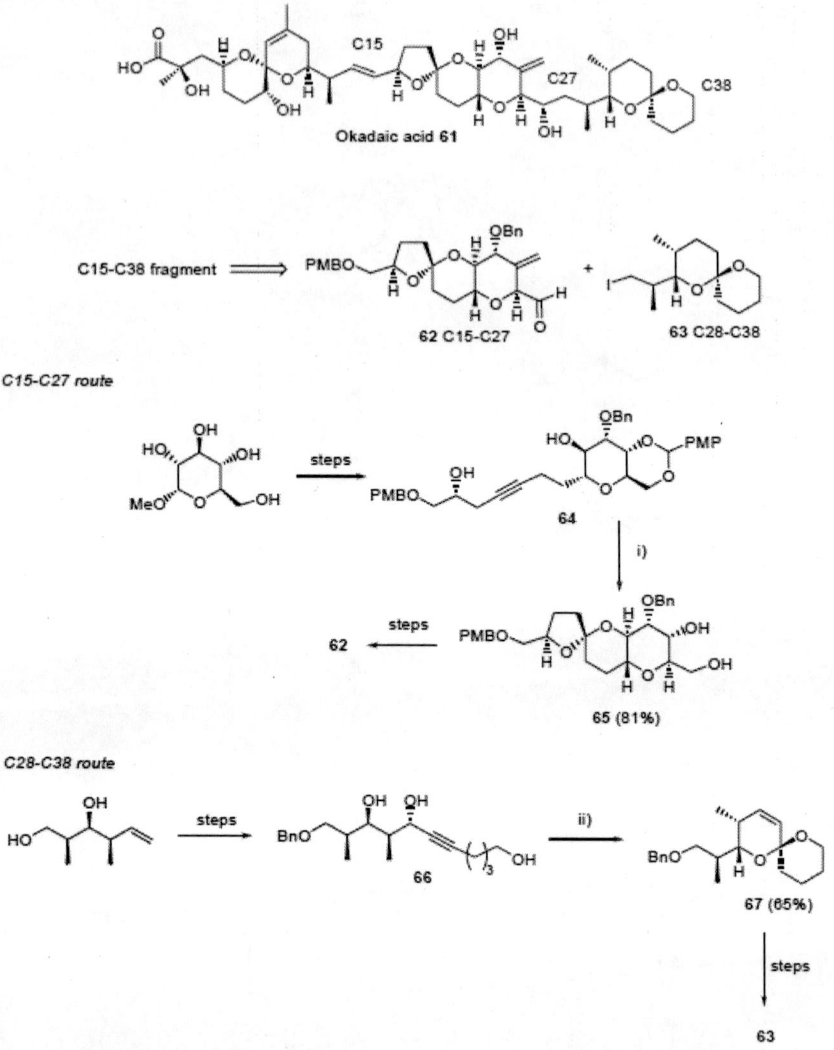

Okadaic acid 61

C15-C38 fragment ⟹ 62 C15-C27 + 63 C28-C38

C15-C27 route

steps → 64 → i) → 65 (81%) → steps → 62

C28-C38 route

steps → 66 → ii) → 67 (65%) → steps → 63

Scheme 19 Synthesis of C15-C27 and C28-C38 fragments of okadaic acid

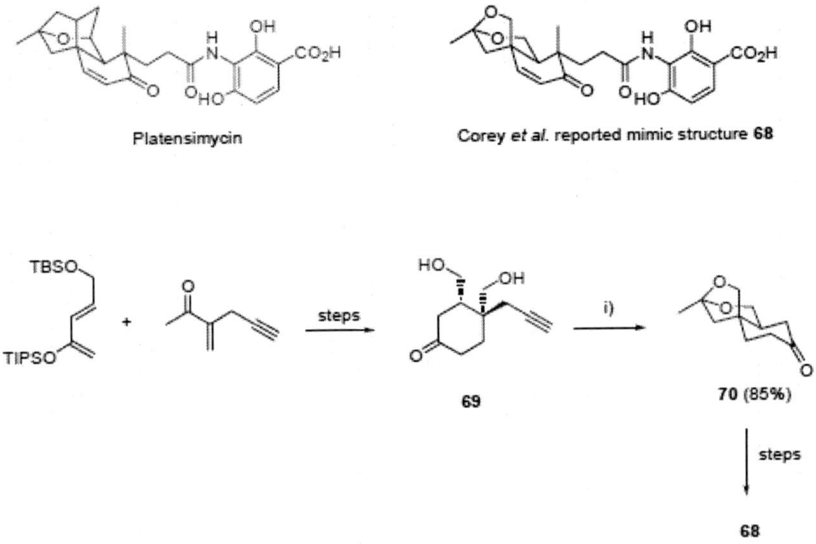

Platensimycin

Corey *et al.* reported mimic structure **68**

TBSO TIPSO + steps **69** i) **70** (85%) steps **68**

Scheme 20 Synthesis of platensimycin-derived structure

CONCLUSIONS

In this overview we have collected the most recent advances in gold-catalyzed cycloisomerization of alkynol and alkynediol-based systems for the preparation of natural products and derivatives. This type of process has become an established methodology for accessing a large number of both carbocyclic and heterocyclic structures, containing different sized skeletons. Three to seven-membered carbon rings, such as furan, pyrans, piperidines, and different ketal and spiroketal systems are therefore accessible through this strategy. The reactions discussed herein demonstrate the high synthetic potential of alkynol-based compounds undergoing gold catalyzed cyclization. On the other hand, the efficiency of gold salts and gold complexes have been also documented, allowing mild reaction conditions and great functional group compatibility, specially compared to related thermal or basic rearrangements. In addition, the extremely large number of natural bioactive compounds containing these type of structural motifs, readily available through gold-catalyzed conditions, will certainly provide a renewed and continuous topic of investigation in this field.

ACKNOWLEDGMENTS

Support for this work by the DGI-MICINN (Project CTQ2009-09318), Conunidad Autónoma de Madrid (CAM) (Project S2009/PPQ-1752), and SANTANDER-UCM (Project GR35/10-A) are gratefully acknowledged. J. M. A. thanks CAM and Fondo Social Europeo for a postdoctoral contract.

REFERENCES

1. Hashmi, A.S.K.; Rudolph, M. Gold Catalysis in Total Synthesis. Chem. Soc. Rev. 2008, 37, 1766-1775.
2. Fürstner, A. Gold and Platinum Catalysis-A Convenient Tool for Generating Molecular Complexity. Chem. Soc. Rev. 2009, 28, 3208-3221.
3. Alcaide, B.; Almendros, P.; Alonso, J.M. Gold Catalyzed Oxycyclizations of Alkynols and Alkynediols. Org. Biomol. Chem. 2011, 9, 4405-4416.
4. Zeni, G.; Larock, R.C. Synthesis of Heterocycles via Palladium n-Olefin and n-Alkyne Chemistry. Chem. Rev. 2004, 104, 2285-2309.
5. Alonso, F.; Beletskaya, I.P.; Yus, M. Transition-Metal-Catalyzed Addition of Heteroatom- Hydrogen Bonds to Alkynes. Chem. Rev. 2004, 104, 3079-3159.
6. Michael, P.; Gerd, R.; Theo, S. (Eli Lilly Co.) Imidazoline Derivatives for the Treatment of Diabetes, Especially Type II Diabetes. WO Patent 2000078726 A1, 28 December 2000.
7. Tagami, K.; Yoshimura, H.; Nagai, M.; Hibi, S.; Kikuchi, K.; Sato, T.; Okita, M.; Okamoto, Y.; Nagasaka, Y.; Kobayashi, N.; Hida, T.; Tai, K.; Tokuhara, N.; Kobayashi, S. (Eisai Co., Ltd.) Preparation of Fused-ring Carboxylic Acid Compounds as Retinoic Acid Receptor Agonists. WO Patent 9734869 A1, 25 September 1997.
8. Masaki, S.; Mitsunori, K.; Yuhei, M.; Masakuni, K. (Takeda Pharmaceutical) Amide Compound. WO Patent 2010018874 A1, 18 February 2010.
9. Hashmi, A.S.K.; Yang, W.; Rominger, F. Gold(I)-Catalyzed Formation of Benzo[b]furans from 3-Silyloxy-1,5-enynes. Angew. Chem. Int. Ed. 2011, 50, 5762-5765.
10. Henkel, T.; Brunne, R.M.; Reichel, F.; Muller, H. Statiscal Investigation into the Structural Complementarity of Natural Products and

Synthetic Compounds. Angew. Chem. Int. Ed. 1999, 38, 643-647.

11. Pearce, C. Biologically Active Fungal Metabolites. Adv. Appl. Microbiol. 1997, 44, 1-80.
12. For isolation and structural determination of guanacastepene, see: Brady, S.F.; Singh, M.P.; Janso, J.E.; Clardy, J. Guanacastepene, a Fungal-Derived Diterpene Antibiotic with a New Carbon Skeleton. J. Am. Chem. Soc. 2000, 122, 2116-2117.
13. For selected studies about the mentioned pathogens, see: Neu, H.C. The Crisis in Antibiotic Resistance. Science 1992, 257, 1064-1073.
14. Swartz, M.N. Hospital-Acquired Infections: Diseases with Increasingly Limited Therapies. Proc. Natl. Acad. Sci. USA 1994, 91, 2420-2427.
15. Gagosz, F. Unusual Gold(I)-Catalyzed Isomerization of 3-Hydroxylated 1,5-Enynes: Highly Substrate-Dependent Reaction Manifolds. Org. Lett. 2005, 7, 4129-4132.
16. Ohloff, G.; Strickler, H.; Willhalm, B.; Borer, C.; Hinder, M. En-Syntheses with Singlet Oxygen. II. Dye-Sensitized Photooxygenation of (−)-Thujopsene and Stereochemistry of the Prepared Thujopsanols. Helv. Chim. Acta 1970, 53, 623-637.
17. Hatsui, T.; Suzuki, N.; Takeshita, H. Dicyanoanthracene-Sensitized Photooxygenation of Thujopsene. Chem. Lett. 1985, 639-642.
18. Fehr, C.; Vuagnoux, M.; Buzas, A.; Arpagaus, J.; Sommer, H. Gold- and Copper-Catalyzed Cycloisomerizations towards the Synthesis of Thujopsanone-Like Compounds. Chem. Eur. J. 2011, 17, 6214-6220.
19. Graening, T.; Schmalz, H.-G. Total Synthesis of Colchicine in Comparison: A Journey through 50 Years of Synthetic Organic Chemistry. Angew. Chem. Int. Ed. 2004, 43, 3230-3256.
20. Boyé, O.; Brossi, A. The Alkaloids; Brossi, A., Cordell, G.A., Eds.; Academic Press: New York, NY, USA, 1992; Volume 41, p. 125.
21. Jordan, M.A.; Wilson, L. Microtubules as a Target for Anticancer Drugs. Nat. Rev. Cancer 2004, 4, 253-265.
22. Vorogushin, A.V.; Wulff, W.D.; Hansen, H.-J. Central-to-Axial Chirality Transfer in the Benzannulation Reaction of Optically Pure Fischer Carbene Complexes in the Synthesis of Allocolchicinoids. Tetrahedron 2008, 64, 949-968.
23. Besong, G.; Jarowski, K.; Kocienski, P.J.; Sliwinski, E.; Boyle, F.T. Synthesis of (S)-(−)-NAcetylcolchinol Using Intramolecular Biaryl Oxidative Coupling. Org. Biomol. Chem. 2006, 4, 2193-2207.
24. Leblanc, M.; Fagnou, K. Allocolchicinoid Synthesis via Direct Arylation. Org. Lett. 2005, 7, 2849-2852.

25. Büttner, F.; Bergemann, S.; Guénard, D.; Gust, R.; Seitz, G.; Thoret, S. Two Novel Series of Allocolchicinoids with Modified Seven Membered B-Rings: Design, Synthesis, Inhibition of Tubulin Assembly and Citotoxicity. Bioorg. Med. Chem. 2005, 13, 3497-3511.

26. Wu, T.R.; Chong, J.M. Asymmetric Synthesis of Propargylamines via 3,3′-Disubstituted Binaphtol-Modified Alkynylboronates. Org. Lett. 2006, 8, 15-18.

27. Vorogushin, A.V.; Predeus, A.V.; Wulff, W.D.; Hansen, H.-J. Diels-Alder Reaction- Aromatization Approach towards Functionalized Ring C Allocolchicinoids. Enantioselective Total Synthesis of (−)-7S-Allocolchicine. J. Org. Chem. 2003, 64, 5826-5831.

28. For a reported synthesis of 16 see: Boyer, F.-D.; Hanna, I. Synthesis of Allocolchicines Using Sequential Ring-Closing Enyne Methathesis-Diels-Alder Reactions. Org. Lett. 2007, 9, 715-718.

29. Boyer, F.-D.; LeGoff, X.; Hanna, I. Gold(I)-Catalyzed Cycloisomerization of 1,7- and 1,8-Enynes: Application to the Synthesis of a New Allocolchicinoid. J. Org. Chem. 2008, 74, 5163-5166.

30. Carbone, M.; Yan, L.; Irace, C.; Mollo, E.; Castelluccio, F.; Di Pascale, A.; Cimino, G.; Santamaria, R.; Guo, Y.-W.; Gavagnin, M. Structure and Cytotoxicity of Phidianidines A and B: First Finding of 1,2,4-Oxadiazole System in a Marine Natural Product. Org. Lett. 2011, 13, 2516-2519.

31. Sunderhaus, J.D.; Sherman, D.H.; Williams, R.M. Studies on the Biosynthesis of the Stephadin and Notoamide Natural Products: A Stereochemical and Genetic Coundrum. Israel J. Chem. 2011, 51, 442-452.

32. Wu, M.; Wu, P.; Xie, H.; Wu, G.; Wei, X. Monoterpenoid Indole Alkaloid Mediating DNA Strand Scission from Turpina Arguta. Planta Medica 2011, 77, 284-286.

33. Ozcelik, B.; Kartal, M.; Orhan, I. Cytotoxicity, Antiviral and Antimicrobial Activities of Alkaloids, Flavonoids and Phenolic Acids. Pharm. Biol. 2011, 49, 396-402.

34. Yap, W.-S.; Gan, C.-Y.; Low, Y.-Y.; Choo, Y.-M.; Etoh, T.; Hayashi, M.; Komiyama, K.; Kam, T.-S. Grandilodines A-C, Biologically Active Indole Alkaloids from Kopsia Grandifolia. J. Nat. Prod. 2011, 74, 1309-1312.

35. Chung, Y.-M.; Lan, Y.-H.; Hwang, T.-L.; Leu, Y.-L. Anti-Inflammatory and Antioxidant Components from Hygroyza Aristata. Molecules 2011, 16, 1917-1927.

36. Foo, K.; Newhouse, T.; Mori, I.; Takayama, H.; Baran, P.S. Total Synthesis Guided Structure Elucidation of (+)-Psychotetramine. Angew. Chem. Int. Ed. 2011, 50, 2716-2719.

37. Finlayson, R.; Pearce, A.; Norrie, A.; Page, M.J.; Kaiser, M.; Bourguet-Kondracki, M.-L.; Harper, J.; Webb, V.; Copp, B. Didemnidines A and B, Indole Spermidine Alkaloids from the New Zealand Ascidian Didemnum sp. J. Nat. Prod. 2011, 74, 888-892.

38. Nakadate, S.; Nozawa, K.; Horie, H.; Fuji, Y.; Yaguchi, T. New Type Indole Diterpene, Eujindoles, From Eupenicillium Javanicum. Heterocycles 2011, 83, 351-356.

39. Ruiz-Sanchis, P.; Savina, S.; Albericio, F.; Alvarez, M. Structure, Bioactivity and Synthesis of Natural Products with Hexahydropyrrolo[2,3-b]Indole. Chem. Eur. J. 2011, 17, 1388-1408.

40. Yamada, Y.; Kitajima, M.; Kogure, N.; Wongseripipatana, S.; Takayama, H. Seven New Monoterpenoid Indole Alkaloids from Gelsemium Elegans. Chem. Asian J. 2011, 6, 166-173.

41. Palmisano, G.; Penoni, A.; Sisti, M.; Tiblietti, F.; Tollari, S.; Nicholas, K. Synthesis of Indole Derivatives with Biological Activity by Reactions between Unsaturated Hydrocarbons and N-Aromatic Precursors. Curr. Org. Chem. 2010, 14, 2409-2441.

42. Nakajima, T.O.; Satoshi, Y.; Fukuyama, T. Total Synthesis of (−)-Mersicarpine. J. Am. Chem. Soc. 2010, 132, 1236-1237.

43. Zhang, L. Tandem Au-catalyzed 3,3-Rearrangement-[2+2] Cycloadditions of Propargylic Esters: Expeditious Access to Highly Functionalized 2,3-Indoline-Fused Cyclobutanes. J. Am. Chem. Soc. 2005, 127, 16804-16805.

44. Ferrer, C.; Amijs, C.H.M.; Echavarren, A.M. Intra- and Intermolecular Reactions of Indoles with Alkynes Catalyzed by Gold. Chem. Eur. J. 2007, 13, 1358-1373.

45. u, Y.; Du, X.; Jia, X.; Liu, Y. Gold Catalyzed Intermolecular Reactions of (Z)-Enynols with Indoles for the Construction of Dihydrocyclohepta[b] indole Skeletons through a Cascade Friedel-Crafts/Hydroarylation Sequence. Adv. Syth. Catal. 2009, 351, 1517-1522.

46. Raffa, G.; Belot, S.; Balme, G.; Monteiro, N. Iodocyclization versus Diiodination in the Reaction of 3-Alkynyl-4-methoxycoumarins with Iodine: Synthesis of 3-Iodofuro[2,3-b]chromones. Org. Biomol. Chem. 2011, 9, 1474-1478.

47. Wang, L.; Peng, S.; Wang, J. Palladium-Catalyzed Cascade Reactions of Coumarins with Alkynes: Synthesis of Highly Substituted

Cyclopentadiene Fused Chromones. Chem. Commun. 2011, 47, 5422-5424.

48. Zhao, J.; Zhao, Y.; Fu, H. Transition-Metal-Free Intramolecular Ullmann-Type O-Arylation: Synthesis of Chromone Derivatives. Angew. Chem. Int. Ed. 2011, 50, 3769-3773.
49. Renault, J.; Qian, Z.; Uriac, P.; Goualt, N. Electrophilic Carbon Transfer in Gold Catalysis: Synthesis of Substituted Chromones. Tetrahedron Lett. 2011, 52, 2476-2479.
50. For a review, see: Boumendjel, A. Aurones: A Subclass of Flavone with Promising Biological Potential. *Curr. Med. Chem.* 2003, *10*, 2621-2330.
51. Brooks, C.J.; Watson, D.G. Phytoalexins. *Nat. Prod. Rep.* **1985**, 427-459.
52. Morimoto, M.; Fukumoto, H.; Nozoe, T.; Hagiwara, A.; Komai, K. Synthesis and Insect Antifeedant Activity of Aurones against Spodoptera Litura Larvae. *J. Agric. Food. Chem.* 2007, *55*, 700-705.
53. Okombi, S.; Rival, D.; Bonnet, S.; Mariotte, A.-M.; Perrier, E.; Boumendjel, A. Discovery of Benzylidenezofuran-3(2H)-one (Aurones) as Inhibitors of Tyrosinase Derived from Human Melanocytes. *J. Med. Chem.* 2006, *49*, 329-333.
54. Venkateswarlu, S.; Panchagnula, G.K.; Subbaraju, G.V. Synthesis and Antioxidative Activity of 3′,4′,6,7-tetrahydroxyaurone, a Metabollite of Biden Frondosa. *Biosci. Biotechnol. Biochem.* 2004, *68*, 2183-2185.
55. Auf′mkolk, M.; Koerhle, J.; Hesch, R.D.; Cody, V. Inhibition of Rat Liver Iodothyronine Deiodinase. Interaction of Aurones with the Iodothyronine Ligand-Binding Site. *Biol. Chem.* 1986, *261*, 11623-11630.
56. Donnelly, J.A.; Fox, M.J.; Sharma, T.C. α-Halo Ketones. XI Generation of the Wheeler Aurone Synthesis. *Tetrahedron* 1979, *35*, 875-879.
57. Bose, G.; Mondal, E.; Khan, A.T.; Bordoloi, M.J. An Environmentally Benign Synthesis of Aurones and Flavones from 2′-Acetoxychalcones using n-Tetrabutylammonium Tribromide. *Tetrahedron Lett.* 2001, *42*, 8907-8909.
58. Lévai, A.; Tökés, A.L. Synthesis of Aurones by the Oxidative Rearrangement of 2′-Hydroxychalcones with Thallium(III) Nitrate. *Synth. Commum.* 1982, *12*, 701-707.
59. Imafuku, K.; Honda, M.; McOmie, J.F.W. Cyclodehydrogenation of 2′-Hydroxychalcones with 2,3-Dichloro-5,6-dicyano-p-benzoquinone: A Simple Route for Flavones and Aurones. *Synthesis* 1987, 199-201.
60. Sekizaki, H. Synthesis of 2-Benzylidene-3(2H)-benzofuran-3-ones (Aurones) by Oxidation of 2′- Hydroxychalcones with Mercury(II) Acetate. *Bull. Chem. Soc. Jpn.* 1988, *61*, 1407-1409.
61. Thakkar, K.; Cushman, M. A Novel Oxidative Cyclization of

2′-Hydroxychalcones to 4,5- Dialkoxyaurones by Thallium(III) Nitrate. *J. Org. Chem.* 1995, *60*, 6499-6510.

62. An, Z.-W.; Catellani, M.; Chiusoli, G.P. Palladium-catalyzed Synthesis of Aurone from Salicyloyl Chloride and Phenylacetylene. *J. Organomet. Chem.* 1990, 397, 371-373.

63. Harkat, H.; Blanc, A.; Weibel, J.-M.; Pale, P. Versatile and Expeditious Synthesis of Aurones via AuI-Catalyzed Cyclization. *J. Org. Chem.* 2008, *73*, 1620-1623.

64. Atta-ur-Rahman; Choudhary, M.I.; Hayat, S.; Khan, A.; Ahmed, A. Two New Aurones from Marine Brown Alga Spatoglossum Variable. *Chem. Pharm. Bull.* 2001, *49*, 105-107.

65. Kobayashi, S.; Miyase, T.; Noguchi, H. Polyphenolic Glycosides and Oligosacharide Multiesters from the Roots of Polygala Dalmaisiana. *J. Nat. Prod.* 2002, *65*, 319-328.

66. Trost, B.M.; Dong, G. Total Synthesis of Bryostatin 16 Using a Pd-Catalyzed Diyne Coupling as Macrocyclization Method and Synthesis of C20-epi-Bryostatin 7 as a Potent Anticancer Agent. *J. Am. Chem. Soc.* 2010, *132*, 16403-16416.

67. For more information of bryostatin 16, see: Pettit, G.R.; Gao, F.; Blumberg, P.M.; Herald, C.L.; Coll, J.C.; Kamano, Y.; Lewin, N.E.; Schmidt, J.M.; Chapuis, J.-C. Antineoplastic Agents. 340. Isolation and Structural Elucidation of Bryostatins 16-18. J. Nat. Prod. 1996, 59, 286-289.

68. Hale, K.J.; Hummersome, M.G.; Manaviazar, S.; Frigerio, M. The Chemistry and Biology of the Bryostatin Antitumour Macrolides. Nat. Prod. Rep. 2002, 19, 413-453.

69. Newman, D.J.; Cragg, G.M. Marine Natural Products and related Compounds in Clinical and Advanced Preclinical Trials. J. Nat. Prod. 2004, 67, 1216-1238.

70. Hale, K.J.; Manaviazar, S. New Approaches to the Total Synthesis of Bryostatin Antitumor Macrolides. Chem. Asian J. 2010, 5, 704-754.

71. Manaviazar, S.; Hale, K.J. Total Synthesis of Bryostatin 1: A Short Route. Angew. Chem. Int. Ed. 2011, doi:10.1002/anie.201101562

72. Serrano, M. The Tumour Suppressor Protein p16INK4a. Exp. Cell. Res. 1997, 237, 7-13.

73. Pettit, G.R.; Inoue, M.; Kamano, Y.; Herald, D.L.; Arm, C.; Dufresne, C.; Christie, N.D.; Schmidt, J.M.; Doubek, D.L.; Krupa, T.S. Antineoplastic Agent. 174. Isolation and Styructure of the Cytostatic Depsipeptide Dolastatin 13 from the Sea Hare Dolabella Auricularia. J. Am. Chem. Soc. 1989, 110, 2006-2007.

74. Jeong, J.U.; Sutton, S.C.; Kim, S.; Fuchs, P.L. Biomimetic Total Syntheses of (+)-Cephalostatin 12, and (+)-Ritterazine K. J. Am. Chem. Soc. 1995, 117, 10157-10158.

75. LaCour, T.G.; Guo, C.; Bhandatu, S.; Fuchs, P.L.; Boyd, M.R. Interphylal Product Splicing: The First Total Syntheses of Cephalostatin 1, the North Hemisphere of Ritterazine G, and the High Active Hybrid Analog, Ritterostatin GN1N. J. Am. Chem. Soc. 1998, 120, 692-707.

76. Jeong, J.U.; Guo, C.; Fuchs, P.L. Synthesis of the South Unit of Cephalostatin. 7. Total Syntheses of (+)-Cephalostatin 7, (+)-Cephalostatin 12, and (+)-Ritterazine K. J. Am. Chem. Soc. 1999, 121, 2071-2084.

77. Kim, S.; Sutton, S.C.; Guo, C.; LaCour, T.G.; Fuchs, P.L. Synthesis of the North 1 Unit of the Cephalostatin Family from Hecogenin Acetate. J. Am. Chem. Soc. 1999, 121, 2056-2070.

78. Lee, S.; Fuchs, P.L. The First Total Synthesis of (Corrected) Ritterazine M. Org. Lett. 2002, 4, 317-318.

79. Fortner, K.C.; Kato, D.; Tanaka, Y.; Shair, M.D. Enantioselective Synthesis of (+)-Cephalostatin 1. J. Am. Chem. Soc. 2010, 132, 275-280.

80. Ueda, M.; Sato, A.; Ikeda, Y.; Miyoshi, T.; Naito, T.; Miyata, O. Direct Synthesis of Trisubstituted Isoxazoles through Gold-Catalyzed Domino Reaction of Alkynyl Oxime Ethers. Org. Lett. 2010, 11, 2594-2597.

81. He, W.; Li, C.; Zhang, L. An Efficient [2+2+1] Synthesis of 2,5-Disubstututed Oxazoles via Gold-Catalyzed Intermolecular Alkyne Oxidation. J. Am. Chem. Soc. 2011, 133, 8482-8485.

82. Hashmi, A.S.K.; Schuster, A.M.; Zimmer, M.; Rominger, F. Synthesis of 5-Halo-4H-1,3-oxazine- 6-amines by Copper-Mediated Domino Reaction. Chem. Eur. J. 2011, 17, 5511-5515.

83. Gao, X.; Pan, Y.-M.; Li, M.L.; Chen, L.; Zhan, Z.-P. Facile One-Pot Synthesis of Three Different Substituted Thiazoles from Propargylic Alcohols. Org. Biomol. Chem. 2010, 8, 3259-3266.

84. Yoshimatsu, M.; Matsui, M.; Yamamoto, T.; Sawa, A. Convinient Preparation of 4-Arylmethyland 4-Hetarylmethyl Thiazoles by Regioselective Cycloaddition Reactions of 3-Sulfanyl- and Selenylpropargyl Alcohols. Tetrahedron Lett. 2010, 66, 7975-7987.

85. Asanuma, Y.; Fujiwara, S.-I.; Shin-Ike, T.; Kambe, N. Selenoimidoylation of Alcohols with Selenium and Isocyanides and its Application to the Synthesis of Selenium-Containing Heterocycles. J. Org. Chem. 2004, 69, 4845-4848.

86. Wilckens, K.; Uhlemann, M.; Czekelius, C. Gold-Catalyzed endo-Cyclizations of 1,4-Diynes to Seven-Membered Ring Heterocycles. Chem. Eur. J. 2009, 15, 13323-13326.

87. Yan, B.; Zhou, Y.; Zhang, H.; Chen, J.; Liu, Y. Highly Efficient Synthesis of Functionalized Indolizines and Indolizinones by Copper-Calayzed Cycloisomerization of Propargylic Pyridines. J. Org. Chem. 2007, 72, 7783-7786.

88. Bunnelle, E.M.; Smith, C.R.; Lee, S.K.; Singaram, S.W.; Rhodes, A.J.; Sarpong, R. Pt-Catalyzed Cyclization/Migration of Propargylic Alcohols for the Synthesis of 3(2H)-Furanones, Pyrrolones, Indolizines, and Indolizinones. Tetrahedron Lett. 2008, 64, 7008-7014.

89. For some examples of non-metal catalyzed cycloisomerizations see: Ji, K.-G.; Zhu, H.-T.; Yang, F.; Shu, X.-Z.; Zhao, S.-C.; Liu, X.-Y.; Shaukat, A.; Liang, Y.-M. A Novel Iodine- Promoted Tandem Cyclization: An Efficient Synthesis of Substituted 3,4-Diiodoheterocyclic Compounds. Chem. Eur. J. 2010, 16, 6151-6154.

90. Zang, X.; Teo, W.T.; Chan, S.W.H.; Chan, P.W.H. Brønsted Acid Catalyzed Cyclization of Propargylic Alcohols wiyh Thioamides. Facile Synthesis of Di- and Trisubstituted Thiazoles. J. Org. Chem. 2010, 75, 6290-6293.

91. Kothandaraman, P.; Rao, W.; Foo, S.J.; Chan, P.W.H. Gold-Catalyzed Cycloisomerization Reaction of 2-Tosylamino-phenylprop-1-yn-3-ols via a Versatile Approach for Indole Synthesis. Angew. Chem. Int. Ed. 2010, 49, 4619-4623.

92. Jung, H.H.; Floreancig, P.E. Gold-Catalyzed Synthesis of Oxygen- and Nitrogen-Containing Heterocycles from Alkynyl Ethers: Application to the Total Synthesis of Andrachcinidine. J. Org. Chem. 2007, 72, 7359-7366.

93. Mill, S.; Hootelé, C. Alkaloids of Andrachne Aspera. J. Nat. Prod. 2000, 63, 762-764.

94. Forsyth, C.J. Asymmetric Synthesis, 2nd ed.; Christmann, M., Braese, S., Eds.; Wiley-VCH Verlag Gmbh & Co.: Weinheim, Germany, 2008; pp. 271-276.

95. Ballini, R.; Petrini, M. Nitroalkanes as Key Building Blocks for the Synthesis of Heterocyclic Derivatives. ARKIVOC 2008, 9, 195-223.

96. Forsyth, C.J. Asymmetric Synthesis; Christmann, M., Braese, S., Eds.; Wiley-VCH Verlag Gmbh & Co.: Weinheim, Germany, 2007; pp. 256-261.

97. Sherry, B.D.; Maus, L.; Laforteza, B.N.; Toste, D.F. Gold(I)-Catalyzed Synthesis of Dihydropyrans. J. Am. Chem. Soc. 2006, 128, 8132-8133.

98. Alcaide, B.; Almendros, P.; Carrascosa, R.; Torres, M.R. Gold/Acid-Cocatalyzed Regiodivergent Preparation of Bridged Ketals via Direct Bis-Oxycyclization of Alkynic Acetonides. Adv. Synth. Catal. 2010, 352, 1277-1283.

99. Perron, F.; Albizati, K.F. Chemistry of Spiroketals. Chem. Rev. 1989, 89, 1617-1661.

100. Aho, J.E.; Pihko, P.M.; Rissa, T.K. Nanometric Spiroketals in Natural Products: Structures, Sources, and Synthetic Strategies. Chem. Rev. 2005, 105, 4406-4440.

101. Rama Raju, B.; Saikia, A.K. Asymmetric Synthesis of Naturally Ocurring Spiroketals. Molecules 2008, 13, 1942-2038.

102. Blunt, J.W.; Copp, B.R.; Hu, W.P.; Munro, M.H.G.; Northcote, P.T.; Prinsep, M.R. Marine Natural Products. Nat. Prod. Rep. 2009, 26, 170-244.

103. Brimble, M.A.; Bryant, C.J. Synthesis of the Spiroketal-Containing Anti-Helicobacter Pylori Agents CJ-12,954 and CJ-13,014. Chem. Commun. 2006, 4506-4508.

104. Izquierdo-Cubero, I.; Plaza Lopez-Espinosa, M.T.; Kari, N. Synthesis of Optically Active Chalcogran from L-Sorbose. Carbohydr. Res. 1994, 261, 231-242.

105. Antoniotti, S.; Genin, E.; Michelet, V.; Genêt, J.-P. Highly Efficient Access to Strained Bicyclic Ketals via Gold-catalyzed Cycloisomerization of Bis-Homopropargylic Diols. J. Am. Chem. Soc. 2005, 127, 9976-9977.

106. Liu, L.-P.; Hammond, G.B. Highly Efficient and Tunable Synthesis of Dioxabicyclo[4.2.1] Ketals and Tetrahydropyrans via Gold-Catalyzed Cycloisomerization of 2-Alkynyl-1,5-diols. Org. Lett. 2009, 11, 5090-5092.

107. Liu, B.; De Brabander, J.K. Metal-catalyzed Regioselective Oxy-Functionalization of Internal Alkynes: An Entry into Ketones, Acetals, and Spiroketals. Org. Lett. 2006, 8, 4907-4910.

108. Hashmi, A.S.K.; Bührle, M.; Wölfe, M.; Rudolph, M.; Wieteck, M.; Rominger, F.; Frey, W. Gold Catalysis: Tandem reactions of Diyne-Diols and External Nuchleophiles as an Easy Access to Tricyclic Cage-Like Structures. Chem. Eur. J. 2010, 16, 9846-9854.

109. Zhang, Y.; Xue, J.; Xin, Z.; Xie, Z.; Li, Y. Gold-Catalyzed Double Intramolecular Alkyne Hydroalkoxylation: Synthesis of the Bisbenzannulated Spiroketal Core of Rubromycins. Synlett 2008, 6, 0940-0944.

110. Brockmann, H.; Lenk, W.; Scwantje, G.; Zeeck, A. Robromycins. Tetrahedron Lett. 1966, 3525-3530.

111. Goldman, M.E.; Salituro, G.S.; Bowen, J.A.; Williamson, J.M.; Zink, L.; Scleif, W.A.; Emini, E.A. Inhibition of Human Immunodefiency Virus-1 reverse Transcriptase Activity by Robromycins: Competitive Interaction at the Template Primer Site. Mol. Pharmacol. 1990, 38, 20.

112. Trani, A.; Dallanoce, C.; Pranzone, G.; Ripamonti, F.; Goldstein, B.P.; Cibiatti, R. Semisynthetic Derivatives of Purpuromycin as Potential Topical Agents for Vaginal Infections. J. Med. Chem. 1997, 40, 967-971.

113. Chino, M.; Nishikawa, K.; Umekia, M.; Hayashi, C.; Yamazaki, T.; Tsuchida, T.; Sawa, T.; Hamada, M.; Takeuchi, T. Heliquinomycin, A New Inhibitor of DNA Helicase, Produced by Streptomyces sp. MJ929-SF2. Taxonomy, Production, Isolation, Physicochemical Properties and Biological Activities. J. Antibiot. 1996, 49, 752.

114. Li, X.; Yao, Y.; Zheng, Y.; Satter, I.; Lin, W. Cephalosporolides H and I, Two Novel Lactones from a Marine-Derived Fungus, Penicillium sp. Arch. Phram. Res. 2007, 30, 812-815.

115. Penning, T.M. Inhibition of 5β-Dihydrocortisone Reduction in Rat Liver Cytosol: A Rapid Spectrophotometric Screen for Nonsteroidal Antiinflammatory Drug Potency. Pharm. Sci. 1985, 74, 651-654.

116. Sami, F.; Dudley, G.B. Stereocontrol of 5,5-Spiroketals in the Synthesis of Cephalosporolide H Epimers. Org. Lett. 2010, 12, 4698-4701.

117. Sami, F.; Dudley, G.B. A Gold-Catalyzed Alkyne-Diol Cycloisomerization for the Synthesis of Oxygenated 5,5-Spiroketals. Beilstein J. Org. Chem. 2011, 7, 570-577.

118. McMahon, I.; Silke, J. Winter Toxicity of Unknown Aetiology in Mussels. Harmful Algae News 1996, 14, 2.

119. Ito, E.; Satake, M.; Ofuji, N.; Kurita, N.; McMahon, T.; James, K.; Yasumoto, T. Multiple Organ Damage Caused by a New Toxin Azaspiracid, Isolated from Mussels Produced in Ireland. Toxicon 2000, 38, 917-930.

120. Satake, M.; Ofuji, K.; Naoki, H.; James, K.J.; Furey, A.; McMahon, T.; Silke, J.; Yasumoto, T. Azaspiracid, a New Marine Toxin Having Unique Spiro Ring Assemblies, Isolated from Irish Mussels, Myttilus Edulis. J. Am. Chem. Soc. 1998, 120, 9967-9968.

121. Nicolau, K.C.; Kotifs, T.V.; Vyskocil, S.; Petrovic, G.; Ling, T.; Yamamda, Y.M.A.; Tang, W.; Frederick, M.O. Structural Revision and Total Synthesis of Azaspiracid-1, part 2: Definition of the ABCD Domain and Total Synthesis. Angew. Chem. Int. Ed. 2004, 43, 4318-4324.

122. Inoki, S.; Mukaiyama, T. A Convinient Method for the Stereoselective Preparation of trans-(2- Hydroxymethyl)tetrahydrofurans by the Oxidative Cyclization of 5-Hydroxy-1-alkenes with Molecular Oxygen Catalyzed by Cobalt(II) Complex. Chem. Lett. 1990, 67-70.

123. Li, Y.; Zhou, F.; Forsyth, C.J. Gold(I)-Catalyzed Bis-Spiroketalization: Synthesis of the Trioxadispiroketal-Containing A−D Rings of Azaspiracid. Angew. Chem. Int. Ed. 2007, 46, 279-282.

124. Tachibana, K.; Scheuer, P.J.; Tsukitamni, Y.; Kikuchi, H.; Van Engen, D.; Clardy, J.; Gopichand, Y.; Schimtz, F.J. Okadaic Acid, A Cytotoxic Ployether from Two Marine Sponges of the Genus Halichondria. J. Am. Chem. Soc. 1981, 103, 2469-2471.

125. Dounay, A.B.; Forsyth, C.J. Okadaic Acid: The Archetypal Serine/ Threonine Protein Phosphatase Inhibitor. Curr. Med. Chem. 2002, 9, 1939-1980.

126. Scheuer, P. Marine Natural Products Research: A Look into the Dive Bag J. Nat. Prod. 1995, 58, 335-343.

127. Yasumoto, T.; Murata, M.; Oshima, Y.; Sano, M.; Matsumoto, G.K.; Clardy, J. Diarrhetic Shellfish Toxins. Tetrahedron 1985, 41, 1019-1022.

128. Bialojan, C.; Takai, A. Inhibitory Effect of Marine-Sponge Toxin, Okadaic Acid, on Protein Phosphatases. Specifity and Kinetics. Biochem. J. 1988, 256, 283-290.

129. Fang, C.; Pang, Y.; Forsyth, C.J. Formal Total Synthesis of Okadaic Acid via Regiocontrolled Gold(I)-Catalyzed Spiroketalizations. Org. Lett. 2010, 12, 4528-4531.

130. Wang, J.; Soisson, S.M.; Young, K.; Shoop, W.; Kodali, S.; Galgoci, A.; Painter, R.; Parthasarathy, G.; Tang, Y.S.; Cummings, R.; et al. Platensimycin is a Selective FabF Inhibitor with Potent Antibiotic Properties. Nature 2006, 441, 358-361.

131. Singh, S.B.; Jayasuriya, H.; Ondeyka, J.G.; Herath, K.B.; Zhang, C.; Zink, D.L.; Tsou, N.N.; Ball, R.G.; Basilio, A.; Genilloud, O.; Diez, M.T.; Vicente, F.; Pelaez, F.; Young, K.; Wang, J. Isolation, Structure, and Absolute Stereochemistry of Platensimycin, a Broad Spectrum Antibiotic Discovered Using an Antisense Differential Sensitivity Strategy. J. Am. Chem. Soc. 2006, 128, 11916-11920.

132. Häbich, D.; von Nussbaum, F. Platensimycin, A New Antibiotic and "Superbug Challenger" from Nature. ChemMedChem 2006, 1, 951-954.

133. Yeung, Y.-Y.; Corey, E.J. A Simple, Efficient, and Enantiocontrolled Synthesis of a Near- Structural Mimic of Platensimycin. Org. Lett. 2008, 17, 3877-3878.

134. Nicolau, K.C.; Tang, Y.; Wang, Y.; Stepan, A.F.; Li, A.; Montero, A. Total Synthesis and Antibacterial Properties of Carbaplatensimycin. J. Am. Chem. Soc. 2007, 129, 14850-14851.

135. Nicolau, K.C.; Lister, T.; Denton, R.M.; Montero, A.; Edmons, D.J. Adamantaplatensimycin: A Bioactive Analogue of Platensimycin. Angew. Chem. Int. Ed. 2007, 46, 4712-4714.

Chapter 8

SOLID-PHASE SYNTHESIS OF METHYL N-(PYRIMIDIN-2-YL) GLYCINATE

Denis S. Ermolat'ev and Eugene V. Babaev*

Chemistry Department, Moscow State University, Moscow 119899, Russia

INTRODUCTION

Aminoacid derivatives, in particular N-substituted α-aminoacids, are important synthetical precursors and building blocks in combinatorial chemistry. It is sometimes very difficult to obtain such N-substituted acids, especially for the cases when a heterocyclic substituent is attached to the amino group. Attempting to prepare ethyl N-(pyrimidin-2-yl)glycinate via reaction of 2-chloropyrimidine and ethyl glycinate we observed predominant formation of diketopiperazine. An alternative strategy - alkylation of the endocyclic nitrogen of 2-aminopyrimidine with chloroacetic acid (or its esters) followed by Dimroth rearrangement (which usually is a safe route to 2-alkylaminopyrimidines) – also failed due to intramolecular self-condensation between the 2-amino group and the acetic acid fragment. An alternative strategy is therefore required to prepare such a target.

RESULTS AND DISCUSSION

Our aim was to synthesize N-(pyrimidin-2-yl)glycinate (I). This compound can be formally considered as a 2-alkylaminopyrimidine, and there are two main synthetic routes for this class of compounds (Scheme 1).

Scheme 1

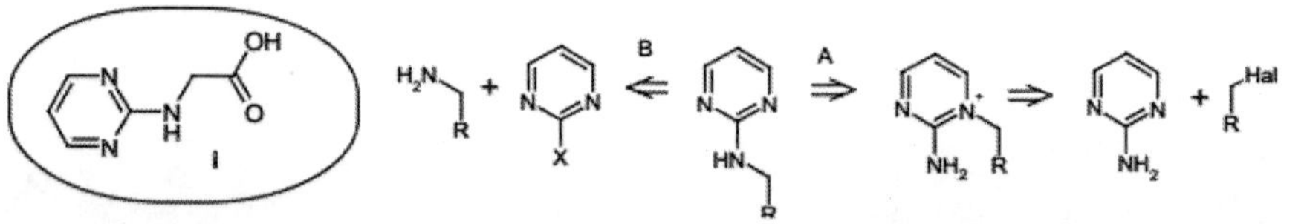

The first route (A) is the Dimroth rearrangement of 1-alkylpyrimidinium salts, which in turn can be easily prepared by alkylation of the endocyclic nitrogen atom of 2-aminopyrimidine (one would expect that for the target molecule I such an alkylating agent might correspond to an α-halogenated acetic acid derivative). The second strategy (B) is a nucleophilic substitution in a pyrimidine ring with a good leaving group in the alpha position by amines.

We have attempted to apply strategy A by alkylating 2-aminopyrimidine with ethyl bromoacetate and sodium chloroacetate. Both salts were obtained, and their structures were proven by NMR. However, our attempts to perform the Dimroth rearrangement gave undesirable compounds with deep coloration. It was assumed that the expected intramolecular self-condensation between the 2-amino group and the acetic acid fragment occurred, probably followed by further oxidative dimerization under basic conditions (Scheme 2):

Scheme 2

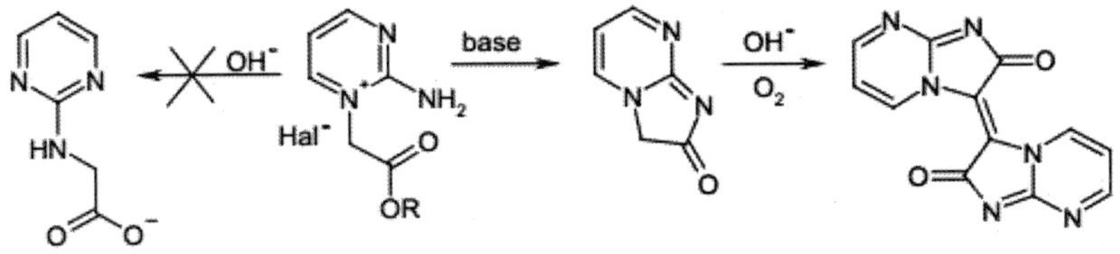

The second route to prepare the target compound I is the nucleophilic substitution in the pyrimidine ring bearing a good leaving group in the alpha position by amines. Such nucleophilic substitutions in pyrimidine rings are well-studied reactions [1]. We tried to apply this reaction to 2-chloropyrimidine using ethyl glycinate chlorohydrate in the presence of strong base (Scheme 3).

Unfortunately, the expected substitution proceeds quite slowly, in contrast to the quick self-reaction of ethyl glycinate, and consequently we observed predominant formation of diketopiperazine.

Scheme 3

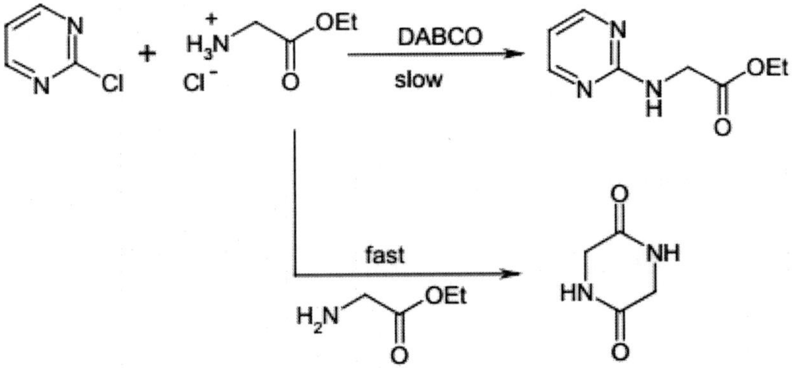

Attempting to resolve the problem, we have applied a solid-phase synthetic approach (Scheme 4). At first Boc-protected glycine (in the form of its cesium salt) was attached onto Merrifield resin under standard conditions [2] in order to prevent possible self-condensation. Solid phase supported glycine was deprotected and successfully involved in the reaction with 2-chloropyrimidine. The degree of conversion was monitored by the ninhydrin test. Finally, the product was cleaved from the Merrifield resin in the presence of sodium methoxide in the methanol/tetrahydrofuran solution [3].

Scheme 4

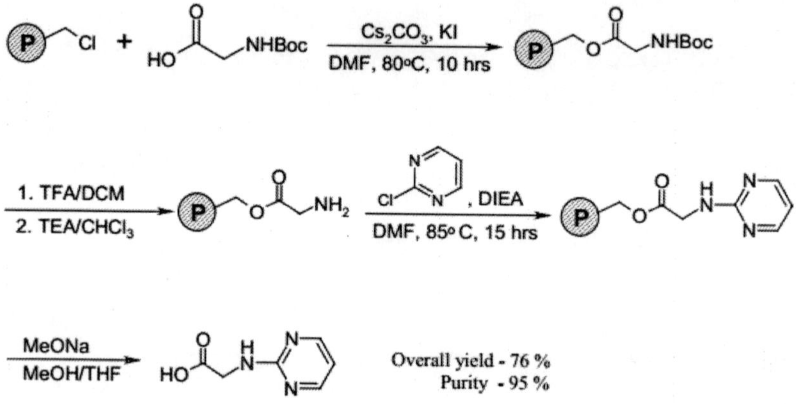

P–Cl + HO–C(O)–CH₂–NHBoc $\xrightarrow[\text{DMF, 80°C, 10 hrs}]{Cs_2CO_3\text{, KI}}$ P–O–C(O)–CH₂–NHBoc

$\xrightarrow[\text{2. TEA/CHCl}_3]{\text{1. TFA/DCM}}$ P–O–C(O)–CH₂–NH_2 $\xrightarrow[\text{DMF, 85°C, 15 hrs}]{\text{2-chloropyrimidine, DIEA}}$ P–O–C(O)–CH₂–NH–(pyrimidin-2-yl)

$\xrightarrow[\text{MeOH/THF}]{\text{MeONa}}$ HO–C(O)–CH₂–NH–(pyrimidin-2-yl)

Overall yield - 76 %
Purity - 95 %

Conclusions

We have found a solid-phase synthetic strategy to be advantageous for synthesis of the target compound I, compared to usual liquid phase processes.

EXPERIMENTAL

General

1H-NMR spectra were obtained using a Bruker AC 400 NMR spectrometer and were recorded at 360 MHz. Merrifield resin (2% crosslinked with divinylbenzene) was obtained from Reanal. All reagents were obtained from ACROS.

SYNTHETIC PROCEDURE

Merrifield resin (1 g, 1.0 mmol) was swelled in DMF solution (6 mL) containing of Boc-protected glycine (0.525 g, 3 mmol), cesium carbonate (0.49 g, 1.5 mmol) and potassium iodide (0.17 g, 1 mmol). The mixture was constantly stirred and heated at 80oC for about 10 hours. The resin was washed successively with DMF, MeOH, DCM and dried in vacuo. Immobilized glycine was deprotected by suspending the resin (1.15 g) in a 1:1 mixture of TFA and DCM at room temperature for 0.5 hours [4]. The resin was washed sequentially with 1:1 DCM-DMF, DCM, MeOH, 1:10 TEA CHCl3, and MeOH. In the next step dried resin (1.06 g) was heated in DMF solution containing 2-chloropyrimidine (1.137 g, 10 mmol) and DIEA (0.55 mL, 4 mmol) at 90oC for at least 15 hours.

The resin turned to brown color and gave a negative ninhydrin test. After washing by DMF, DCM and drying we obtained 1.13 g of dark colored resin. Then this amount of resin was refluxed for 6 hours in a 4:1 mixture of THF-MeOH (5 mL) in the presence of sodium methoxide (0.5 mL of 0.2 M MeONa solution in MeOH). The resin was filtered and washed with THF (15 mL) and MeOH (10 mL, twice).

After evaporation of solvents a solid residue was obtained (0.126 g) and analyzed without further purification. The overall yield was 76% and the purity of the product was evaluated as 95%. The structure was proven by its 1H-NMR spectra (CDCl3, 360 MHz), δ: 8.21-8.25 (2CH, d, J=7.6 Hz, 2H); 6.49-6.62 (CH, t, J=7.5 Hz, 1H); 5.45-5.60 (NH, bs, 1H); 4.13-4.16 (CH2CO2, d, J=5.2 Hz, 2H); 3.70 (CH3, s, 3H).

REFERENCES

1. Illuminati G., Nucleophilic heteroaromatic substitution. Adv. Heterocycl. Chem. 1964, Vol. 3, 275-283

2. Merrifield R.B.. Solid phase peptide synthesis. I. The synthesis of a tetrapeptide J. Am. Chem. Soc. 1963, 85, 2149-2154

3. Kobayashi, S.; Akiyama, R.; Kitagawa, H. Polymer-Supported α-Imino Acetates. Versatile Reagents for the Synthesis of alpha-Amino Acid Libraries J. Comb. Chem. 2000, 2, 438-440

4. Mukaiyama T., Oxidation-Reduction Condensation a New Method for Peptide Synthesis. Synth. Comm. 1972, 2, 243-265

Chapter 9

ENANTIOSPECIFIC SYNTHESIS OF [2.2]PARACYCLOPHANE-4-THIOL AND DERIVATIVES

Gareth J. Rowlands[1] and Richard J. Seacome[2]

[1]Institute of Fundamental Sciences, Massey University, Private Bag 11 222, Palmerston North, New Zealand

[2]Chemistry Division, Department of Chemistry and Biochemistry, University of Sussex, Falmer, Brighton, BN1 9QJ, UK

ABSTRACT

This paper describes a simple route to enantiomerically enriched [2.2] paracyclophane-4-thiol via the stereospecific introduction of a chiral sulfoxide to the [2.2]paracyclophane skeleton. The first synthesis of an enantiomerically enriched planar chiral benzothiazole is also reported.

INTRODUCTION

[2.2]Paracyclophane (**1**; R = H) is a fascinating compound comprising of two eclipsing benzene rings that are held in place by two ethyl bridges at the para positions (Figure 1). The close proximity of the arene moieties results in strong electronic and structural interactions between the two rings and between substituents appended to each layer [1,2]. The resulting unique properties have led to derivatives of [2.2]paracyclophane being employed in a wide range of disciplines including polymer, material and electronic chemistry [3-9]. Whilst enantiomerically pure derivatives have been utilised in chiral catalysis [10,11] and as probes for biological recognition processes [12-14], the full potential of these systems has not been realised due to the difficulties encountered when trying to access enantiomerically pure [2.2]paracyclophane derivatives [15].

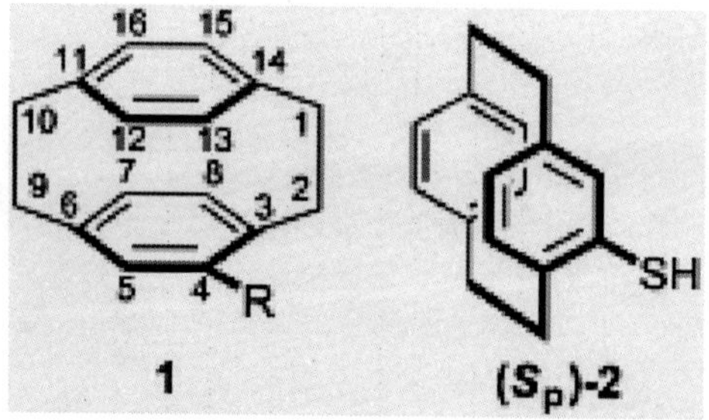

Figure 1: [2.2]Paracyclophane (**1**) showing standard numbering and [2.2]paracyclophane-4-thiol (**2**).

Chiral sulfur [2.2]paracyclophane derivatives are beginning to attract attention due to the great potential such compounds exhibit [16-18]. Non-cyclophane-based thiophenol derivatives have been employed in the nucleophilic addition of thioacetals to suitable electrophiles [19,20], sigmatropic rearrangements [21] and as either thiyl radical precursors [22] or as a source of hydrogen in radical chemistry [23]. With the appropriate sulfur derivative, stereoselective variants of all these transformations can be envisaged.

Currently, there are few examples of sulfur containing [2.2] paracyclophane compounds; aryl sulfonylation and the related sulfenylation facilitates the synthesis of sulfonic acids, sulfonamides and protected thiols [24-26] whilst directed metallation has allowed the formation of various sulfides [27-30]. Very few methodologies allow the synthesis of simple chiral monosubstituted thiols such as [2.2]paracyclophane-4-thiol **2** (Figure 1); the first reported preparations of racemic **2** were the conversion of 4-hydroxy[2.2] paracyclophane to the desired compound *via* a Newman-Kwart reaction or the direct reaction of 4-lithio[2.2]paracyclophane with sulfur [17]. Use of enantiomerically pure 4-hydroxy[2.2] paracyclophane or application of our own sulfoxide-metal exchange protocol [31] would permit enantiospecific variants of either route, but neither has been reported. An elegant entry to a variety of racemic alkyl sulfides and sulfoxides by an S_EAr reaction mediated by a sulfonium salt has recently been divulged [16]. The only reported synthesis of enantiomerically pure [2.2]paracyclophane-4-thiol entails the palladium-mediated addition of triisopropylsilanethiol to a triflate formed from previously resolved (*R*)-4-hydroxy[2.2] paracyclophane [18]. We are developing a 'tool-box' for the synthesis of enantiomerically enriched [2.2]paracyclophane derivatives based on the chemistry of [2.2]paracyclophane sulfoxides [15,31-33]. This methodology has allowed us to develop routes to enantiomerically pure 4-monosubstituted [2.2]paracyclophanes[31] along with a range of disubstituted derivatives [33]. The basis of the strategy is the stereospecific introduction of a sulfoxide to [2.2]paracyclophane to give readily separable diastereoisomers, thus resolving the planar chirality[34]. The sulfoxide moiety is used to direct further elaboration of the [2.2]paracyclophane framework or is displaced *via* sulfoxide-metal exchange. We were interested in modifying this methodology to permit the synthesis of enantiomerically enriched [2.2]paracyclophane-4-thiol and related compounds.

RESULTS AND DISCUSSION

The synthesis of (S_p)-**2** is depicted in Scheme 1. Key to the success of this strategy was the resolution of the planar chirality of [2.2] paracyclophane by incorporation of the *tert*-butylsulfinyl moiety to give the diastereoisomers (S_p,R_S)-**5** and (R_p,R_S)-**5**. Standard iron-catalysed bromination of **1** gave (±)-4-bromo[2.2]paracyclophane **3** in good yield [35,36]. Halogen-lithium exchange and addition to

Ellman's (*R*)-*tert*-butyl *tert*-butanethiosulfinate [37] **4**furnished a 1 : 1.4 mixture of (S_p,R_S)-**5** and (R_p,R_S)-**5** in a combined 72% yield [33]. The two diastereoisomers are readily separable by standard column chromatography, with the diastereoisomer (S_p,R_S)-**5** being eluted first. As (*R*)-**4** was prepared with an *ee* of 80%, as judged by optical rotation, we assume that each diastereoisomer displays an *ee* of 80%. As the sulfinylation reaction proceeds with inversion at sulfur, the two diastereoisomers only differ by the chirality of the [2.2] paracyclophane therefore allowing the facile resolution of the planar chirality. The assignment of configuration is based on a combination of X-ray studies [33,38], formation of all stereoisomers and analogy to our previous tolylsulfinyl chemistry [31,39].

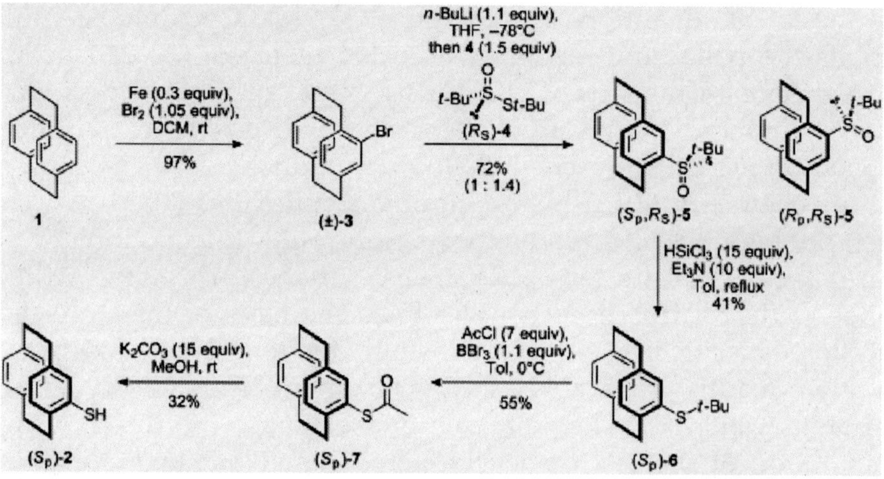

Scheme 1: Conversion of [2.2]paracyclophane to enantiomerically enriched [2.2]paracyclophane-4-thiol.

Unlikethepreviouslyprepared4-tolylsulfinyl[2.2]paracyclophane [31], direct sulfoxide-metal exchange was not possible with the *tert*-butyl derivative [38]; presumably, the *tert*-butyl group and the lower ring of the [2.2]paracyclophane moiety shield the sulfur from attack. As a result a stepwise procedure for the conversion of (S_p,R_S)-5 to [2.2]paracyclophane-4-thiol (S_p)-2 was investigated (Scheme 1). The first step, the reduction of (S_p,R_S)-5 to sulfide (S_p)-6, proved the most problematic; use of a large excess of trichlorosilane and triethylamine resulted in deoxygenation in moderate yield after recrystallisation

[40]. By comparison, reduction of the less hindered aryl sulfoxide, (R_p,S_S)-4-bromo-13-*p*-tolylsulfinyl[2.2]paracyclophane, occurs efficiently in 98% yield suggesting the *tert*-butyl group is the source of the problem [39]. Exchange of the *tert*-butyl group for an acetyl group was achieved by reaction of a mixture of (S_p)-6 and acetyl chloride in toluene with boron tribromide [41]. The resulting thioacetic acid *S*-[2.2]paracyclophane ester (S_p)-7 readily undergoes simple base-catalysed hydrolysis to give the desired (*S*)-(+)-[2.2]paracyclophane-4-thiol (S_p)-2.

There are two advantages to our methodology compared to the previously reported syntheses of [2.2]paracyclophane thiols; the first is that resolution of the planar chirality is complicit in the addition of the sulfur moiety and does not require resolution of any precursors. Secondly, the sulfinyl moiety permits further functionalisation of the [2.2]paracyclophane skeleton [33]. It is the latter reason that prompted the synthesis of the thiol *via* the *tert*-butyl derivative and not by direct sulfoxide-metal exchange; whilst this route would have delivered**2** more rapidly it would not have permitted elaboration of the [2.2]paracyclophane framework. The utility of the *tert*-butyl derivative is demonstrated in the synthesis of the planar chiral benzothiazole (R_p)-**10** (Scheme 2).

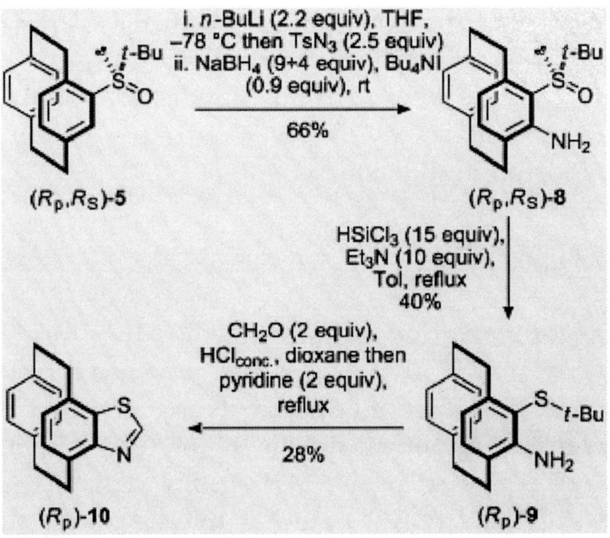

Scheme 2: Synthesis of [2.2](4,7)benzo[*d*]thiazoloparacyclophane (R_p)-**10**.

Benzothiazoles are important heterocycles having found use as dyes, pharmaceuticals and ligands in catalysis[42]. Planar chiral heterocycles are still rare but show considerable potential as probes in stereocontrolled recognition processes in biological systems as highlighted by Gmeiner [12-14] and as ligands or catalysts [43-47]. We have previously prepared planar chiral benzimidazoles [32] and wanted to extend the range of heterocycles that could be accessed.

Diastereoisomer (R_p,R_S)-**5** was functionalised by sulfinyl-directed *ortho* lithiation with *n*-butyllithium followed by reaction with tosyl azide. The resulting azo[2.2]paracyclophane was reduced *in situ* to give the amine (R_p,R_S)-**8** in good yield for the two steps (Scheme 2). Trichlorosilane-mediated deoxygenation proceeded uneventfully to furnished amino sulfide (R_p)-**9**. Simultaneous sulfide deprotection and thiazole formation was achieved by treating (R_p)-**9** with concentrated hydrochloric acid, paraformaldehyde and pyridine [48]. Although the yield of (R_p)-**10** is not yet satisfactory, it shows the potential of our methodology for the formation of these valuable heterocycles.

In conclusion, we have developed a straightforward method for the synthesis of enantiomerically enriched [2.2]paracyclophane-4-thiol that does not rely on the resolution of precursors to the introduction of the sulfur moiety. Furthermore, we have shown that this methodology has the potential to produce a wide-range of thiol derivatives and this has permitted the first synthesis of a planar chiral benzothiazole, [2.2](4,7)benzo[*d*]thiazoloparacyclophane. The use of these thiols in asymmetric synthesis is currently being investigated and will be reported in due course.

EXPERIMENTAL

NMR spectra were recorded on a Bruker 400 MHz , Bruker 300 MHz, Varian 500 MHz or Varian 400 MHz using residual isotopic solvent as internal reference. Infrared spectra were recorded on a Perkin-Elmer 1600 Fourier Transform spectrometer. Mass spectra and exact mass data were recorded by Dr. Ali Abdul-Sada at the University of Sussex or by the EPSRC national mass spectrometry service, Swansea. Melting points were recorded on a Gallenkamp melting point apparatus and are uncorrected. Optical rotation was recorded on a Perkin Elmer 241 polarimeter using a sodium lamp emitting at 589 nm. All samples were measured in chloroform (c = 1) in a 10 cm cell and an average taken of 10 readings; average temperature was 27 °C. Glassware was oven dried and reactions were performed under an inert atmosphere of nitrogen or argon where applicable. Chromatography refers to flash column chromatography on Merck Kieselgel 60 (230-400 mesh) or Fischer Davisil 60 silica gel unless otherwise stated. TLC refers to analytical thin-layer chromatography performed using pre-coated glass-backed plates (Merck Kieselgel 60 F_{254}) and visualised with ultraviolet light, iodine, acidic ammonium molybdate (IV), acidic ethanolic vanillin, aqueous potassium manganate(VII), ninhydrin or acidic anisaldehyde as appropriate. Petrol refers to redistilled petroleum ether (60–80 °C), and ether to diethyl ether. Ether and THF were distilled from sodium-benzophenone ketyl, toluene from 4Å molecular sieves or calcium chloride. Dioxane was stored over sodium wire and DMF was stored over 4Å molecular sieves.

(±)-4-Bromo[2.2]paracyclophane (3)

All stages of this reaction were performed in the dark by covering the flasks with aluminium foil. Bromine (7.8 mL, 0.15 mol, 1.05 equiv) was dissolved in DCM (1.5 L). 10% of the solution (150 mL) was transferred to a flask containing iron filings (2.4 g, 0.04 mol, 0.3 equiv) and stirred at rt for 1.5 h. A solution of [2.2]paracyclophane (30.0 g, 0.14 mol, 1.0 equiv) in DCM (2.8 L) was added and the suspension stirred for 20 min. The remaining solution of bromine in DCM (1350 mL) was added via cannula over 30 minutes. After 5 min TLC indicated complete reaction and saturated aqueous NH_4Cl solution (5.0 L) was added. The aqueous phase was extracted with DCM (3 × 1 L) and the combined organic extracts washed with aqueous $Na_2S_2O_3$ solution (10% w/v; 1 L) then dried ($MgSO_4$). The solvent was removed to give **3** as a white powder (40.2 g, 97%); mp = 134–135 °C (lit. [35]: 136–138 °C); ν_{max} (film) 3055, 2987, 1422, 1265, 896, 739 and 705 cm^{-1}; δ_H (300 MHz, $CDCl_3$) 7.15 (1H, d, J = 9.0 Hz, H-13), 6.56 (1H, d, J = 9.0 Hz, H-8), 6.51–6.46 (4H, m, H-5, H-7, H-15 and H-16), 6.44 (1H, d, J = 9.0 Hz, H-12), 3.46 (1H, ddd, J = 10.2, 7.7, 2.1 Hz, H-2 *endo*), 3.24–3.15 (1H, m, H-1 *endo*), 3.98–3.13 (5H, m, H-1 *exo*, 2 × H-9 and 2 × H-10), 2.95–2.77 (1H, m, H-2 *exo*); δ_C (75 MHz,

$CDCl_3$) 142.0, 139.7, 139.5, 137.6, 135.4, 133.7, 133.4, 133.3, 132.6, 131.8, 129.1, 127.4, 36.2, 35.9, 35.2, 33.8.

($\boldsymbol{R_p,R_S}$)-(−)-4-*tert*-Butylsulfinyl[2.2]paracyclophane [($\boldsymbol{R_p,R_S}$)-5] and ($\boldsymbol{S_p,R_S}$)-(−)-4-*tert*-Butylsulfinyl[2.2] paracyclophane [($\boldsymbol{S_p,R_S}$)-5]

To a solution of (±)-4-bromo[2.2]paracyclophane (5.50 g, 19.16 mmol, 1.0 equiv) in THF (180 mL) at −78 °C was added *n*-BuLi (2.5 M in hexanes; 8.5 mL, 21.08 mmol, 1.1 equiv) dropwise over 15 min. After 45 min, (*R*)-*tert*-butyl*tert*-butanethiosulfinate (*R*)-**4** (80% *ee*; 5.57 g, 28.74 mmol, 1.5 equiv) was added as a solid and the reaction stirred at rt overnight. The solvent was removed and the resulting residue purified by chromatography (Et_2O/*n*-heptane gradient) to yield (R_p,R_S)-**5** (1.79 g, 30%) and (S_p,R_S)-**5** (2.51 g, 42%).

($\boldsymbol{R_p,R_S}$)-(−)-4-*tert*-Butylsulfinyl[2.2]paracyclophane [($\boldsymbol{R_p,R_S}$)-5]

mp 124–126 °C; $[\alpha]_D$ −39.6 (*c* 1, $CHCl_3$) (assumed 80% *ee* see text); ν_{max} (film) 2962, 2926, 1585, 1456, 1473, 1500, 1170, 1054, 1024, 908 and 847 cm^{-1}; δ_H (500 MHz, $CDCl_3$) 7.02 (1H, s, H-5), 6.83 (1H, d, *J* = 7.5 Hz, H-13), 6.62 (1H, d, *J* = 7.5 Hz, H-7), 6.54 (1H, d, *J* = 8.0 Hz, H-12), 6.52 (2H, s, H-15, H-16), 6.48 (1H, d, *J* = 8.0 Hz, H-8), 3.54 (1H, ddd, *J* = 13.5, 12.3, 2.5 Hz, H-2 *endo*), 3.27 (1H, ddd, *J* = 13.0, 9.1, 5.5 Hz, H-1 *endo*), 3.16–3.06 (5H, m, H-1 *exo*, 2 × H-9 & 2 × H-10), 2.89 (1H, ddd, *J* = 10.0, 8.9, 5.5 Hz, H-2 *exo*), 1.05 (9H, s, *t*-Bu); δ_C (125 MHz, $CDCl_3$) 140.7 (C), 139.5 (C), 139.0 (C), 138.9 (C), 138.9 (C), 136.0 (CH), 134.6 (CH), 133.1 (CH), 132.7 (CH), 132.6 (CH), 132.3 (CH), 130.3 (CH), 56.6 (C), 35.2 (CH_2), 35.1 (CH_2), 34.7 (CH_2), 33.6 (CH_2), 22.7 (CH_3); *m/z* (EI+) 256 $[M\text{-}t\text{-}Bu]^+$, 240, 152, 135, 123, 104, 91, 78 (Found: $[M]^+$, 312.1539. $C_{20}H_{24}OS$ requires $[M]^+$, 312.1542).

($\boldsymbol{S_p,R_S}$)-(+)-4-*tert*-Butylsulfinyl[2.2]paracyclophane [($\boldsymbol{S_p,R_S}$)-5]

mp = 122–124 °C; $[\alpha]_D$ +151.4 (*c* 1, $CHCl_3$) (assumed 80% *ee* see text); ν_{max} (film) 2970, 2927, 2852, 1587, 1474, 1459, 1432, 1410, 1175, 1039,

904, 847 and 805 cm^{-1}; δ_H (500 MHz, $CDCl_3$) 6.93 (1H, d, *J* = 10.5 Hz, H-13), 6.58 (1H, d, *J* = 10.0 Hz, H-12), 6.54–6.47 (5H, m, H-5, H-7, H-8, H-15, H-16), 4.35 (1H, t, *J* = 14.5 Hz, H-2 *endo*), 3.37 (1H, ddd, *J* = 12.5, 13.0, 7.0 Hz, H-1 *endo*), 3.22–2.98 (5H, m, H-1 *exo*, 2 × H-9 and 2 × H-10), 2.82–2.77 (1H, m, H-2 *exo*), 1.05 (9H, s, *t*-Bu); δ_C (125 MHz, $CDCl_3$) 142.2 (C), 140.7 (C), 139.3 (C), 139.0 (C), 137.7 (CH), 135.7 (CH), 134.2 (CH), 133.3 (CH), 133.0 (CH), 132.7 (CH), 132.6 (CH), 132.5 (CH), 56.5 (C), 36.1 (CH_2), 35.2 (CH_2), 35.0 (CH_2), 34.2 (CH_2), 23.1 (CH_3); *m/z* (EI+) 256 [M − *t*-Bu]$^+$, 240, 207 [[2.2]paracyclophane]$^+$, 152, 136, 123, 104, 91, 78 (Found: [M]$^+$, 312.1545. $C_{20}H_{24}OS$ requires [M]$^+$, 312.1542).

(S_p)-(+)-4-*tert*-Butylsulfanyl[2.2]paracyclophane [(S_p)-6]

Triethylamine (11.92 mL, 85.58 mmol, 10 equiv) was added to a solution of trichlorosilane (17.9 mL, 128.36 mmol, 15 equiv) and (S_p,R_S)-(+)-4-*tert*-butylsulfinyl[2.2]paracyclophane (2.67 g, 8.56 mmol, 1.0 equiv) in toluene (41 mL). The reaction was heated to reflux for 18 hours. After cooling to 0 °C a solution of aqueous NaOH (3.0 M; 200 mL) was added carefully. The aqueous phase was extracted with Et_2O (3 × 100 mL) and the combined organic phases dried ($MgSO_4$). After removal of the solvent, the residue was purified by chromatography (5% Et_2O/hexane) followed by trituration of the yellow semi-solid with petrol gave (S_p)-**6** as a white solid (1.0 g, 41.0%); mp = 53 °C; $[\alpha]_D$ + 89.5 (*c* 1, $CHCl_3$) (assumed 80% *ee* see text); ν_{max} (film) 2957, 2926, 2894, 1471, 1455, 1433, 1411, 1389, 1362 and 1165 cm^{-1}; δ_H (500 MHz, $CDCl_3$) 6.68 (1H, s, H-5), 6.68 (1H, d, *J* = 10.0 Hz, H-13), 6.54 (2H, dd, *J* = 8.0, 3.5 Hz, H-7, H-8), 6.50 (2H, s, H-15, H-16), 6.45 (1H, d, *J* = 8.0 Hz, 12-H), 3.84 (1H, t, *J* = 11.0 Hz, H-2 *endo*), 3.20–3.14 (1H, m, H-1 *endo*), 3.12–2.97 (5H, m, H-1 *exo*, 2 × H-9, 2 × H-10), 2.83 (1H, ddd, *J* = 12.0, 5.5, 3.7 Hz, H-2 *exo*), 1.17 (9H, s, *t*-Bu); δ_C (125 MHz, $CDCl_3$) 145.7 (C), 144.4 (C), 139.9 (C), 139.4 (C), 139.1 (C), 134.2 (CH), 133.5 (CH), 133.0 (CH), 132.9 (CH), 132.8 (CH), 132.5 (CH), 131.0 (CH), 46.3 (C), 35.4 (CH_2), 35.4 (CH_2), 34.8 (CH_2), 34.7 (CH_2), 30.9 (CH_3); *m/z* (EI+) 296 [M]$^+$, 240 [M – *t*-Bu]$^+$, 207 [M – S*t*-Bu]$^+$, 136, 104, 91, 78 (Found: [M]$^+$, 296.1591. $C_{20}H_{24}S$ requires [M]$^+$, 296.1593).

(S_p)-(+)-Thioacetic acid *S*-[2.2]paracyclophan-4-yl ester [(S_p)-7]

Boron tribromide (0.36 mL, 3.77 mmol, 1.1 equiv) was added to a solution of (S_p)-(+)-4-*t*-butylsulfanyl[2.2]paracyclophane (S_p)-**6** (1.0 g, 3.43 mmol, 1.0 equiv) and acetyl chloride (1.7 mL, 24.01 mmol, 7.0 equiv) in toluene (34.3 mL) at rt. The reaction was stirred for 1 hour then poured into a solution of ice cold saturated aqueous NH_4Cl (200 mL). Aqueous phase was extracted with Et_2O (3 × 50 mL). The combined organic phase was washed with aqueous $Na_2S_2O_3$ (10% w/v) (3 × 50 ml), dried ($MgSO_4$) and concentrated. Purification by chromatography (10% Et_2O/hexane) afforded (S_p)-**7** as a white solid (0.53 g, 55%); mp = 134–136 °C; $[\alpha]_D$ + 84.3 (*c* 1, $CHCl_3$) (assumed 80% *ee* see text); ν_{max} (film) 2920, 2850, 1694, 1498, 1476, 1447, 1432, 1408, 1113, 954, 906, 850 and 792 cm^{-1}; δ_H (500 MHz, $CDCl_3$) 6.68 (1H, d, *J* = 8.0 Hz, H-13), 6.60 (2H, s, H-15, H-16), 6.52 (3H, m, H-5, H-7, H-8), 6.42 (1H, d, *J* = 7.5 Hz, H-12), 3.41–3.35 (1H, m, H-2 *endo*), 3.13–2.97 (5H, m, H-1 *endo*, 2 × H-9, 2 × H-10), 2.99 (1H, t, *J* = 10.2 Hz, H-1 *exo*), 2.92–2.86 (1H, m, H-2 *exo*), 2.36 (3H, s, Me); δ_C (125 MHz, $CDCl_3$) 190.7 (CO), 143.1 (C), 140.4 (C), 139.4 (C), 139.1 (C), 135.1 (CH), 135.0 (CH), 134.5 (CH), 133.2 (CH), 133.1 (CH), 132.3 (CH), 130.2 (CH), 129.2 (CH), 35.4 (CH_2), 34.9 (CH_2), 34.6 (CH_2), 34.4 (CH_2), 30.2 (CH_3); *m/z*(EI+) 282 $[M]^+$, 240 $[M+H-Ac]^+$, 207 $[[2.2]paracyclophane]^+$, 178, 136, 104, 91, 78, 43 (Found: $[M]^+$, 282.1075. $C_{18}H_{18}OS$ requires $[M]^+$, 282.1073).

(S_p)-(+)-[2.2]Paracyclophane-4-thiol [(S_p)-2]

To a solution of (S_p)-**7** (0.51 g, 1.80 mmol, 1.0 equiv) in methanol (18 mL) was added K_2CO_3 (3.71 g, 26.84 mmol, 15 equiv) and the reaction stirred for 3 hours at rt. The reaction was poured into saturated aqueous NH_4Cl solution (100 mL) and extracted with Et_2O (3 × 100 mL). The combined organic layers were dried ($MgSO_4$) and concentrated before purification by chromatography (1% Et_2O/hexane) gave (S_p)-**2** as a white solid (0.14 g, 32%); mp = 144–146 °C; $[\alpha]_D$ + 164.0 (*c* 1, $CHCl_3$) (assumed 80% *ee* see text); ν_{max} (film) 3011, 2929, 2848, 2558, 1587, 1548, 1499, 1480, 1449, 1432, 1411, 1060, 938, 897, 849, 804 and 791 cm^{-1}; δ_H (500 MHz, $CDCl_3$) 7.21 (1H, d, *J* = 7.5 Hz, H-13), 6.57 (1H, d, *J* = 7.5 Hz, H-8), 6.47 (1H, d, *J* = 7.5 Hz, H-7),

6.43 (1H, d, *J* = 7.5 Hz, H-12), 6.40 (2H, t,*J* = 7.5 Hz, H-15, H-16), 6.22 (1H, s, H-5), 3.41 (1H, t, *J* = 12.0 Hz, H-2 *endo*), 3.26 (1H, ddd, J = 13.0, 6.0, 3.9 Hz, H-1 *endo*), 3.13 (1H, s, S-H), 3.11–3.01 (4H, m, 2 × H-9, 2 × H-10), 2.90–2.87 (1H, t, *J* = 9.0 Hz, H-1 *exo*), 2.83–2.77 (1H, m, H-2 *exo*); δ_C (125 MHz, $CDCl_3$) 140.4 (C), 139.3 (C), 139.1 (C), 138.5 (C), 135.8 (CH), 134.8 (CH), 133.4 (CH), 132.8 (CH), 131.9 (CH), 131.5 (CH), 130.5 (CH), 127.7 (CH), 35.4 (CH_2), 34.9 (CH_2), 34.6 (CH_2), 33.1 (CH_2); *m/z* (EI+) 240 $[M]^+$, 207 $[[2.2]paracyclophane]^+$, 136, 104, 91, 78 (Found: $[M]^+$, 240.0969. $C_{16}H_{16}S$ requires $[M]^+$, 240.0967).

(R_p,R_S)-(–)-4-*tert*-Butylsulfinyl-5-amino[2.2]paracyclophane [(R_p,R_S)-8]

To a solution of (R_p,R_S)-(–)-4-*t*-butylsulfinyl[2.2]paracyclophane [(R_p,R_S)-**5**] (5.80 g, 18.6 mmol, 1.0 equiv) in THF (350 mL) at 0 °C was added *n*-BuLi (2.5M in hexanes; 16.5 mL, 41.25 mmol, 2.2 equiv) dropwise over 30 min to give an orange solution. After 1 h tosyl azide (9.20 g, 46.70 mmol, 2.5 equiv) was added and the reaction warmed to rt over 18 h. $NaBH_4$ (6.45 g, 171 mmol, 9 equiv) and *tetra-n*-butyl ammonium iodide (6.31 g, 17.1 mmol, 0.9 equiv) were added and the reaction stirred for a further 24 h at rt whereupon a further portion of $NaBH_4$ (2.80 g, 74.0 mmol, 4.0 equiv) was added. After further stirring at rt for 5 days the reaction was poured into saturated aqueous NH_4Cl (250 mL) causing effervescence. The aqueous phase was extracted with Et_2O (500 mL + 200 mL) and the combined organic extracts dried ($MgSO_4$) and the solvent removed. The residue was purified by chromatography (neutralized silica gel 40% Et_2O/*n*-heptane) to yield **8** as a pale yellow solid powder, which was recrystalised from $CHCl_3$ / heptane (4.00 g, 66%); mp = 130–132 °C; $[\alpha]_D$ −118.0 (*c* 1, $CHCl_3$) (assumed 80% *ee* see text); ν_{max} (film) 3430, 3055, 2987, 1637, 1421, 1265, 896, 739 and 705 cm^{-1}; δ_H (500 MHz, $CDCl_3$) 7.15 (1H, d, *J* = 8.0 Hz, H-13), 6.89 (1H, d, *J* = 7.5 Hz, H-16), 6.63 (1H, d, *J* = 8.0 Hz, H-15), 6.42 (1H, d, *J* = 8.0 Hz, H-7), 6.35 (1H, d, *J* = 7.5 Hz, H-12), 6.01 (1H, d, *J* = 7.5 Hz, H-8), 5.62 (2H, s, broad, NH_2), 3.46 (1H, t, *J* = 12.0 Hz, H-2 *endo*), 3.23–3.18 (1H, m, H-1 *endo*), 3.12–3.01 (4H, m, H-1 *exo*, H-9 *endo*, 2 × H-10), 2.71–2.62 (2H, m, H-2 *exo*, H-9 *exo*), 1.19 (9H, s, *t*-Bu); δ_C (125 MHz, $CDCl_3$) 150.6 (C), 141.6 (C), 138.5 (C), 138.1 (C), 133.1 (C), 132.1 (C), 131.9 (CH), 129.7 (CH), 127.1 (CH), 126.5 (CH), 123.0 (CH), 114.7 (CH), 60.19 (C), 34.1 (CH_2), 33.9 (CH_2), 32.8

(CH_2), 30.8 (CH_2), 23.6 (CH_3);*m/z* (ESI+) 350.1547 $[M+Na]^+$ (Found: $[M+Na]^+$, 350.1549. $C_{20}H_{25}OSNNa$ requires $[M+Na]^+$, 350.1546).

(*R*$_p$)-(–)-4-*tert*-Butylsulfanyl-5-amino[2.2]paracyclophane [(*R*$_p$)-9]

Triethylamine (1.47 mL, 10.53 mmol, 10 equiv) followed by trichlorosilane (2.60 mL, 25.76 mmol, 15.0 equiv) were added carefully to a solution of (R_p,R_S)-4-(–)-*t*-butylsulfinyl-5-amino[2.2] paracyclophane [(R_p,R_S)-**8**] (0.50 g, 1.61 mmol, 1.0 equiv) in toluene (8 mL,) at 0 °C and the reaction heated to reflux for 16 h. The reaction was cooled to 0 °C and aqueous NaOH (3M; 100 mL) was added. The aqueous phase was extracted with Et_2O (3 × 50 mL), dried ($MgSO_4$) and concentrated to give a pale yellow solid. Purification by chromatography (neutralized silica, 20% Et_2O/hexane) gave (R_p)-**9** as a pale yellow solid (0.19 g, 40%); mp = 72–74 °C; $[\alpha]_D$ −192.3 (*c* 1, $CHCl_3$) (assumed 80% *ee* see text); ν_{max} (film) 3472, 3364, 2931, 2856, 1592, 1464, 1456, 1429, 1408, 1155, 876, 802, 741 and 717 cm^{-1}; δ_H (500 MHz, $CDCl_3$) 7.0 (1H, d, *J* = 9.5 Hz, H-13), 6.60 (1H, d, *J* = 10.0 Hz, H-7), 6.47 (1H, d, *J* = 10.0 Hz, H-12), 6.41 (1H, d, *J* = 9.5 Hz, H-8), 6.33 (1H, d, *J* = 9.5 Hz, H-15), 6.23 (1H, d, *J* = 9.5 Hz, H-16), 4.47 (2H, br s, NH_2), 3.70 (1H, ddd, *J* = 14.0, 9.6, 3.2 Hz, H-2 *endo*), 3.11–2.96 (5H, m, H-1 *endo*, 2 × H-9, 2 × H-10), 2.78–2.65 (2H, m, H-1 *exo*, H-2 *exo*), 1.21 (9H, s, *t*-Bu); δ_C (125 MHz, $CDCl_3$) 149.4 (C), 147.3 (C), 139.1 (C), 138.3 (C), 135.7 (CH), 133.0 (CH), 132.2 (CH), 129.7 (CH), 126.3 (CH), 124.3 (C), 123.0 (CH), 117.9 (C), 48.1 (C), 35.3 (CH_2), 34.1 (CH_2), 32.8 (CH_2), 32.3 (CH_2), 30.9 (CH_3); *m/z* (EI+) 311 $[M]^+$, 255 $[M+H-t\text{-}Bu]^+$, 207 $[[2.2]paracyclophane]^+$, 151, 106 (Found: $[M+H]^+$, 312.1791. $C_{20}H_{26}NS$ requires $[M+H]^+$, 312.1780).

(*R*$_p$)-(+)-[2.2](4,7)Benzo[*d*]thiazoloparacyclophane [(*R*$_p$)-10]

To a solution of **9** (141 mg, 0.45 mmol, 1.0 equiv) and paraformaldehyde (54.5 mg, 1.82 mmol, 2.0 equiv) in dioxane (5.0 mL) and water (1.0 mL) was added aqueous $HCl_{conc.}$ (9.0 M; 0.108 mL). Pyridine (0.19 mL, 1.82 mmol, 2.0 equiv) was added and the reaction mixture heated to reflux for 48 h. The reaction mixture was poured into aqueous NaOH (3.0 M; 20 mL) and the aqueous phase extracted with Et_2O (3

× 30 mL). The combined organic phases were dried ($MgSO_4$) and the solvent removed; purification by chromatography (neutralized silica gel, 50% Et_2O/heptane) furnished (R_p)-**10** as a pale yellow powder (36.3 mg, 28%); mp = 144–146 °C; $[\alpha]_D$ +67.4 (*c* 1, $CHCl_3$) (assumed 80% *ee* see text); ν_{max} (film) 3467, 3005, 2978, 2934, 2873, 1448, 1384, 1351, 1217 and 1111 cm^{-1}; δ_H (500 MHz, $CDCl_3$; paracyclophane numbering) 8.89 (1H, s, thiazole CH), 6.81 (1H, d, *J* = 7.5 Hz, H-13), 6.68 (1H, d, *J* = 7.5 Hz, H-12), 6.52 (1H, d, *J* = 8.0 Hz, H-16), 6.46 (1H, d, *J* = 7.5 Hz, H-15), 6.18 (1H, d, *J* = 7.5 Hz, H-7), 5.85 (1H, d, *J* = 7.5 Hz, H-8), 3.99–3.93 (1H, m, H-2 *endo*), 3.27–3.21 (1H, m, H-1 *endo*), 3.17–3.11 (1H, m, H-9 *endo*), 3.06–2.99 (5H, m, H-1 *exo*, H-2 *exo*, H-9 *exo*, 2 × H-10); δ_C (125 MHz, $CDCl_3$) 154.4 (CH), 151.8 (C), 139.2 (C), 137.2 (C), 134.8 (C), 134.4 (C), 132.8 (C), 132.3 (CH), 131.7 (CH), 130.5 (CH), 126.2 (CH), 124.9 (CH), 35.0 (CH_2), 34.5 (CH_2), 33.5 (CH_2), 32.4 (CH_2); *m/z* (EI+) 265 $[M]^+$, 161, 104, 78 (Found: $[M]^+$, 265.0921. $C_{17}H_{15}NS$ requires $[M]^+$, 265.0920).

ACKNOWLEDGEMENTS

The authors would like to thank Massey University and the University of Sussex for financial support.

REFERENCES

1. Gleiter, R.; Hopf, H., Eds. *Modern Cyclophane Chemistry;* Wiley-VCH: Weinheim, Germany, 2004. Return to citation in text: [1]
2. Vögtle, F. *Cyclophane Chemistry;* Wiley: Chichester, U.K., 1993. Return to citation in text: [1]
3. Amthor, S.; Lambert, C. *J. Phys. Chem. A* **2006,** *110,* 3495–3504. doi:10.1021/jp055098o Return to citation in text: [1]
4. Ball, P. J.; Shtoyko, T. R.; Krause Bauer, J. A.; Oldham, W. J.; Connick, W. B. *Inorg. Chem.* **2004,** *43,* 622–632.doi:10.1021/ic0348648 Return to citation in text: [1]
5. Bazan, G. C. *J. Org. Chem.* **2007,** *72,* 8615–8635. doi:10.1021/jo071176n Return to citation in text: [1]
6. Morisaki, Y.; Chujo, Y. *Angew. Chem., Int. Ed.* **2006,** *45,* 6430–6437. doi:10.1002/anie.200600752 Return to citation in text: [1]

7. Valentini, L.; Mengoni, F.; Taticchi, A.; Marrocchi, A.; Kenny, J. M. *J. Mater. Chem.* **2006,** *16,* 1622–1625.doi:10.1039/b600333h Return to citation in text: [1]
8. Valentini, L.; Mengoni, F.; Taticchi, A.; Marrocchi, A.; Landi, S.; Minuti, L.; Kenny, J. M. *New J. Chem.* **2006,** *30,*939–943. doi:10.1039/b601535b Return to citation in text: [1]
9. El-Shaieb, K. M.; Mourad, A. F. E.; Hopf, H. *ARKIVOC* **2006,** No. ii193–200. Return to citation in text: [1]
10. Gibson, S. E.; Knight, J. D. *Org. Biomol. Chem.* **2003,** *1,* 1256–1269. doi:10.1039/b300717k Return to citation in text: [1]
11. Rozenberg, V.; Sergeeva, E.; Hopf, H. Cyclophanes as Templates in Stereoselective Synthesis. In *Modern Cyclophane Chemistry;* Gleiter, R.; Hopf, H., Eds.; Wiley-VCH: Weinheim, Germany, 2004; pp 435–462. doi:10.1002/3527603964.ch17 Return to citation in text: [1]
12. Ortner, B.; Hübner, H.; Gmeiner, P. *Tetrahedron: Asymmetry* **2001,** *12,* 3205–3208. doi:10.1016/S0957-4166(01)00558-4 Return to citation in text: [1] [2]
13. Ortner, B.; Waibel, R.; Gmeiner, P. *Angew. Chem., Int. Ed.* **2001,** *40,* 1283–1285. doi:10.1002/1521-3773(20010401)40:7<1283::AID-ANIE1283>3.0.CO;2-# Return to citation in text: [1] [2]
14. Schlotter, K.; Boeckler, F.; Hübner, H.; Gmeiner, P. *J. Med. Chem.* **2006,** *49,* 3628–3635.doi:10.1021/jm060138d Return to citation in text: [1] [2]
15. Rowlands, G. J. *Org. Biomol. Chem.* **2008,** *6,* 1527–1534. doi:10.1039/b800698a Return to citation in text: [1] [2]
16. Lohier, J.-F.; Foucoin, F.; Jaffres, P.-A.; Garcia, J. I.; Sopková-de Oliveira Santos, J.; Perrio, S.; Metzner, P.*Org. Lett.* **2008,** *10,* 1271–1274. doi:10.1021/ol800161m Return to citation in text: [1] [2]
17. Kane, V. V.; Gerdes, A.; Grahn, W.; Ernst, L.; Dix, I.; Jones, P. G.; Hopf, H. *Tetrahedron Lett.* **2001,** *42,* 373–376. doi:10.1016/S0040-4039(00)01992-4 Return to citation in text: [1] [2]
18. Kreis, M.; Bräse, S. *Adv. Synth. Catal.* **2005,** *347,* 313–319. doi:10.1002/adsc.200404299 Return to citation in text: [1] [2]
19. Nakamura, S.; Ito, Y.; Wang, L.; Toru, T. *J. Org. Chem.* **2004,** *69,* 1581–1589. doi:10.1021/jo035558e Return to citation in text: [1]
20. Vargas-Diaz, M. E.; Lagunas-Rivera, S.; Joseph-Nathan, P.; Tamariz, J.; Zepeda, L. G. *Tetrahedron Lett.* **2005,***46,* 3297–3300. doi:10.1016/j.tetlet.2005.03.104 Return to citation in text: [1]
21. Gonda, J. *Angew. Chem., Int. Ed.* **2004,** *43,* 3516–3524. doi:10.1002/anie.200301718 Return to citation in text: [1]

22. Dondoni, A. *Angew. Chem., Int. Ed.* **2008,** *47,* 89958997. doi:10.1002/anie.200802516 Return to citation in text: [1]

23. Majumdar, K. C.; Debnath, P. *Tetrahedron* **2008,** *64,* 9799–9820. doi:10.1016/j.tet.2008.07.107 Return to citation in text: [1]

24. van Lindert, H. C. A.; van Doorn, J. A.; Bakker, B. H.; Cerfontain, H. *Recl. Trav. Chim. Pays-Bas* **1996,** *115,*167–178. Return to citation in text: [1]

25. van Lindert, H. C. A.; Koeberg-Telder, A.; Cerfontain, H. *Recl. Trav. Chim. Pays-Bas* **1992,** *111,* 379–388. Return to citation in text: [1]

26. Braddock, D. C.; MacGilp, I. D.; Perry, B. G. *Adv. Synth. Catal.* **2004,** *346,* 1117–1130.doi:10.1002/adsc.200404065 Return to citation in text: [1]

27. Menichetti, S.; Faggi, C.; Lamanna, G.; Marrocchi, A.; Minuti, L.; Taticchi, A. *Tetrahedron* **2006,** *62,* 5626–5631.doi:10.1016/j.tet.2006.03.101 Return to citation in text: [1]

28. Pelter, A.; Mootoo, B.; Maxwell, A.; Reid, A. *Tetrahedron Lett.* **2001,** *42,* 8391–8394. doi:10.1016/S0040-4039(01)01808-1 Return to citation in text: [1]

29. Marchand, A.; Maxwell, A.; Mootoo, B.; Pelter, A.; Reid, A. *Tetrahedron* **2000,** *56,* 7331–7338.doi:10.1016/S0040-4020(00)00644-X Return to citation in text: [1]

30. Hou, X.-L.; Wu, X.-W.; Dai, L.-X.; Cao, B.-X.; Sun, J. *Chem. Commun.* **2000,** 1195–1196. doi:10.1039/b002679o Return to citation in text: [1]

31. Hitchcock, P. B.; Rowlands, G. J.; Parmar, R. *Chem. Commun.* **2005,** 4219–4221. doi:10.1039/b507394d Return to citation in text: [1] [2] [3] [4] [5]

32. Hitchcock, P. B.; Hodgson, A. C. C.; Rowlands, G. J. *Synlett* **2006,** 2625–2628. doi:10.1055/s-2006-950432 Return to citation in text: [1] [2]

33. Hitchcock, P. B.; Rowlands, G. J.; Seacome, R. J. *Org. Biomol. Chem.* **2005,** *3,* 3873–3876.doi:10.1039/b509994c Return to citation in text: [1] [2] [3] [4] [5]

34. Reich, H. J.; Yelm, K. E. *J. Org. Chem.* **1991,** *56,* 5672–5679. doi:10.1021/jo00019a039 Return to citation in text: [1]

35. Ernst, L.; Wittkowski, L. *Eur. J. Org. Chem.* **1999,** *1999,* 1653–1663. doi:10.1002/(SICI)1099-0690(199907)1999:7<1653::AID-EJOC1653>3.0.CO;2-R Return to citation in text: [1] [2]

36. Reich, H. J.; Cram, D. J. *J. Am. Chem. Soc.* **1969,** *91,* 3534–3543. doi:10.1021/ja01041a018 Return to citation in text: [1]

37. Weix, D. J.; Ellman, J. A. *Org. Lett.* **2003,** *5,* 1317–1320. doi:10.1021/ol034254b Return to citation in text: [1]

38. Seacome, R. J. The synthesis and methodology development of novel [2.2]paracyclophane-based compounds. D.Phil. Thesis, University of Sussex, U.K., 2008. Return to citation in text: [1] [2]

39. Parmar, R. New methodology for the synthesis of enantiopure [2.2] paracyclophane derivatives, D.Phil. Thesis, University of Sussex, U.K., 2006. Return to citation in text: [1] [2]

40. Mancheño, O. G.; Priego, J.; Cabrera, S.; Arrayás, R. G.; Llamas, T.; Carretero, J. C. *J. Org. Chem.* **2003,** *68,*3679–3686. doi:10.1021/jo0340657 Return to citation in text: [1]

41. Stuhr-Hansen, N. *Synth. Commun.* **2003,** *33,* 641–646. doi:10.1081/SCC-120015820 Return to citation in text: [1]

42. Ulrich, H. Product Class 19: 1,2- and 1,3-Thiaphospholes and Benzoannulated Analogues. In *Hetarenes and Related Ring Systems: Five-Membered Hetarenes with One Chalcogen and One Additional Heteroatom;*Schaumann, E., Ed.; Science of Synthesis, Vol. 11; Georg Thieme Verlag: Stuttgart - New York, 2002; pp 835–912. Return to citation in text: [1]

43. Fu, G. C. *Acc. Chem. Res.* **2006,** *39,* 853–860. doi:10.1021/ar068115g Return to citation in text: [1]

44. Fu, G. C. *Acc. Chem. Res.* **2004,** *37,* 542–547. doi:10.1021/ar030051b Return to citation in text: [1]

45. Bräse, S.; Dahmen, S.; Hoefener, S.; Lauterwasser, F.; Kreis, M.; Ziegert, R. E. *Synlett* **2004,** 2647–2669.doi:10.1055/s-2004-836029 Return to citation in text: [1]

46. Fu, G. C. Planar-chiral Heterocycles as Enantioselective Organocatalysts. In *Asymmetric Synthesis - The Essentials;* Christmann, M.; Bräse, S., Eds.; Wiley-VCH: Weinheim, Germany, 2007; pp 195–199. Return to citation in text: [1]

47. Bräse, S. Planar Chiral Ligands Based on [2.2]Paracyclophanes. In *Asymmetric Synthesis - The Essentials;*Christmann, M.; Bräse, S., Eds.; Wiley-VCH: Weinheim, Germany, 2007; pp 67–71. Return to citation in text: [1]

48. Royer, R.; Lechartier, J. P.; Demerseman, P. *Bull. Soc. Chim. Fr.* **1973,** *11,* 3017–3018. Return to citation in text: [1]

Chapter 10

A NOVEL AND ONE-POT SYNTHESIS OF 6-ARYLPYRIMIDIN-4-OL

Jitendra M. Gajera, Shital N. Tondlekar, and Laxmikant A. Gharat

Drug Discovery Chemistry, Glenmark Research Centre, Glenmark Pharmaceuticals Limited, MIDC, Mahape, Navi Mumbai 400709, India

ABSTRACT

We have developed a novel and one-pot synthesis of 6-arylpyrimidine-4-ol by reacting commercially available alkyl 3-oxo-3-arylpropanoate with formamide in the presence of stoichiometric amount of ammonium acetate.

INTRODUCTION

Pyrimidine derivatives are very well known for their various therapeutic applications. Pyrimidine derivatives are used as anticancer [1], anti-HIV [2], antibacterial [3], antimalarial [4], antihypertensive [5], sedative, hypnotics [6], anticonvulsant [7], antithyroid [8], antihistaminic agents [9], and antibiotics [10]. A very recent review describes the significance of pyrimidine derivatives as anti-inflammatory agents [11]. 2-thiopyrimidine derivatives possess potent activity against inflammation and immune disorders [12]. Recently, various pyrimidine derivatives have been reported as vanilloid receptor antagonists [13].

After looking at the diverse properties of pyrimidine derivatives, we selected 4,6-disubstituted pyrimidines as a part of our pharmacophore to synthesize novel anti-inflammatory agents. Literature survey revealed that 4,6-disubstituted pyrimidines can be prepared using either Biginelli approach [14] or reaction of β-iminoesters with formamide [15] or reaction of 4,6-dichloropyrimidines with appropriate boronic acids [16]. The only competitive method similar to our approach described in the literature uses formamidine in DMF and affords the product in only 14% after a reaction time of 3 days [17]. During the course of our research work on synthesis on various 4,6-disubstituted pyrimidines, we developed a novel and one-pot method for the synthesis of 6-arylpyrimidine-4-ol using commercially available raw materials. The method comprises reaction between 3-oxo-3-arylpropanoate, stoichiometric amount of ammonium acetate, and formamide at elevated temperature.

RESULTS AND DISCUSSION

Various 6-arylpyrimidin-4-ols 3a-h were prepared by reacting methyl-3-oxo-3-arylpropanoates 1a-hwith formamide in the presence of ammonium acetate with a yield of 50–70%.

The reaction of methyl-3-oxo-3-arylpropanoates 1 with the in situ generated ammonia gives the intermediate methyl 3-amino-3-arylacrylate 2, which subsequently reacts with formamide to give 6-arylpyrimidin-4-ols 3 (Scheme 1). One of the intermediates methyl 3-amino-3-phenylacrylate was isolated in 80% yield and characterized.

Isolation and characterization of 2a confirm the reaction pathway (Table 1). The procedure described in the experimental part provides a novel and one-pot approach for the synthesis of 6-arylpyrimidin-4-ol.

Table 1: Yield and melting point of compounds 3a-h and 2a.

Compound	R_1	R_2	Yield (%)	M. P. (°C)
3a	H	H	65	271-272
3b	Cl	H	70	213–215
3c	Br	H	68	311–313
3d	F	H	61	313–315
3e	CH_3	H	57	291-292
3f	OCH_3	H	53	282–284
3g	OCH_3	OCH_3	50	252–254
3h	H	Cyclopentyloxy	55	260–264
2a	H	H	80	—

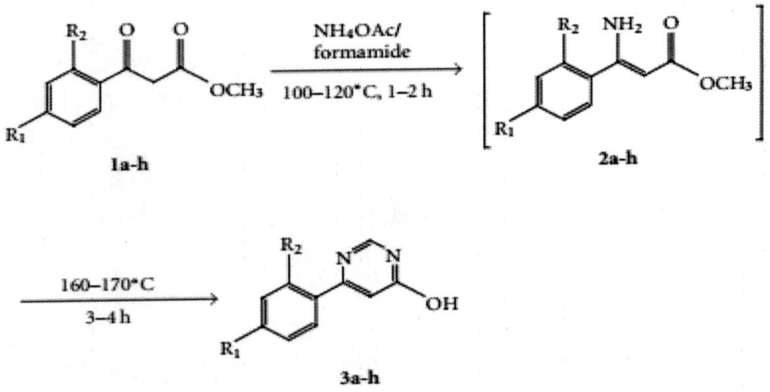

SCHEME 1

Scheme 1

EXPERIMENTAL

Commercial solvents and reagents were used without further purification. 1H NMR spectra were recorded on a Varian 300 MHz spectrometer. Melting points are uncorrected. Elemental analysis was performed on a Perkin-Elmer analyzer. Mass spectra were recorded on Thermo Finnigan LCQ DECA XP MAX (ION TRAP) mass spectrometer using atmospheric pressure chemical ionization (APCI) source in positive/negative mode at capillary voltage 3.14 V and capillary temperature 250°C.

Methyl 3-Amino-3-Phenylacrylate (2a) To a stirred solution of methyl 3-oxo-3-phenylpropanoate (1 mmol) in formamide (5.0 mL) was added ammonium acetate (5 mmol) at ambient temperature. Reaction mixture was then heated to 110–120°C over a period of 1 hour and then held at 110–120°C for 1 more hour. It was then cooled to room temperature; diluted with cold water; and extracted with diethyl ether. The residue obtained after removal of diethyl ether was purified through silica gel column using ethyl acetate: petroleum ether (8:2) as an eluent to give methyl 3-amino-3-arylacrylate 2a as thick oil. IR (KBr): 3445, 1622, 1491, 1320 cm− 1. 1H NMR (DMSO-*d*6): δ 3.56 (S, 3H); 4.78 (s, 1H); 7.34–7.59 (m, 5H); 7.95 (brs, 2H). [M−1]−:176.23. Anal. calcd for $C_{10}H_{11}NO_2$: C, 67.78; H, 6.26; N, 7.90. Found: C, 67.65; H, 6.24; N, 7.91.

General Procedure for Preparation of 6-Arylpyrimidine-4-Ol (3a-h) A stirred solution of appropriate methyl 3-oxo-3-arylpropanoate 1 (1 mmol) in formamide (50 mol) was added ammonium acetate (5 mmol) and heated to 100–120°C over a period of 1 hour and held at 110–120°C for 1 more hour. It was then stirred for 4-5 hours at 160–170°C. Reaction mixture was cooled to room temperature and diluted with cold water. The precipitated material was extracted with ethyl acetate. The solid product obtained after removal of ethyl acetate was washed with diethyl ether to get pure product.

6-Phenylpyrimidine-4-Ol (3a) This compound was obtained as light yellow solid. IR (KBr): 3435, 1668, 1592, 1252, 1024 cm^{-1}. 1H NMR (DMSO- *d*6): δ 6.87 (s, 1H); 7.45–7.47 (m, 3H); 8.02-8.03 (m, 2H); 8.25 (s, 1H); 12.51 (brs, 1H). MS [M + 1]+:172.32. Anal. calcd for $C_{10}H_8N_2O$: C, 69.76; H, 4.68; N, 16.27. Found: C, 69.74; H, 4.69; N, 16.31.

6-(4-Chlorophenyl) pyrimidine-4-Ol (3b) This compound was obtained as light yellow solid. IR (KBr): 3339, 1667, 1594, 1243, 1014 cm− 1. 1H NMR (DMSO- *d6*): δ 6.91 (s, 1H); 7.52 (d, 2H, *J* = 8 . 4 Hz); 8.05 (d, 2H, *J* = 8 . 7 Hz); 8.26 (s, 1H); 12.54 (brs, 1H). MS [M + 1]+:207.39. Anal. calcd for $C_{10}H_7ClN_2O$: C, 58.13; H, 3.41; N, 13.58. Found: C, 58.20; H, 3.42; N, 13.61.

6-(4-Bromophenyl)pyrimidine-4-Ol (3c) This compound was obtained as light yellow solid. IR (KBr) 3445, 1685, 1589, 1257, 1010 cm− 1. 1H NMR (DMSO- *d6*): δ 6.91 (s, 1H); 7.66 (d, 2H, *J* = 9 Hz); 7.98 (d, 2H, *J* = 8 . 4 Hz); 8.26 (s, 1H); 12.54 (bs, 1H). MS [M + 1]+:251.39. Anal. calcd for $C_{10}H_7BrN_2O$: C, 47.84; H, 2.81; N, 11.16. Found: C, 47.89; H, 2.80; N, 11.15.

6-(4-Fluorophenyl)pyrimidine-4-Ol (3d) This compound was obtained as light yellow solid. IR (KBr) 3444, 1672, 1600, 1242, 1035 cm− 1. 1H NMR (DMSO- *d6*): δ 6.88 (s, 1H); 7.26–7.31 (m, 2H); 8.07–8.11 (m, 2H); 8.25 (s, 1H); 12.52 (brs, 1H). [M + 1]+:190.97. Anal. calcd for $C_{10}H_7FN_2O$: C, 63.16; H, 3.71; N, 14.73. Found: C, 63.12; H, 3.70; N 14.76.

6-(4-Methylphenyl)pyrimidine-4-Ol (3e) This compound was obtained as light yellow solid. IR (KBr): 3422, 1664, 1592, 1254, 1175, 1037 cm− 1. 1HNMR (DMSO- *d6*): δ 2.34 (s, 3H); 6.81 (s, 1H); 7.26 (d, 2H, *J* = 7 . 8 Hz); 7.92 (d, 2H, *J* = 7 . 8 Hz); 8.22 (s, 1H); 12.20 (brs, 1H). [M + 1]+:186.39. Anal. calcd for $C_{11}H_{10}N_2O$: C, 70.95; H, 5.41; N, 15.04. Found: C, 71.11; H, 5.43; 15.00.

6-(4-Methoxyphenyl)pyrimidine-4-Ol (3f) This compound was obtained as light yellow solid. IR (KBr) 3397, 1666, 1605, 1245, 1177 cm− 1. 1H NMR (DMSO- *d6*): δ 3.80 (s, 3H); 6.77 (s, 1H); 7.00 (d, 2H, *J* = 8 . 7 Hz); 8.00 (d, 2H, *J* = 9 Hz); 8.21 (s, 1H); 12.42 (brs, 1H). [M + 1]+:203.37. Anal. calcd for $C_{11}H_{10}N_2O_2$: C, 65.34; H, 4.98; N, 13.85. Found: C, 65.38; H, 4.97; N, 13.88.

6-(2,4-Dimethoxyphenyl)pyrimidine-4-Ol (3g) This compound was obtained as light yellow solid. IR (KBr) 3414, 1673, 1610, 1270, 1170 cm− 1. 1H NMR (DMSO- *d6*): δ 3.81 (d, 6H, *J* = 4 . 5 Hz); 6.84 (s, 1H); 7.02 (d, 1H, *J* = 8 . 4 Hz); 7.58 (s, 1H); 7.65 (d, 1H, *J* = 9 Hz); 8.20 (s, 1H); 12.32 (brs, 1H). [M + 1]+:233.36. Anal. calcd for $C_{12}H_{12}N_2O_3$: C, 62.06; H, 5.21; N, 12.06. Found: C, 62.20; H, 5.23; N, 12.04.

6-(2-(cyclopentyloxy)phenyl)pyrimidine-4-Ol (3h) This compound was obtained as light yellow solid. IR (KBr): 3313, 1643,

1574, 1279, 1242 cm− 1. 1H NMR (DMSO- *d*6): δ 1.63–1.76 (m, 6H); 1.90–1.99 (m, 2H); 4.95 (m, 1H); 6.93 (s, 1H); 7.00 (t, 1H, *J* = 7 . 8 Hz); 7.11 (d, 1H, *J* = 8 . 4 Hz); 7.38 (t, 1H, *J* = 6 . 9 Hz); 7.93 (d, 1H, *J* = 6 . 3 Hz); 8.20 (s, 1H); 12.42 (brs, 1H). [M + 1]+:257.13. Anal. calcd for $C_{15}H_{16}N_2O_2$: C, 70.29; H, 6.29; N, 10.93. Found: C, 70.19; H, 6.27; N, 10.91.

ACKNOWLEDGMENT

The authors are thankful to analytical department for analytical support.

REFERENCES

1. E. C. Taylor and B. Liu, "A new and efficient synthesis of pyrrolo[2,3-d] pyrimidine anticancer agents: alimta (LY231514, MTA), homo-alimta, TNP-351, and some aryl 5-substituted pyrrolo[2,3-d]pyrimidines," The Journal of Organic Chemistry, vol. 68, no. 26, pp. 9938–9947, 2003. View at Publisher · View at Google Scholar · View at PubMed
2. P. Herdewijn, J. Balzarini, M. Baba, et al., "Synthesis and anti-HIV activity of different sugar-modified pyrimidine and purine nucleosides," Journal of Medicinal Chemistry, vol. 31, no. 10, pp. 2040–2048, 1988. View at Publisher · View at Google Scholar
3. J. Matsumoto and S. Minami, "Pyrido[2,3-d]pyrimidine antibacterial agents. 3. 8-alkyl- and 8-vinyl-5,8-dihydro-5-oxo-2-(1-piperazinyl) pyrido[2,3-d]pyrimidine-6-carboxylic acids and their derivatives," Journal of Medicinal Chemistry, vol. 18, no. 1, pp. 74–79, 1975. View at Publisher ·View at Google Scholar
4. A. Agarwal, K. Srivastava, S. K. Puri, and P. M. S. Chauhan, "Antimalarial activity and synthesis of new trisubstituted pyrimidines," Bioorganic & Medicinal Chemistry Letters, vol. 15, no. 12, pp. 3130–3132, 2005. View at Publisher · View at Google Scholar · View at PubMed
5. H. Maeda, T. Akaike, Y. Miyamoto, and M. Yoshida, "Use of 4-amino-6-hydroxypyrazolo[3,4-d]pyrimidine for the manufacture of an antihypertensive agent," European patent no. EP07959298, 1997.
6. K. S. Jain, T. S. Chitre, P. B. Miniyar, et al., "Biological and medicinal significance of pyrimidines,"Current Science, vol. 90, no. 6, pp. 793–803, 2006.

7. A. K. Gupta, Sanjay, H. P. Kayath, A. Singh, G. Sharma, and K. C. Mishra, "Anticonvulsant activity of pyrimidine thiols," Indian Journal of Pharmacology, vol. 26, no. 3, pp. 227–228, 1994.

8. J.-F. Lagorce, F. Comby, A. Rousseau, J. Buxeraud, and C. Raby, "Synthesis and antithyroid activity of pyridine, pyrimidine and pyrazine derivatives of thiazole-2-thiol and 2-thiazoline-2-thiol," Chemical and Pharmaceutical Bulletin, vol. 41, no. 7, pp. 1258–1260, 1993.

9. V. Alagarsamy, V. R. Solomon, and M. Murugan, "Synthesis and pharmacological investigation of novel 4-benzyl-1-substituted-4H-[1,2,4]triazolo[4,3-a]quinazolin-5-ones as new class of H1-antihistaminic agents," Bioorganic & Medicinal Chemistry, vol. 15, no. 12, pp. 4009–4015, 2007.View at Publisher · View at Google Scholar · View at PubMed

10. R. L. Tolman, R. K. Robins, and L. B. Townsend, "Pyrrolo[2,3-d] pyrimidine nucleoside antibiotics. Total synthesis and structure of toyocamycin, unamycin B, vengicide, antibiotic E-212, and sangivamycin (BA-90912)," Journal of the American Chemical Society, vol. 90, no. 2, pp. 524–526, 1968. View at Publisher · View at Google Scholar

11. M. Amir, S. Javed, and H. Kumar, "Pyrimidine as antiinflammatory agent: a review," Indian Journal of Pharmaceutical Sciences, vol. 69, no. 3, pp. 337–343, 2007.

12. R. V. Bonnert, P. A. Cage, S. F. Hunt, I. J. S. Walters, and R. P. Austin, "Pteridinone derivatives as modulators of chemokine receptor activity," PCT International Patent Application, WO/2003/024966, 2003.

13. X. Wang, P. P. Chakrabarti, V. I. Ognyanov, et al., "Trisubstituted pyrimidines as transient receptor potential vanilloid 1 (TRPV1) antagonists with improved solubility," Bioorganic & Medicinal Chemistry Letters, vol. 17, no. 23, pp. 6539–6545, 2007. View at Publisher · View at Google Scholar · View at PubMed

14. J. Lu, Y. Bai, Z. Wang, B. Yang, and H. Ma, "One-pot synthesis of 3,4-dihydropyrimidin-2(1H)-ones using lanthanum chloride as a catalyst," Tetrahedron Letters, vol. 41, no. 47, pp. 9075–9078, 2000. View at Publisher · View at Google Scholar

15. F. Zurmühlen, "Process for preparing 5-chloro-4-hydroxypyrimidines," US patent no. 5523404, 1996.

16. E. M. Doherty, J. Zhu, M. Stec, et al., "Amino-pyridine, -pyridine and pyridazine derivatives for use as vanilloid receptor ligands for the treatment of pain," PCT International Patent Application, WO 099284, 2003.

17. G. T. Wang, S. Wang, R. Gentles, et al., "Amino-substituted heterocycles as isosteres of trans-cinnamides: design and synthesis of heterocyclic biaryl sulfides as potent antagonists of LFA-1/ICAM-1 binding," Bioorganic & Medicinal Chemistry Letters, vol. 15, no. 1, pp. 195–201, 2005.View at Publisher · View at Google Scholar View at PubMed

Citations

CHAPTER 1

Chen, D.-K.; Kang, Q.; Wu, T.R. Modular Synthesis of Polyphenolic Benzofurans, and Application in the Total Synthesis of Malibatol A and Shoreaphenol.Molecules 2010, 15, 5909-5927.

CHAPTER 2

Samuël Demin, Dirk Van Haver and Pierre J. De Clercq*; Asymmetric Synthesis of (7aS)-7a-Methyl-4,5,7,7a-tetrahydro-1H-indene-2,6-dione and Useful Derivatives Thereof; doi:10.3390/11090707

CHAPTER 3

Jonathan Clayden and Paul MacLellan Asymmetric synthesis of tertiary thiols and thioethers doi:10.3762/bjoc.7.6

CHAPTER 4

Harshita Sachdeva, Rekha Saroj, Sarita Khaturia, and Diksha Dwivedi, "Environ-Economic Synthesis and Characterization of Some New 1,2,4-Triazole Derivatives as Organic Fluorescent Materials and Potent Fungicidal Agents," Organic Chemistry International, vol. 2013, Article ID 659107, 19 pages, 2013. doi:10.1155/2013/659107

CHAPTER 5

Kieron M. G. O'Connell, Monica Díaz-Gavilán§, Warren R. J. D. Galloway and David R. Spring, Two-Directional Synthesis as a Tool for Diversity-Oriented Synthesis: Synthesis of Alkaloid Scaffolds, doi:10.3762/bjoc.8.95.

CHAPTER 6

Harshita Sachdeva, Rekha Saroj, Sarita Khaturia, and Diksha Dwivedi, "Environ-Economic Synthesis and Characterization of Some New 1,2,4-Triazole Derivatives as Organic Fluorescent Materials and Potent Fungicidal Agents," Organic Chemistry International, vol. 2013, Article ID 659107, 19 pages, 2013. doi:10.1155/2013/659107

CHAPTER 7

Benito Alcaide, Pedro Almendros, and José M. Alonso, Gold-Catalyzed Cyclizations of Alkynol-Based Compounds: Synthesis of Natural Products and Derivatives, ISSN 1420-3049, DOI: 10.3390.

CHAPTER 8

Ermolat'ev, D.S.; Babaev, E.V. Solid-Phase Synthesis of Methyl N-(pyrimidin-2-yl)glycinate. Molecules 2003, 8, 467-471. doi:10.3390/80600467

CHAPTER 9

Gareth J. Rowlands and Richard J. Seacome Enantiospecific synthesis of [2.2]paracyclophane-4-thiol and derivatives doi:10.3762/bjoc.5.9

CHAPTER 10

Jitendra M. Gajera, Shital N. Tondlekar, and Laxmikant A. Gharat, "A Novel and One-Pot Synthesis of 6-arylpyrimidin-4-ol," Research Letters in Organic Chemistry, vol. 2008, Article ID 810678, 3 pages, 2008. doi:10.1155/2008/810678

INDEX

D

E

F

G

H

I

L

M

N

P

R

S

T